Metal-free Functionalized Carbons in Catalysis
Synthesis, Characterization and Applications

Catalysis Series

Series editors:
Bert Klein Gebbink, *Utrecht University, The Netherlands*
Jose Rodriguez, *Brookhaven National Laboratory, USA*

Titles in the series:
1: Carbons and Carbon Supported Catalysts in Hydroprocessing
2: Chiral Sulfur Ligands: Asymmetric Catalysis
3: Recent Developments in Asymmetric Organocatalysis
4: Catalysis in the Refining of Fischer–Tropsch Syncrude
5: Organocatalytic Enantioselective Conjugate Addition Reactions: A Powerful Tool for the Stereocontrolled Synthesis of Complex Molecules
6: N-Heterocyclic Carbenes: From Laboratory Curiosities to Efficient Synthetic Tools
7: *P*-Stereogenic Ligands in Enantioselective Catalysis
8: Chemistry of the Morita–Baylis–Hillman Reaction
9: Proton-Coupled Electron Transfer: A Carrefour of Chemical Reactivity Traditions
10: Asymmetric Domino Reactions
11: C–H and C–X Bond Functionalization: Transition Metal Mediation
12: Metal Organic Frameworks as Heterogeneous Catalysts
13: Environmental Catalysis Over Gold-Based Materials
14: Computational Catalysis
15: Catalysis in Ionic Liquids: From Catalyst Synthesis to Application
16: Economic Synthesis of Heterocycles: Zinc, Iron, Copper, Cobalt, Manganese and Nickel Catalysts
17: Metal Nanoparticles for Catalysis: Advances and Applications
18: Heterogeneous Gold Catalysts and Catalysis
19: Conjugated Linoleic Acids and Conjugated Vegetable Oils
20: Enantioselective Multicatalysed Tandem Reactions
21: New Trends in Cross-Coupling: Theory and Applications
22: Atomically-Precise Methods for Synthesis of Solid Catalysts
23: Nanostructured Carbon Materials for Catalysis
24: Heterocycles from Double-Functionalized Arenes: Transition Metal Catalyzed Coupling Reactions
25: Asymmetric Functionalization of C–H Bonds

26: Enantioselective Nickel-catalysed Transformations
27: N-Heterocyclic Carbenes: From Laboratory Curiosities to Efficient Synthetic Tools, 2nd edition
28: Zeolites in Catalysis: Properties and Applications
29: Biocatalysis: An Industrial Perspective
30: Dienamine Catalysis for Organic Synthesis
31: Metal-free Functionalized Carbons in Catalysis: Synthesis, Characterization and Applications

How to obtain future titles on publication:
A standing order plan is available for this series. A standing order will bring delivery of each new volume immediately on publication.

For further information please contact:
Book Sales Department, Royal Society of Chemistry, Thomas Graham House, Science Park, Milton Road, Cambridge, CB4 0WF, UK
Telephone: +44 (0)1223 420066, Fax: +44 (0)1223 420247
Email: booksales@rsc.org
Visit our website at www.rsc.org/books

Metal-free Functionalized Carbons in Catalysis

Synthesis, Characterization and Applications

Edited by

Alberto Villa
Università degli Studi di Milano, Italy
Email: alberto.villa@unimi.it

and

Nikolaos Dimitratos
Cardiff University, UK
Email: DimitratosN@cardiff.ac.uk

ROYAL SOCIETY OF CHEMISTRY

Catalysis Series No. 31

Print ISBN: 978-1-78262-863-7
PDF ISBN: 978-1-78801-311-6
EPUB ISBN: 978-1-78801-460-1
ISSN: 1757-6725

A catalogue record for this book is available from the British Library

The Royal Society of Chemistry is a charity, registered in England and Wales, Number 207890, and a company incorporated in England by Royal Charter (Registered No. RC000524), registered office: Burlington House, Piccadilly, London W1J 0BA, UK, Telephone: +44 (0) 207 4378 6556.

For further information see our web site at www.rsc.org

Printed in the United Kingdom by CPI Group (UK) Ltd, Croydon, CR0 4YY, UK

Preface

Carbon-based materials show unique chemico-physical properties in their various allotropic forms and they have been successfully used in many catalytic processes including the production of chemicals and energy. Carbons have been traditionally used as supports for metals. However, carbon itself can act as a potential catalyst with recent pioneering studies showing it can replace the use of metals or metal-oxide-based catalysts in the gas phase oxidative dehydrogenation (ODH) of alkanes, such as ethylbenzene and alkenes. Metal-free carbons were successfully applied in several catalytic processes including photocatalysis, electrocatalysis, and liquid-phase oxidation processes. Compared to traditional metal-based catalysts, carbon materials have many advantages, such as their high surface area, unique electronic properties, high resistance to acids and alkalis, and thermal stability. In particular, the discovery of fullerene, carbon nanotubes, and graphene offered materials with higher activity and stability compared to classical carbon black and activated carbon. Their well-defined structures present great opportunities to advance the understanding of carbon materials chemistry. The catalytic activities of carbon materials are intimately related to their defects, structures, and surface chemistry. The introduction of defects and heteroatoms (N, B, P, S) alters the electronic properties of the surfaces, often increasing the reactivity of the surface of nanocarbons. The inclusion of hetero-elements in the honeycomb carbon structure induces a deep redistribution of the electronic properties. An ever-growing number of functionalized carbons (preferentially containing nitrogen) are nowadays employed with success as metal-free catalysts. Indeed, nitrogen-functionalized carbons present different surface functional groups, and they can be used as multifunctional catalysts, either through their electronic or nucleophilic

Catalysis Series No. 31
Metal-free Functionalized Carbons in Catalysis: Synthesis, Characterization and Applications
Edited by Alberto Villa and Nikolaos Dimitratos

Published by the Royal Society of Chemistry, www.rsc.org

properties, or ability to form additional H bonds with substrates. The scope of this book is to provide an overview of the preparation, characterization and application of metal-free functionalized carbons. The first three chapters will cover the preparation of the most common functionalized carbon utilized in metal-free catalysis, including carbon functionalized by covalent methods (Chapter 1) and non-covalent methods (Chapter 2). Chapter 3 provides an insight into the preparation of carbon nitrides (g-C_3N_4) and covalent triazine frameworks (CTF). Chapters 4 and 5 describe the most useful spectroscopy techniques to characterize the carbon surface, such as Raman, infrared spectroscopy, inelastic neutron scattering (INS) and X-ray photoelectron spectroscopy (XPS). The last part of the book provides a description of the different applications of metal-free functionalized carbon including liquid-phase reactions (Chapter 6), gas-phase reactions (Chapter 7), electrocatalysis (Chapter 8) and photocatalysis (Chapter 9). Finally, Chapter 10 describes the utilization of metal-free doped carbons for electroanalytical sensors. As editors, we believe that all the contents of this book constitute a valuable contribution to understanding the fundamental chemistry of metal-free carbon materials and will inspire both researchers already involved in the subject and those who would like to approach this field.

Alberto Villa
Nikolaos Dimitratos

Contents

Section I: Synthesis

Catalysis Series No. 31
Metal-free Functionalized Carbons in Catalysis: Synthesis, Characterization and Applications
Edited by Alberto Villa and Nikolaos Dimitratos

Published by the Royal Society of Chemistry, www.rsc.org

Section II: Characterization

Section III: Applications

Section I

Synthesis

CHAPTER 1

Covalent Methods for Functional Carbons' Synthesis

J.-P. TESSONNIER*[a], R. G. RAO[a], G. GIAMBASTIANI*[b,c] AND G. TUCI[b,d]

[a]Iowa State University, Department of Chemical and Biological Engineering, 617 Bissell Road, Ames, IA 50011-1098, United States; [b]National Research Council, Institute of Chemistry of OrganoMetallic Compounds ICCOM-CNR, Via Madonna del Piano 10, 50019, Sesto Fiorentino, Florence, Italy; [c]Kazan Federal University, 420008 Kazan, Russian Federation; [d]Università di Firenze, Dipartimento di Chimica "Ugo Schiff", Via della Lastruccia, 3-13, 50019, Sesto Fiorentino, Florence, Italy
*E-mail: tesso@iastate.edu, giuliano.giambastiani@iccom.cnr.it

1.1 Towards a Molecular Understanding of the Reactivity of Carbon Surfaces

1.1.1 Role of Curvature and Topological Defects on Surface Reactivity

The unique chemico–physical properties of carbon-based materials in their various allotropic forms have made them valuable candidates for application in many catalytic processes. Traditionally, carbon materials have been employed as supports for metal-based multi-phase catalysts in heterogeneous processes although their role (innocent or non-innocent)

Catalysis Series No. 31
Metal-free Functionalized Carbons in Catalysis: Synthesis, Characterization and Applications
Edited by Alberto Villa and Nikolaos Dimitratos

Published by the Royal Society of Chemistry, www.rsc.org

in the catalytic performance remains largely questionable. The last few years have witnessed a *technological renaissance* that has boosted the exploitation of carbon-based nanomaterials for running a number of key industrial transformations with good to excellent catalytic outcomes. This renaissance is in line with seminal papers published in the 1960s by Donnet, Boehm, and Coughlin,[1–3] who showed that the catalytic performance (activity and selectivity) of carbon catalysts is related to two largely interconnected features: the chemistry and electronic properties of their outer surfaces.

Many efforts in the scientific community have been devoted to the understanding of molecular-level phenomena that govern carbon material reactivity. The discovery of well-defined carbon nanomaterials in the 1980s and 1990s, namely fullerene and carbon nanotubes, offered unprecedented opportunities to advance our understanding of carbon materials' rich chemistry. Works on defect-free fullerenes and single-walled carbon nanotubes (SWCNTs) revealed notable connections between the curvature and the reactivity of carbon surfaces. For instance, for fullerenes, the reactivity of sp^2 hybridized carbons was demonstrated to be a result of the strain induced by the curvature.[4] Indeed, any C-sp^2 carbon atom within a high curvature environment (radius of curvature from 5 to 10 Å) will undergo a strain-induced pyramidalization of its orbitals. This pyramidalization strongly facilitates addition reactions, thus breaking C=C bonds and inducing atoms' relaxation due to rehybridization to C-sp^3.[5] Chirality ordinarily present in SWCNTs creates additional strain on C=C bonds due to the "twisting" of the rolled graphene sheet that leads to a misalignment of the p_z orbitals.[6,7] The high reactivity of strained carbon surfaces is not limited to well-defined nanocarbons like fullerene and SWCNTs, but it actually occurs on any site with a high local curvature. In the case of Stone–Wales defects (Figure 1.1), the presence of C_5 and C_7 rings would induce curvature in the carbon materials.[4,8–11]

1.1.2 Surface Modification through Heteroatom Insertion

Typically, topological defects are reactive sites and their interaction with oxygen leads to modification of surface properties. Their oxidation generates a variety of oxygen-containing functional groups (carboxylic acid, anhydride, hydroxyl, lactone, ether), which alter both the polarity and acid–base properties of the carbon surface. Deep oxidation eventually results in the formation of CO_2 and the creation of a vacancy. This reactivity explains why carbons present a quilted surface consisting of defect-free graphitic domains interconnected by regions with amorphous carbon (sp^3), point defects, and vacancies. The interaction of defects with oxygen also introduces oxygen-doping functionalities on the surface of carbon materials, for example around vacancies as well as at the edge/prismatic plane carbon atoms as these sites are far more reactive towards oxidation than that of basal plane carbons.[12–15] For instance, carbon atoms on the edges

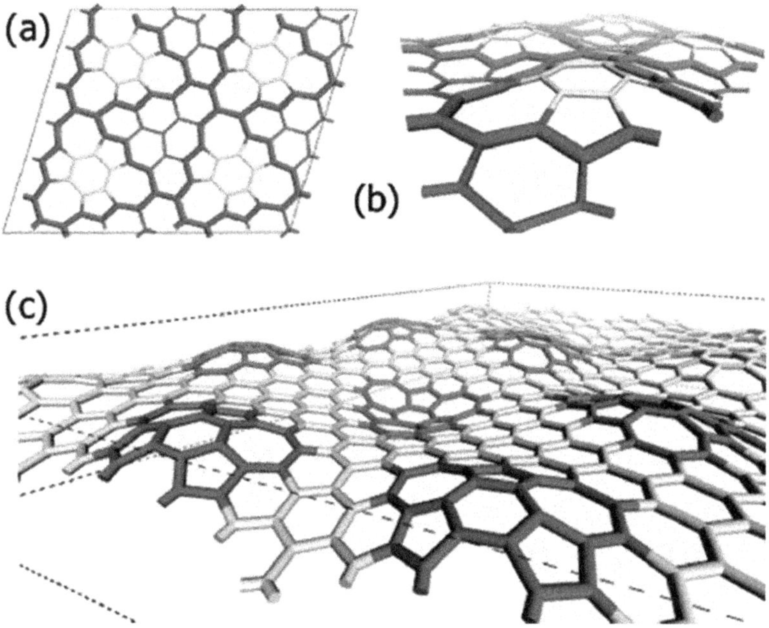

Figure 1.1 Top (a) and side views (b, c) of a graphene sheet presenting regions with high local curvature induced by C_5 and C_7 rings. Reprinted figure with permission from M. T. Lusk and L. D. Carr, *Phys. Rev. Lett.*, **100**, 175503, 2008. Copyright (2008) by the American Physical Society.[8]

of graphene can lead to activation of O_2 to form epoxides that migrate or hop on the surface of the basal plane.[16,17] Therefore, the presence of edges or defects is essential for introducing heteroatoms in the basal plane of carbons.

Defects are generally introduced in carbon materials during synthesis as a result of fast reaction kinetics compared to the thermodynamically-favored formation of a defect-free honeycomb structure.[18] Alternatively, the post-synthetic introduction of new defects typically relies on the use of strong oxidants able to disrupt carbon–carbon double bonds, such as the mixture of nitric acid and potassium chlorate as proposed by B. C. Brodie in 1859 for the oxidation of graphite.[19] Similarly, the works of Staudenmaier, Hummers, and Offeman utilized potassium permanganate and sulfuric acid for further improving the process of graphene oxide (GO) synthesis.[20,21] The introduction of defects and heteroatoms disrupts the delocalized electron cloud and alters the electronic properties of the surfaces, allowing the grafting of functional groups on the surface of nanocarbons.

In recent years, the heteroatom-doping (N, B, P, S) of carbon nanomaterials has emerged as an effective way to manipulate the electronic properties of nanocarbons. The inclusion of hetero-elements in the honeycomb carbon structure induces a deep redistribution of the electronic properties. An ever-growing number of light hetero-doped carbons (preferentially containing nitrogen) are nowadays employed with success as metal-free catalysts. Selected systems from this series can offer higher catalytic activity and long-term operational stability than those measured with benchmark state-of-the-art metal-based systems.[22–24] Post-synthetic chemical and thermal treatments (typically with strong and oxidizing acids or bases) have also been used to tune the materials' surface chemistry. The control of porosity and defect-sites' surface density along with the introduction of dangling (acidic or basic) functional groups is the new contemporary approach to prepare a family of new powerful catalysts.[25–29] However, both heteroatom-doping and post-synthetic treatments generally suffer from moderate control on the nature of the chemical surface groups. Therefore, many efforts have been made to identify the active functionalities for a specific catalytic transformation and establish clear-cut structure-composition–reactivity relationships.

1.2 Strategies for the Synthesis of Functional Carbon Catalysts

1.2.1 Catalytic Chemical Vapor Deposition

The post-synthetic chemical functionalization of carbon materials typically requires multiple steps, making it a tedious process. In the case of carbon nanotubes (CNTs), many efforts have, therefore, focused on the introduction of heteroatoms directly during the synthesis of the nanocarbon using modified catalytic chemical vapor deposition (CCVD) techniques.[30–32] Typically, the CCVD process consists of a hydrocarbon gas source that is heated in the presence of a transition metal catalyst (Fe, Co, Ni).[18] This technique allows an easy incorporation of heteroatoms in the structure by altering the reaction mixture using precursors such as ammonia,[33] aniline,[34] phthalocyanine,[30,35] melanine,[32] acetonitrilie,[36] pyridine,[32,36] triazine,[32] among others. N-doped CNTs have shown promising results in the high temperature (400 °C) oxidative dehydrogenation (ODH) of propane.[34] Graphene containing N-atoms synthesized by adding ammonia gas in the CCVD process has also been used for applications in oxygen reduction reactions (ORR) as a metal-free electrocatalyst.[37] The electrocatalytic applications demonstrated by the N-doped carbon nanomaterials in ORR and fuel cells are due to the modification of electronic properties caused by the electron donating/withdrawing nature of the heteroatoms which induce charge redistribution on the surface.[35,37,38]

In addition to nitrogen, other heteroatoms such as phosphorus and silicon, can also be doped in nanocarbons using precursors such as triphenylphosphine[39,40] and methoxytrimethylsilane.[39] Boron and phosphorus atoms have also been added in the CNT structure using CVD in the presence of triphenylborane,[38] triethylborate,[41] diborane,[42] and triphenylphosphine.[41] The as-prepared materials showed promising results in ORR and the oxidation of cyclohexane.

Nanoporous graphene materials co-doped with sulfur and nitrogen using pyridine and thiophene have recently shown activity for the hydrogen evolution reaction (HER) comparable to 2D MoS_2, recognized as the best Pt-free HER catalyst.[43] The key reason for the performance of the catalyst in the HER was demonstrated to be the alteration of its electronic properties and, thereby, of the Gibbs free energy of hydrogen binding, a key descriptor for this reaction. These examples provide evidence that the electronic properties of nanocarbons can be tuned using heteroatom doping and, thereby, their performance can be varied by altering the heteroatom amount, its nature, and chemical environment.

While being an efficient strategy for introducing heteroatoms in the carbon backbone, CCVD suffers from several drawbacks. Notably, the high temperatures required for growing nanocarbons do not offer opportunities to control the nature of the inserted moieties. For instance, in the case of nitrogen, high temperatures favor the formation of pyridinic and quaternary nitrogen moieties, thus limiting the applications of these materials for reactions catalyzed by strong Brønsted bases. It should also be noted that the heteroatoms are typically distributed throughout the nanomaterial, meaning that only a fraction of the inserted heteroatoms are present on the surface and actively participate in catalytic reactions. Finally, it has also been shown that the addition of heteroatom-containing precursors in the gas feed during synthesis alters the growth process. For example, the addition of ammonia in the feed yielded bamboo-shaped CNTs instead of multiwalled carbon nanotubes (MWCNTs). The added arches close the inner channels, leading to a significant decrease in surface area and pore volume compared to undoped MWCNTs.

1.2.2 Post-synthetic Functionalization

The vast majority of covalently functionalized carbons are produced through post-synthetic treatments. The methods employed are typically derived from organic chemistry techniques and consist of tethering heteroatom-containing moieties to the carbon surface through oxidation and coupling reactions.

In the case of oxidations, the key challenges of the post-treatment techniques are the changes in the morphological and textural properties of carbons due to harsh conditions and the lack of control of the nature of active groups introduced on the surface. In the case of GO synthesis from graphite, the use of harsh oxidants leads to highly defective, wrinkled, graphene oxide sheets with difficult-to-control morphology and textural properties. However, modifications of the thermochemical procedures facilitate the introduction of functional groups without altering the structural integrity of carbons. For instance, the gas-phase oxidation of CNTs using HNO_3 vapors reported by Xia *et al.* introduced oxygen functionalities on their surface without deteriorating the material morphology.[44] The challenges associated with the control of the amount and nature of O-containing groups were addressed by partial defunctionalization methods. The knowledge of the thermal decomposition of specific functional groups provided an excellent handle on the tunability of carbon's surface chemistry.[45] The control of surface functionalization was utilized for developing carbocatalysts, for instance by tuning the diketone and quinone groups for the ODH of ethyl benzene.[46]

Apart from tuning the catalytic performance of carbons by controlling the surface O-functionalities, strategies were implemented for introducing other heteroatoms such as nitrogen. Treating oxidized CNTs or graphene with ammonia at elevated temperatures, among various other techniques, facilitated the introduction of nitrogen functional groups and control of acid/base properties of carbons.[47–49] Boron could also be inserted in the carbon matrix by thermally annealing oxidized CNTs or GO in the presence of elemental boron and boron oxide.[50,51] Sulfonate moieties can be incorporated in CNTs and graphite by treatment with oleum.[52,53] The addition of heteroatoms such as boron and nitrogen have been reported to play a role in enhancing carbocatalysts' activity and selectivity for ODH and oxidation reactions in general.[54,55] The doped carbon materials have demonstrated better catalytic performance than noble-metal-based catalysts for oxygen reduction reactions (ORR) along with better control of activity, selectivity and stability by tuning the dopant concentration.[56]

Nitrogen incorporation is also possible using hydrothermal synthesis with melamine as a precursor.[57] Hydrothermal techniques open possibilities related to the use of multiple dopants in the same carbocatalyst. For example, boron and nitrogen doped graphene quantum dots deposited on GO using hydrothermal techniques (Figure 1.2) result in carbocatalysts that demonstrate better ORR activity than the benchmark Pt/C.[58] Further, nitrogen-doped graphene on CNTs, synthesized *via* hydrothermal techniques, successfully performed the metal-free electrocatalytic ORR.[59] The assembly of graphitic C_3N_4 on graphene using hydrothermal techniques also showed a promising performance for the HER reaction, comparable to those of metal catalysts.[60]

Thermal synthesis is another technique by which heteroatoms can be incorporated into nanocarbon materials. This method includes the decomposition of precursors such as melamine[61] or dicyandiamide[62] on the surface of nanocarbons. Recent studies have also utilized food-grade components as starting materials (*i.e.* glucose, citric acid, and ammonium carbonate) as aqueous solutions or physical mixtures for preparing highly N-rich carbon composites in the form of either film coatings for a variety of organic and inorganic macroscopic supports[63,64] or 3D self-standing and shape-controllable open-cell mesoporous C-based foams (Figure 1.3).[65] These materials were successfully applied as metal-free catalysts for the electrochemical ORR, H_2S partial oxidation and direct dehydrogenation (DDH) of ethylbenzene to styrene. Another bottom-up synthesis technique comprised the annealing glucose, urea, and phosphoric acid together to obtain dual-doped nanoporous graphene, which depicted high catalytic activity for the hydrogen evolution reaction (HER), comparable to traditional metallic catalysts.[66]

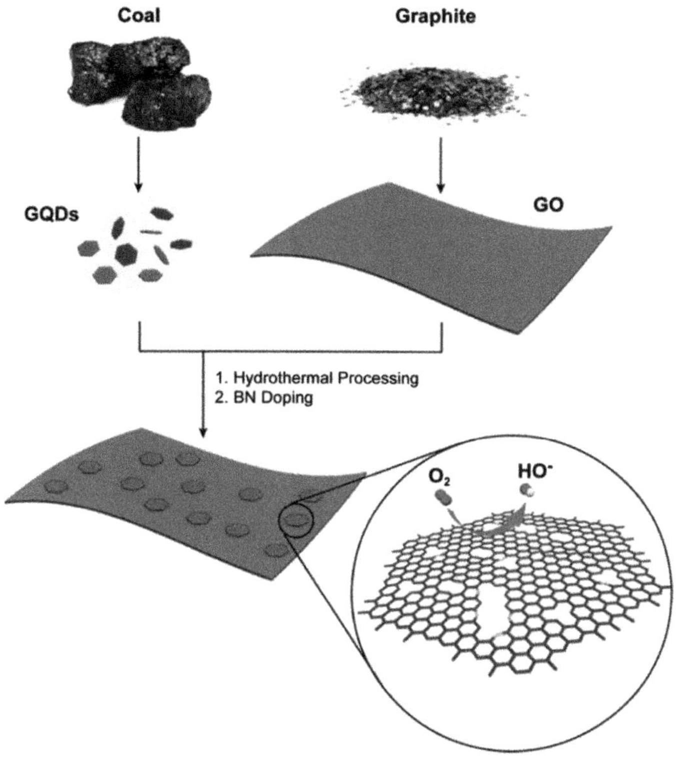

Figure 1.2 Illustration of the preparation procedure for the BN-GQD/G nanocomposite. Reprinted with permission from H. Fei, R. Ye, G. Ye, Y. Gong, Z. Peng, X. Fan, E. L. G. Samuel, P. M. Ajayan and J. M. Tour, *ACS Nano*, 2014, **8**, 10837–10843.[58] Copyright (2014) American Chemical Society.

In addition to the application of carbocatalysts in ODH, ORR, and oxidation reactions in general, these materials also demonstrated activity for hydrogenation reactions when creating frustrated Lewis pairs (FLPs) on the carbon surface. FLPs consist of a Lewis acid and a base moiety in organic molecules, separated by an appropriate distance to maintain the interaction without being neutralized. In general, FLPs have been reported to be capable of hydrogen activation.[67–69] Similar reaction mechanisms have been proposed for the FLP-driven hydrogenation of acetylene and alkene on graphene consisting of oxygen, phosphorus, sulfur, and nitrogen functionalities.[70] GO and reduced graphene oxide (rGO) synthesized by the pyrolysis of alginate have also demonstrated hydrogenation activity for various nitro compounds.[71] Recent work by Sun *et al.* further provided theoretical evidence for the use of boron and nitrogen co-doped bilayer graphene and a graphene nanoribbon for hydrogen activation due to the presence of FLPs.[72] This evidence for the capabilities of carbocatalysts as metal-free hydrogenation catalysts provides new opportunities for the development of sustainable and greener routes for the synthesis of specialty and fine chemicals.

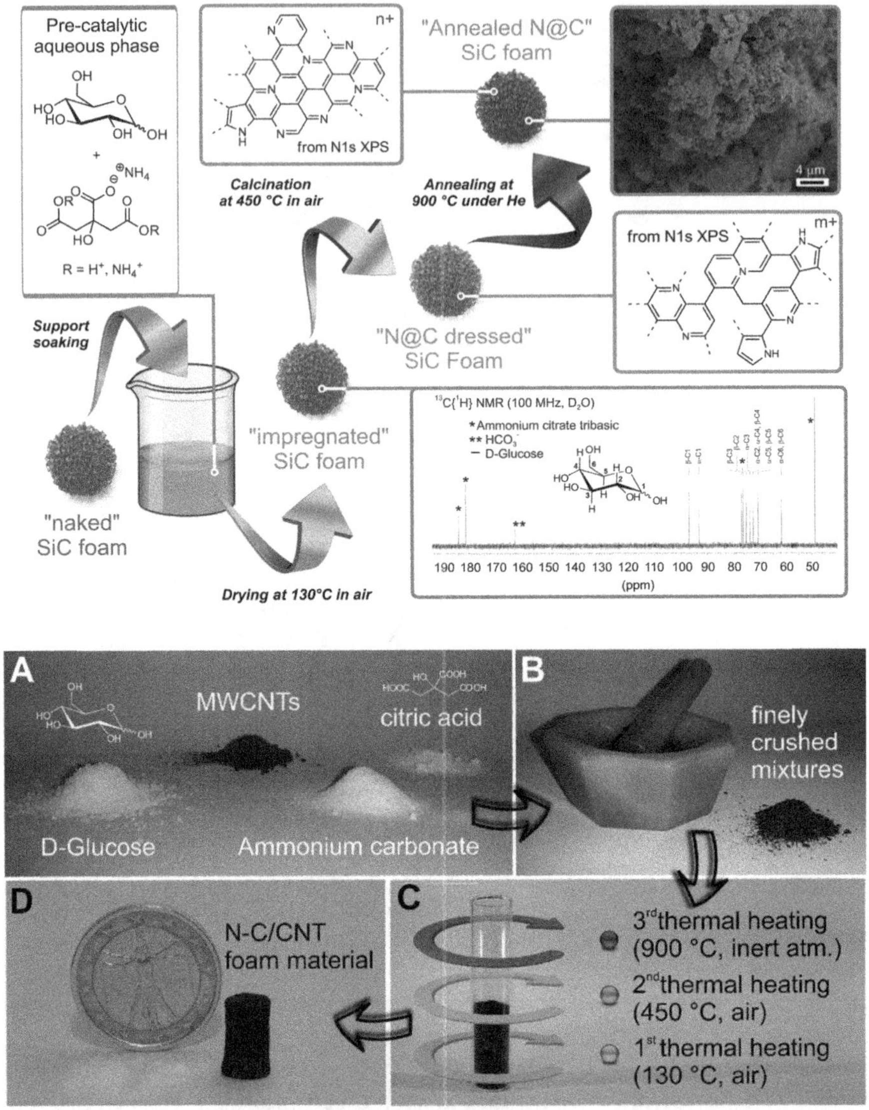

Figure 1.3 Top: synthesis of a highly N-doped carbon-based coating (N@C) on a model macroscopically shaped host matrix (SiC foam) (reproduced from ref. 63 with permission from the Royal Society of Chemistry). Bottom: sequential steps (A → D) for the preparation of N-doped 3D self-standing composites (N–C/CNT) (D), from a physical mixture (wt/wt) of commercially available food-grade components and pristine MWCNTs (reprinted with permission from H. Ba, Y. Liu, L. Truong-Phuoc, C. Duong-Viet, J.-M. Nhut, D. L. Nguyen, O. Ersen, G. Tuci, G. Giambastiani and C. Pham-Huu, *ACS Catal.*, 2016, **6**, 1408–1419.[65] Copyright (2016) American Chemical Society).

1.2.3 Tethering Discrete Active Sites Through Nanomaterials' Surface Functionalization

The covalent grafting of well-defined (organic) functionalities or molecular entities at the nanomaterial surface is of growing interest.[73,74] For many years, functionalization protocols have been optimized considering their foundational character in organic chemistry and material science.[75,76] One of the main tasks driving this optimization is the improvement of the carbon nanomaterial dispersion/deaggregation in liquid media or in polymeric matrices to achieve an easier processing in composites production.[77] Recent findings in the field have demonstrated that chemical functionalization holds huge potential for a rational bottom-up design of complex carbon nanostructures with tailored chemical and electronic surface characteristics. Indeed, functional groups at the nanomaterial surface play a key role in specific catalytic processes, offering at the same time unique tools for the comprehension of their mechanism of action as metal-free systems. There is a huge number of original research papers and review articles dealing with the covalent functionalization of carbon nanomaterials. In the following sections, we will review the most representative studies in relation to their direct contribution to the catalytic performance of the nanomaterial as a metal-free catalyst.

1.3 Oxidized Carbon Nanomaterials in Catalysis

Oxidation is certainly the most widely employed methodology for nanomaterial surface grafting with a variety of O-containing groups (carboxyls, epoxides and hydroxyls). Variable distributions and loadings of oxygenated functionalities can be obtained depending on the oxidative protocol. The harsher the oxidation conditions, the higher the functionalization degree. High oxidative degrees typically translate into materials with better processability and dispersibility in various media but they also determine a progressive loss of their inherent electrical conductivity, a fundamental pre-requisite for the samples' exploitation as electrocatalysts.[78]

Recent literature findings have demonstrated the impressive flexibility and performance of oxidized carbon nanomaterials within a very wide variety of catalytic processes. The interest in the sustainable development of single-phase catalysts that replace classical metal or noble-metal-based systems is shown by the number of papers that appeared on the topic. In some cases, controlled oxidation protocols together with a precise surface characterization offer an unambiguous explanation of the role played by O-containing functional groups and surface defects. For all processes outlined below, catalyst recovery and re-use are commonly claimed to be feasible, without appreciable loss of performance. Table 1.1 summarizes all the main catalytic transformations mediated by oxidized carbon nanomaterials and afterwards discussed in Sections 1.3.1–1.3.4.

Table 1.1 Class of reactions catalyzed by variably oxidized carbon nanomaterials.

Class of metal-free catalyzed reaction	Catalyst/co-catalyst or reagent	Reference
Oxidation of alcohols to aldehydes/ketones	GO/O_2	79
Hydration of alkynes to ketones	GO/O_2	79
Electrochemical oxidation of alcohols	ox-CB or ox-MWCNTs	80 and 81
Benzene oxidation to phenol	CMG/H_2O_2	82
Oxidative coupling of amine to imine	GO	83
Oxidative desulfurization (from thiophenes to aromatic sulfones)	rGO/O_2	84
Oxidative alkane dehydrogenation (ODH)	ox-MWCNTs/O_2	85 and 86
Catalytic wet air oxidation (CWAO)	GO, ox-(MWCNTs, CNFs)/O_2	87 and 88
Oxygen evolution reaction (OER)	ox-MWCNTs	89 and 90
Nitroarene reduction to anilines	rGO or H_2O_2 oxidized CNTs/$H_2N–NH_2$	91–93
Photocatalytic water splitting (H_2 production)	GO	94
Hydrogen evolution reaction (HER)	Activated CNTs, ox-MWCNTs	90 and 95
Oxygen reduction reaction (ORR)	ox-MWCNTs	90, 96, 97
Friedel–Crafts alkylation reactions	GO, fullerenol	98–100
Michael and aza-Michael reaction	GO, fullerenol	100–102
Aldol reaction	Fullerenol	100
C–H arylation of benzene	GO	103
CO_2 cycloaddition reaction to propylene oxide	$GO/R_4N^+X^-$, fullerenol/KI	104–106
Ring-opening polymerization (ROP) and olefin polymerization	GO	107 and 108
Electrochemical polymerization	rGO	109
Epoxide ring opening	GO	110

1.3.1 Oxidation Reactions

This is probably one of the main application areas of oxidized nanocarbons in catalysis. Alkanes, alkenes, alcohols, aromatic compounds and amines are common substrates chosen to demonstrate the ability of variably oxidized carbon nanomaterials to convert them into their oxidized counterparts.

In 2010, Bielawski and co-workers have pioneered research in the field of liquid-phase reactions for the oxidation of alcohol by metal-free, carbon-based catalysts. They have shown that the oxygen-group-rich graphene oxide (GO) can selectively catalyze the oxidation of various alcohols to aldehydes and *cis*-stilbene to the corresponding dione. GO has also been proven to be effective in the hydration of various alkynes to their corresponding ketones at ambient conditions (in air).[79] More recently, the Australian team of Zhao and co-workers has reported on oxidized carbon-black (CB)[80] and oxidized MWCNTs[81] as valuable electrocatalysts for alcohol oxidation reactions (AOR of R–OH; R = Me, Et, ^{i}Pr) in alkaline media. Surface-oxidized carbon (*e.g.* ketonic groups) are claimed by these authors to enhance the material's electrocatalytic performance significantly, both in terms of current density (j) and alcohol oxidation onset potentials (E_{on}). Ma *et al.* have recently demonstrated how chemically modified graphene (CMG),[111,112] a variety of rGO, in combination with H_2O_2 as an oxidizing agent, promotes the challenging one-step conversion of benzene into phenol with good activity and excellent selectivity (up to 99%).[82] Although the catalytic role of O-containing surface species in CMG is unclear, the benzene adsorption to the CMG conjugated network and simultaneous H_2O_2 activation by its surface functionalities make this oxidized material an outstanding catalyst for the process.

Chemical, electrochemical, or thermal post-treatments of classically prepared oxidized nanocarbons can change their inherent catalytic behavior. Loh and co-workers have reported on the oxidative coupling of amines to imines catalyzed by GO under aerobic oxidative conditions. The authors have demonstrated how the GO performance can be dramatically enhanced by a sequential basic and acidic treatment.[83]

The chemical processing redraws the material surface chemistry and generates carboxylic groups at the edges of surface defects along with localized unpaired electrons. They work synergistically to trap amines and molecular oxygen with beneficial effects on the catalyst performance (Figure 1.4).

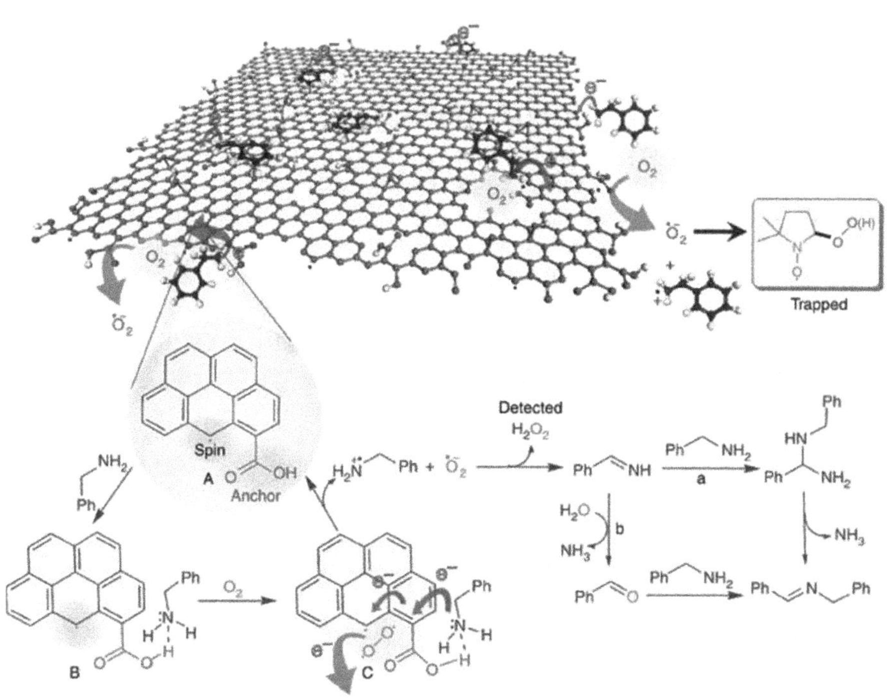

Figure 1.4 Mechanism of the GO-catalyzed oxidative coupling of primary amines. Reprinted by permission from Macmillan Publishers Ltd: *Nature Communications* (ref. 83), copyright (2012).

The combined action of O-containing functional groups (specifically carbonyl) and electronic properties of neighboring carbon sites are also claimed to drive the aerobic oxidative desulfurization of a broad range of sulfur-containing compounds from fuel. Su and co-workers have reported on a variety of reduced graphene oxides and their gas oxidation treatment as a valuable material process for the improvement of the rGO catalytic activity in the process.[84] The same authors have also shown that ketonic carbonyl groups are key functionalities in the nanocarbon-catalyzed oxidative dehydrogenation (ODH) of alkanes to alkenes.[85,113] Recent findings from this team have elegantly contributed to elucidating the catalytic mechanism of action of oxidized nanocarbon in the ODH process by means of conjugated polymers containing ketonic functionalities as model catalysts.[86] Carbon-based materials have also received a great deal of attention as catalysts in industrial wastewater treatment. In the catalytic wet air oxidation (CWAO) of organic pollutants from water, surface functional groups are known to play a pivotal role on the final performance of these metal-free systems. Yang *et al.*[87] and Wang *et al.*[88] have reported on various oxidized carbon samples (GO, rGO, ox-MWCNTs and ox-carbon nanofibers) as excellent metal-free catalysts for the CWAO of phenol in aqueous media. Carboxylic groups at the surface of these materials are claimed to drive the catalyst performance while promoting the production of active radical species throughout the oxidation process.

Oxidized nanocarbons have also answered to the most urgent needs of a clean and sustainable production of energy. In this regard, hydrogen/oxygen production by electrochemical water splitting represents one of the most promising reactions to satisfy the steadily increasing global energy demand.[89] As for oxidation processes, Zhao and collaborators have shown that chemically oxidized MWCNTs followed by controlled hydrothermal and electrochemical activation treatments generate metal-free systems with unprecedented activity in the electrochemical oxygen evolution reaction (OER) in alkaline media.[90,114] The observed OER performance is ascribed to the presence of ketonic groups that activate neighboring carbon surface sites, making them suitable acceptors of water oxidation intermediates.

1.3.2 Reduction Reactions

Chemical reduction processes promoted by oxidized carbon nanomaterials are almost exclusively limited to the activation and conversion of selected aromatic substituents. Ma and Bao have pioneered this research area.[91] In 2011, they showed for the first time that the hydrogenation of nitrobenzene to aniline can be carried out with very high activity and selectivity over rGO at room temperature using hydrazine as the hydrogen source. In this seminal contribution, the authors proposed that carbon atoms at the zig-zag edges of rGO are the active sites for the activation and conversion of the nitro group. Later, Su and co-workers demonstrated that a controlled (and mild) oxidation of carbon nanomaterials by H_2O_2 provides valuable catalytic systems for the process.[92] Experimental evidence suggested that highly oxidized carbon sites at the nanomaterial surface (*i.e.* carboxylic group and anhydride) had only detrimental effects on nitroarene reduction. On the other hand, carbonyl groups seemed to favor the process by fostering hydrazine activation.[93]

Recent studies have shown that oxygen-containing surface groups in nanocarbons hold key roles in reductive photochemical[115] and electrochemical[116] processes at the heart of renewable energy technology,[78] such as the hydrogen evolution reaction (HER) and oxygen reduction reaction (ORR). Under UV or visible light irradiation, GO with an appropriate oxidation level can serve as a photocatalyst for water splitting and H_2 production without employing any charge separation co-catalyst.[94] Chemically activated CNTs *via* acidic oxidation followed by a cathodic electrochemical pre-treatment generate an outstanding class of acid stable metal-free HER electrocatalyst.[95] A hydrothermal method to improve the electrochemical properties of surface-oxidized MWCNTs without reducing their initial oxygen content has recently been proposed by Zhao and collaborators.[90] Such a post-oxidation treatment is found to improve the materials' hydrophilicity, induce transformation of oxygen functional groups and enhance the samples' electrical conductivity. The higher electrocatalytic performance of the hydrothermally treated materials in HER is mainly ascribed to the increased amount of surface groups with stronger electron withdrawing abilities (*i.e.* ketone and carboxylic groups). These authors have also shown that ORR kinetics on MWCNTs are strongly affected by the graphitic structure and surface oxygen functional groups available in the oxidized samples.[90] Similar to carbon samples for HER, hydrothermally treated oxidized MWCNTs offer better electrochemical performance in ORR than their simply oxidized counterparts. Wang's team has demonstrated that oxygen-containing functional groups (*i.e.* C–O and C=O) at the surface of carbon nanomaterials cause similar electronic effects to those observed in nitrogen doped systems.[96] Their surface grafting increases the number of defective carbon atoms and re-draws the material's electronic surface properties. The latter point is claimed to enhance the material oxygen adsorption properties. Lastly, surface functionalities improve the material wettability and lead to a better electrolyte diffusion.[96] As a matter of fact, hydrophilic moieties like carboxylic groups are claimed to impart beneficial effects to the nanomaterials' electrochemical performance in ORR.[97]

1.3.3 C–C and C–N Bond Forming Reactions

Oxidized nanocarbons as metal-free systems catalyze C–C and C–N bond forming reactions, acting as stable and long-term re-usable solid materials. Classically prepared and carefully purified GO was proposed to efficiently catalyze the Friedel–Crafts reaction of indoles with various electron-rich and electron-deficient α,β-unsaturated ketones, giving the corresponding adducts with good yields and without by-products.[98] Song and co-workers have demonstrated that C_{60} fullerenol (a poly-hydroxylated fullerene prepared through a hydrogen peroxide heating method)[117] is an active catalyst for promoting Friedel–Crafts reactions as well as other important C–C bond forming reactions such as Michael addition, Henry reaction, and aldol addition.[100] Among challenging C–C bond forming reactions, Szostak and co-workers have proposed a highly efficient and stereoselective protocol for the coupling reaction between styrenes and alcohols catalyzed by graphene oxides.[99] Mechanistic studies suggest that the reaction proceeds *via tandem* catalysis (Figure 1.5) with both reactants transiently activated at the GO functionalities and properly positioned for a concerted C–C bond forming step. GO is also employed as a simple and efficient phase transfer catalyst for Michael addition reactions in biphasic media, showing similar behavior to that of more classical crown-ethers in the process. Its highly hydrophilic surface along with the hydrophobic nature of the carbon network, make GO a valuable phase transfer candidate for driving potassium cations through two immiscible phases (aqueous and organic).[101] Jain and Khatri found that GO can be a highly efficient, re-usable and cost-effective organocatalyst for the aza-Michael reaction between various amines and electron deficient olefins.[102] These reactions proceed under mild conditions and afford C–N coupling products with excellent yields within short reaction times. The highly hydrophilic character of GO finally makes its separation and re-use after each catalytic run easy by means of simple extractions of the reaction mixture with dichloromethane.

More recently, Ma and co-workers have reported on the GO-catalyzed direct C–H arylation of benzene with various aryl iodides as a heterogeneous, cheap and environmentally friendly metal-free protocol enabling the formation of biaryl compounds.[103] The direct correlation between oxygen content in GO and yield of biaryl compounds has been demonstrated; the specific role of edge- or basal-plane O-containing groups has been clarified by means of oxygenated model molecular catalysts. According to the authors' conclusions, the combination of local charge densities induced by various C–O bond

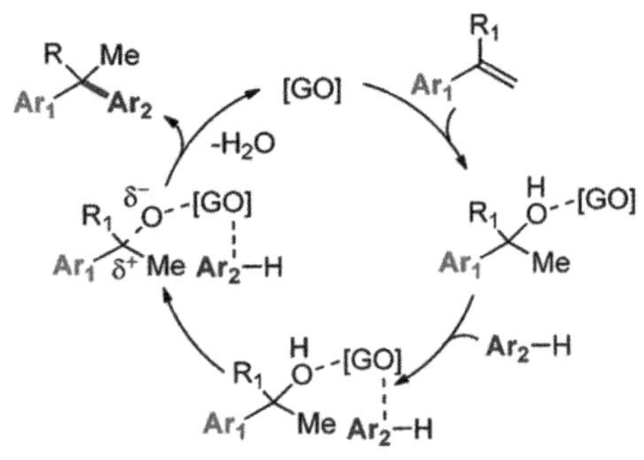

Figure 1.5 Mechanism for the GO-catalyzed alkylation of arenes. Reprinted with permission from F. Hu, M. Patel, F. Luo, C. Flach, R. Mendelsohn, E. Garfunkel, H. He and M. Szostak, *J. Am. Chem. Soc.*, 2015, **137**, 14473.[99] Copyright (2015) American Chemical Society.

polarizations and GO surface topological defects triggers the C–C coupling catalysis.

1.3.4 Cycloaddition and Polymerization Reactions

With the inter- and intra-molecular assistance of ammonium halides, GO also exhibits excellent catalytic performance in the CO_2 cycloaddition to propylene oxide (PO) thus giving the corresponding cyclic carbonate (propylene carbonate, PC) with high yield and selectivity under mild conditions. The reaction can be generalized in terms of epoxides' ring opening *via* CO_2 cycloaddition to provide cyclic carbonates as products. In 2014, Yin and co-workers have shown that GO hydroxyl groups foster the cycloaddition process by activating PO to undergo ammonium-halide (co-catalyst) mediated epoxide ring opening and CO_2 insertion.[104] Later on, the same group has reported on the superior catalytic activity of a properly synthesized multi-functional GO material (containing silanol groups, quaternary ammonium salts and amines) in the cycloaddition process (Figure 1.6).[105]

The authors claim that covalently linked amines act as adsorption sites and activators for CO_2, whereas silanol groups provide activation sites for PO and stabilize the ring opened oxirane before the intramolecular CO_2 addition and subsequent cyclization steps. C_{60} fullerenol was also found to be a highly

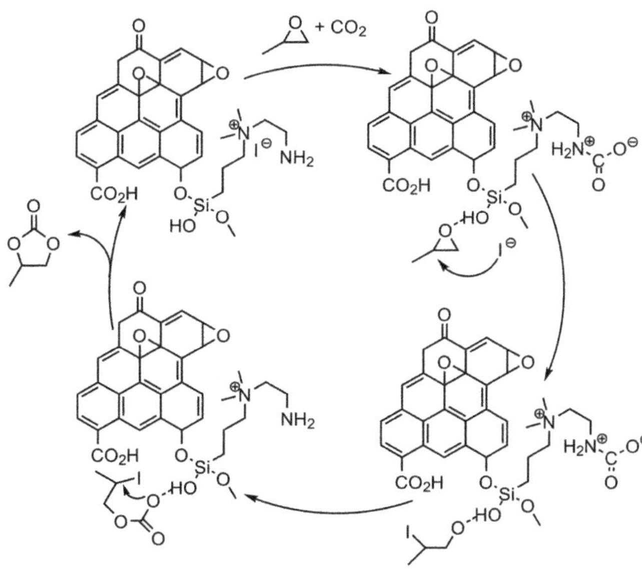

Figure 1.6 Possible reaction mechanism for the CO_2 cycloaddition to PO catalyzed by multifunctional GO. Reprinted from *Carbon*, **93**, D-H. Lan, L. Chen, C.-T. Au, S.-F. Yin, One-pot synthesized multi-functional graphene oxide as a water-tolerant and efficient metal-free heterogeneous catalyst for cycloaddition reaction, 22–31, copyright (2015), with permission from Elsevier.[105]

active, selective and stable catalyst for CO_2 cycloaddition to epoxides to give a variety of cyclic carbonates with excellent yields.[106] The authors propose a solid/liquid interfacial hydrogen-bond assistance of the hydroxyl groups in the cycloaddition mechanism. They act as weak acids for the activation of epoxides; ring opening by a I^- anion followed by CO_2 insertion/cyclization leads to cyclic carbonates. The typical (Brønsted) acidic surface properties of oxidized nanocarbons make them ideal candidates to catalyze acid-mediated transformations. Bielawski's team discovered that GO ($pK_a \sim 3$ in water) catalyzes the efficient ring opening polymerization of various cyclic lactones and lactams to the corresponding polyesters and polyamides, respectively.[107] It can also act as a solid-state initiator for a broader range of acid-catalyzed olefins' polymerization.[108] The excellent activity of GO as an acid catalyst has been also proven by Garcia *et al.* in the ring opening of epoxides using methanol as a solvent and nucleophile at room temperature.[110] On the other hand, rGO seems to be the best candidate to increase the electrochemical polymerization rate of 3-aminophenylboronic acid (ABA). Tan and Feng have systematically compared GO and its reduced counterpart in the ABA electrochemical polymerization, showing that stronger π–π interactions between the monomer and the larger graphitic network in rGO foster the electron transfer from the catalytic material to the adsorbed monomer.[109]

1.4 Tuning Chemical and Electronic Surface Properties of Carbon Nanomaterials in Catalysis *via* Tailored Chemical Functionalization Protocols

Despite the huge versatility of simply oxidized nanocarbons in catalysis and the impressive number of catalytic processes they can promote as metal-free systems, other protocols have been recently applied to the development of covalently functionalized systems with tailored surface groups.

The use of well-defined functionalization procedures meets two fundamental needs: (a) to improve the catalytic performance of the metal-free systems, maximizing the amount of surface functionalities and (b) to provide new insights on the complex structure–reactivity relationship that controls their catalytic performance. Table 1.2 lists all the catalytic transformations promoted by covalently-decorated carbon nanomaterials as metal-free catalysts and afterwards discussed in Sections 1.4.1–1.4.2.

1.4.1 Acid- and Base-catalyzed Reactions with Functionalized Nanocarbons

One of the most relevant applications of chemically functionalized nanocarbons in metal-free catalysis refers to the acidic or basic character of their outer surface. Feng and co-workers have pioneered this research area with the

Table 1.2 Reactions catalyzed by functionalized carbon nanomaterials.

Class of metal-free catalyzed reaction	Covalently linked surface groups	Dominant surface properties	Reference
Esterification and hydrolyses reactions	Aryl sulfonic acids	Acid groups	118–121
Hydroquinone alkylation with ^{t}Bu-OH	Poly(styrene sulfonic acid)		122
Transesterification reactions	Tertiary aliphatic amines	Basic groups	123–125
Knoevenagel condensation	Aliphatic amines		126 and 127
Oxygen reduction reaction (ORR)	Poly(diallyldimethyl-ammonium chloride) (PDDA)	Electronic control	128
	Aziridines		127
	Pyridines		129–131
	Pyrroles		129

development of aryl sulfonic acids chemically grafted to the surface of ordered mesoporous carbons.[118] The authors have demonstrated the effective use and recycling of these solid and highly porous materials as acid catalysts in esterification reactions for biodiesel production.[119] Following these seminal contributions, other related functionalization protocols for the chemical grafting of aryl sulfonic acids have been accomplished with the aim of preparing acidic materials for catalysis. Poly(styrene sulfonic acid)-grafted carbon nanotubes have been presented in 2010 as acidic catalysts for the model alkylation of hydroquinone with *tert*-butanol.[122] Fan and co-workers have also prepared a graphene-based catalyst by direct anchoring of sulfonic acid aryl radicals to reduced graphene.[120] Graphene surfaces undergo irreversible destruction upon direct treatment with sulfuric acid. In addition, the sulfonation reaction normally provides low yields. For these reasons, the use of aryl radicals to decorate graphene with phenyl sulfonic groups has provided an alternative and valuable approach for the preparation of the same kinds of samples. Later, Bitter's group has described a similar functionalization protocol for the surface decoration of carbon nanofibers with aryl sulfonic groups *via* aryl-diazonium salt chemistry.[121] The authors have demonstrated the remarkable catalytic performance of the as-prepared solid acid materials in the transesterification of triolein and they have also shown how this functionalization approach provides samples that outperform benchmark acid catalysts for the same process.

For nanomaterial surface decoration with basic functionalities, Tessonnier *et al.* have proposed the direct alkylation of MWCNTs with tertiary aliphatic amines for application in catalytic biomass conversion. At odds with more traditional and multi-step functionalization paths (*i.e.* MWCNTs' oxidation followed by activation of carboxylic acids as acyl chlorides and amine

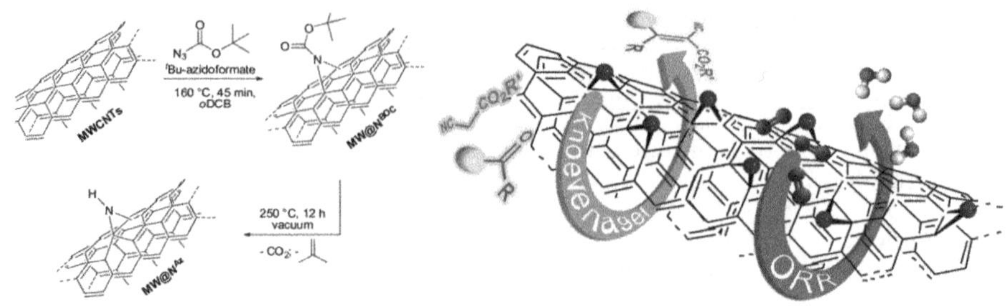

Figure 1.7 Left: [2+1] cycloaddition of ^{t}Bu-azidoformate followed by BOC thermal decomposition for the synthesis of aziridine functionalized MWCNTs. Right: aziridine decorated MWCNTs as a versatile catalytic platform for the Knoevenagel condensation and electrochemical oxygen reduction reaction (ORR). Reprinted with permission from G. Tuci, L. Luconi, A. Rossin, E. Berretti, H. Ba, M. Innocenti, D. Yakhvarov, S. Caporali, C. Pham-Huu and G. Giambastiani, *ACS Appl. Mater. Interfaces*, 2016, **8**, 30099.[127] Copyright (2016) American Chemical Society.

nucleophilic substitution), the authors have proposed an elegant and faster nanotube deprotonation followed by direct amine grafting. The as-prepared metal-free catalyst has been tested with satisfactory outcomes in the trans-esterification of glyceryl tributyrate with methanol as a model reaction for biodiesel production.[123–125]

Later, Cao and Song have reported on the amination reaction of the highly reactive fullerene (C_{60}) using organic amines.[126] The C_{60}-amine functionalized samples have been employed as stable and re-usable solid base catalysts for the Knoevenagel condensation, to give α,β-unsaturated compounds from the reaction between carbon acid compounds and a variety of aliphatic or aromatic carbonyls. Aziridine functionalized MWCNTs offer higher catalytic performance in the Knoevenagel process. Giambastiani *et al.* have proposed a two-step functionalization protocol for nanotube decoration with selected secondary (NH) aziridine groups (Figure 1.7).[127] The [2+1] nitrene cycloaddition of *tert*-butyl azidoformate, followed by a controlled thermal decomposition of the BOC protecting groups, led to the isolation of a highly basic, solid, and re-usable material for the Knoevenagel condensation with superior activity to that of related state-of-the-art N-doped and N-decorated carbon nanomaterials. Besides their basic character, the adopted functionalization protocol has also been used to modulate the surface electronic distribution thus imparting multimodality in catalysis to the as-synthesized samples. Indeed, the close proximity of the N-dopant to the sp^2 C-network creates high energy C-sites that activate small molecules for their conversion under appropriate conditions. The authors have demonstrated that aziridine functionalized MWCNTs act as excellent metal-free electrocatalysts under an alkaline environment for dioxygen activation and subsequent O_2 bond breaking within a prevalent $4e^-$ reduction process.

1.4.2 Chemically Decorated Nanocarbons with Tailored Functional Groups for the Electrochemical Oxygen Reduction Reaction (ORR)

Nitrogen-doped 1D and 2D carbon nanomaterials are known to act as valuable metal-free electrocatalysts for promoting ORR, a fundamental process at the heart of renewable energy technology.[132,133] A relatively high number of N-doped carbon nanostructures with remarkable catalytic performance in ORR have been prepared through classical *in situ* nanomaterial doping (*i.e.* chemical vapor deposition technique).[134] Much less work, though, has been done for the design of catalytically active N-decorated samples by means of *ex situ* (exohedral) organic functionalization techniques.

Dai and co-workers have discovered in 2011 that functionalization of CNTs with polyelectrolyte chains containing positively charged nitrogen moieties creates net positive charges on CNT-carbon atoms *via* intermolecular charge transfer. Such an effect was found to remarkably improve the electrocatalytic activity of the functionalized systems in the ORR, providing metal-free systems with comparable, or higher, electrocatalytic activity and long-term stability to those of commercially available Pt/C electrodes.[128] More recently, Giambastiani and collaborators have proposed the grafting of tailored N-containing heterocycles through an aryl-radical attachment (Tour functionalization protocol) to the surface of purified MWCNTs.[129] This alternative approach offers a unique model system for the elucidation of the role of different N-configurations (pyridinic, pyrrolic, quaternary nitrogen) and that of their electronic surroundings on the O_2 activation mechanism.[130] Giambastiani and Pham-Huu have also demonstrated the existence of a rational trend (coherent with Sabatier's principle) between the electronic charge density distribution at the grafted N-heterocycles and their final control on the overpotential value at which the ORR starts.[131] This trend is conveniently summarized with a volcano plot representation for a quick visual identification of the most suitable N-containing group for the process (Figure 1.8).

Finally, the same group has described the use of a [2+1] nitrene cycloaddition for the preparation of aziridine-decorated carbon-based systems as effective metal-free electrocatalysts for the ORR in an alkaline environment (Section 1.4.1).[127]

1.5 Conclusion

We have provided a snapshot of covalent synthesis techniques for metal-free hetero-doped carbocatalysts. Post-synthetic modifications of graphene, fullerene, and carbon nanotubes using scalable liquid and gas phase techniques have given access to a range of materials active for traditional petrochemical reactions (oxidative dehydrogenation, desulfurization, *etc.*), fine chemicals synthesis (nitroarene reduction, Michael and Friedel–Crafts reactions, *etc.*), and electrocatalysis (OER, ORR). The grafting of desired moieties

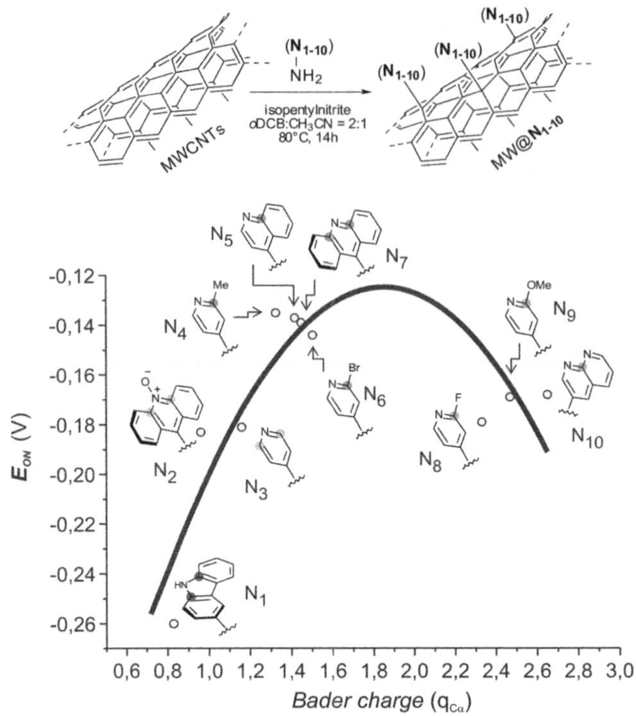

Figure 1.8 Top: MWCNT N-decoration *via* aryldiazonium salt (Tour) protocol.[129,131] Bottom: a volcano plot representation for the "just right" interaction between O_2 and tailored N-groups at the outer surface of chemically decorated MWCNTs engaged in the ORR in an alkaline environment. Reproduced with permission from *CHIMIA*, 2017, **71**, 568.[135] Copyright Swiss Chemical Society.

using C–C coupling reactions has allowed a better understanding of the nature of the active species, notably for electrocatalysis. Future advances in the field will rely on progress in controlling the nature, concentration, and distribution of heteroatom-containing moieties on carbon surfaces using scalable synthesis techniques. The deviation of carbo-catalyzed processes from conventional transition metal- and metal oxide-based catalysis will also require new insights from theory to understand the fundamental steps involved in these new conversions.

Acknowledgements

The Italian MIUR through the PRIN 2015 Project SMARTNESS (2015K7FZLH) “Solar driven chemistry: new materials for photo- and electro-catalysis” is gratefully acknowledged for financial support. GT thanks Fondazione “Ente Cassa di Risparmio di Firenze” for its support to the HORIZON project. This material is also based upon work supported in part by the National Science

Foundation under grant number EEC-0813570. The work is performed within the Russian Government Program of Competitive Growth of Kazan Federal University.

References

1. J. Donnet, *Carbon*, 1968, **6**, 161–176.
2. H. P. Boehm, E. Diehl, W. Heck and R. Sappok, *Angew. Chem., Int. Ed.*, 1964, **3**, 669–677.
3. R. W. Coughlin, *Ind. Eng. Chem. Prod. Res. Dev.*, 1969, **8**, 12–23.
4. R. Haddon, *Science*, 1993, **261**, 1545–1550.
5. S. Niyogi, M. Hamon, H. Hu, B. Zhao, P. Bhowmik, R. Sen, M. Itkis and R. Haddon, *Acc. Chem. Res.*, 2002, **35**, 1105–1113.
6. R. Haddon, *J. Am. Chem. Soc.*, 1990, **112**, 3385–3389.
7. R. C. Haddon, *Acc. Chem. Res.*, 1988, **21**, 243–249.
8. M. T. Lusk and L. D. Carr, *Phys. Rev. Lett.*, 2008, **100**, 175503.
9. P. Allemand, A. Koch, F. Wudl, Y. Rubin, F. Diederich, M. Alvarez, S. Anz and R. Whetten, *J. Am. Chem. Soc.*, 1991, **113**, 1050–1051.
10. R. Haddon, R. Palmer, H. Kroto and P. Sermon, *Philos. Trans. R. Soc., A*, 1993, **343**, 53–62.
11. K. Suenaga, H. Wakabayashi, M. Koshino, Y. Sato, K. Urita and S. Iijima, *Nat. Nanotechnol.*, 2007, **2**, 358–360.
12. S. M. Lee, Y. H. Lee, Y. G. Hwang, J. R. Hahn and H. Kang, *Phys. Rev. Lett.*, 1999, **82**, 217–220.
13. D. Bom, R. Andrews, D. Jacques, J. Anthony, B. Chen, M. S. Meier and J. P. Selegue, *Nano Lett.*, 2002, **2**, 615–619.
14. P. Ajayan, T. Ebbesen, T. Ichihashi, S. Iijima, K. Tanigaki and H. Hiura, *Nature*, 1993, **362**, 522–525.
15. S. C. Tsang, P. J. F. Harris and M. L. H. Green, *Nature*, 1993, **362**, 520–522.
16. L. R. Radovic, A. B. Silva-Tapia and F. Vallejos-Burgos, *Carbon*, 2011, **49**, 4218–4225.
17. L. R. Radovic, A. Suarez, F. Vallejos-Burgos and J. O. Sofo, *Carbon*, 2011, **49**, 4226–4238.
18. J.-P. Tessonnier and D. S. Su, *ChemSusChem*, 2011, **4**, 824–847.
19. B. C. Brodie, *Philos. Trans. R. Soc. London*, 1859, **149**, 249–259.
20. L. Staudenmaier, *Eur. J. Inorg. Chem.*, 1898, **31**, 1481–1487.
21. W. S. Hummers Jr and R. E. Offeman, *J. Am. Chem. Soc.*, 1958, **80**, 1339.
22. X. Liu and L. Dai, *Nat. Rev. Mater.*, 2016, **1**, 16064.
23. P. Tang, G. Hu, M. Li and D. Ma, *ACS Catal.*, 2016, **6**, 6948–6958.
24. D. S. Su, G. Wen, S. Wu, F. Peng and R. Schlögl, *Angew. Chem., Int. Ed.*, 2017, **56**, 936–964.
25. J. L. Figueiredo and M. F. R. Pereira, *Catal. Today*, 2010, **150**, 2.
26. J. Bedia, R. Barrionuevo, J. Rodríguez-Mirasol and T. Cordero, *Appl. Catal., B*, 2011, **103**, 302.
27. E. Raymundo-Piñero, D. Cazorla-Amorós and A. Linares-Solano, *Carbon*, 2003, **41**, 1925.

28. R. S. Ribeiro, A. M. T. Silva, J. L. Figueiredo, J. L. Faria and H. T. Gomes, *Carbon*, 2013, **62**, 97.
29. B. Stöhr, H. P. Boehm and R. Schlögl, *Carbon*, 1991, **29**, 707.
30. M. Yudasaka, R. Kikuchi, Y. Ohki and S. Yoshimura, *Carbon*, 1997, **35**, 195–201.
31. M. Reyes-Reyes, N. Grobert, R. Kamalakaran, T. Seeger, D. Golberg, M. Rühle, Y. Bando, H. Terrones and M. Terrones, *Chem. Phys. Lett.*, 2004, **396**, 167–173.
32. M. Terrones, H. Terrones, N. Grobert, W. Hsu, Y. Zhu, J. Hare, H. Kroto, D. Walton, P. Kohler-Redlich and M. Rühle, *Appl. Phys. Lett.*, 1999, **75**, 3932–3934.
33. C. J. Lee, S. C. Lyu, H.-W. Kim, J. H. Lee and K. I. Cho, *Chem. Phys. Lett.*, 2002, **359**, 115–120.
34. C. Chen, J. Zhang, B. Zhang, C. Yu, F. Peng and D. Su, *Chem. Commun.*, 2013, **49**, 8151–8153.
35. K. Gong, F. Du, Z. Xia, M. Durstock and L. Dai, *Science*, 2009, **323**, 760–764.
36. S. Kundu, T. C. Nagaiah, W. Xia, Y. Wang, S. V. Dommele, J. H. Bitter, M. Santa, G. Grundmeier, M. Bron, W. Schuhmann and M. Muhler, *J. Phys. Chem. C*, 2009, **113**, 14302–14310.
37. L. Qu, Y. Liu, J.-B. Baek and L. Dai, *ACS Nano*, 2010, **4**, 1321–1326.
38. L. Yang, S. Jiang, Y. Zhao, L. Zhu, S. Chen, X. Wang, Q. Wu, J. Ma, Y. Ma and Z. Hu, *Angew. Chem., Int. Ed.*, 2011, **50**, 7132–7135.
39. J. Campos-Delgado, I. O. Maciel, D. A. Cullen, D. J. Smith, A. Jorio, M. A. Pimenta, H. Terrones and M. Terrones, *ACS Nano*, 2010, **4**, 1696–1702.
40. E. Cruz-Silva, D. A. Cullen, L. Gu, J. M. Romo-Herrera, E. Muñoz-Sandoval, F. López-Urías, B. G. Sumpter, V. Meunier, J.-C. Charlier, D. J. Smith, H. Terrones and M. Terrones, *ACS Nano*, 2008, **2**, 441–448.
41. Y. Cao, H. Yu, J. Tan, F. Peng, H. Wang, J. Li, W. Zheng and N.-B. Wong, *Carbon*, 2013, **57**, 433–442.
42. L. Panchakarla, K. Subrahmanyam, S. Saha, A. Govindaraj, H. Krishnamurthy, U. Waghmare and C. Rao, *Adv. Mater.*, 2009, **21**, 4726–4730.
43. Y. Ito, W. Cong, T. Fujita, Z. Tang and M. Chen, *Angew. Chem., Int. Ed.*, 2015, **54**, 2131–2136.
44. W. Xia, C. Jin, S. Kundu and M. Muhler, *Carbon*, 2009, **47**, 919–922.
45. S. Kundu, Y. Wang, W. Xia and M. Muhler, *J. Phys. Chem. C*, 2008, **112**, 16869–16878.
46. M. Pereira, J. Orfao and J. Figueiredo, *Appl. Catal., A*, 1999, **184**, 153–160.
47. R. Arrigo, M. Hävecker, R. Schlögl and D. S. Su, *Chem. Commun.*, 2008, 4891–4893.
48. X. Li, H. Wang, J. T. Robinson, H. Sanchez, G. Diankov and H. Dai, *J. Am. Chem. Soc.*, 2009, **131**, 15939–15944.
49. R. Arrigo, M. Hävecker, S. Wrabetz, R. Blume, M. Lerch, J. McGregor, E. P. J. Parrott, J. A. Zeitler, L. F. Gladden, A. Knop-Gericke, R. Schlögl and D. S. Su, *J. Am. Chem. Soc.*, 2010, **132**, 9616–9630.

50. K. McGuire, N. Gothard, P. L. Gai, M. S. Dresselhaus, G. Sumanasekera and A. M. Rao, *Carbon*, 2005, **43**, 219–227.
51. Z.-H. Sheng, H.-L. Gao, W.-J. Bao, F.-B. Wang and X.-H. Xia, *J. Mater. Chem.*, 2012, **22**, 390–395.
52. F. Liang, J. M. Beach, P. K. Rai, W. Guo, R. H. Hauge, M. Pasquali, R. E. Smalley and W. E. Billups, *Chem. Mater.*, 2006, **18**, 1520–1524.
53. A. Mukherjee, J. Kang, O. Kuznetsov, Y. Sun, R. Thaner, A. S. Bratt, J. R. Lomeda, K. F. Kelly and W. E. Billups, *Chem. Mater.*, 2011, **23**, 9–13.
54. B. Frank, J. Zhang, R. Blume, R. Schlögl and D. S. Su, *Angew. Chem., Int. Ed.*, 2009, **48**, 6913–6917.
55. J. Long, X. Xie, J. Xu, Q. Gu, L. Chen and X. Wang, *ACS Catal.*, 2012, **2**, 622–631.
56. S. Park, Y. Hu, J. O. Hwang, E.-S. Lee, L. B. Casabianca, W. Cai, J. R. Potts, H.-W. Ha, S. Chen, J. Oh, S. O. Kim, Y.-H. Kim, Y. Ishii and R. S. Ruoff, *Nat. Commun.*, 2012, **3**, 638–645.
57. W.-J. Jiang, J.-S. Hu, X. Zhang, Y. Jiang, B.-B. Yu, Z.-D. Wei and L.-J. Wan, *J. Mater. Chem. A*, 2014, **2**, 10154–10160.
58. H. Fei, R. Ye, G. Ye, Y. Gong, Z. Peng, X. Fan, E. L. G. Samuel, P. M. Ajayan and J. M. Tour, *ACS Nano*, 2014, **8**, 10837–10843.
59. P. Chen, T.-Y. Xiao, Y.-H. Qian, S.-S. Li and S.-H. Yu, *Adv. Mater.*, 2013, **25**, 3192–3196.
60. Y. Zhao, F. Zhao, X. Wang, C. Xu, Z. Zhang, G. Shi and L. Qu, *Angew. Chem., Int. Ed.*, 2014, **53**, 13934–13939.
61. Z.-H. Sheng, L. Shao, J.-J. Chen, W.-J. Bao, F.-B. Wang and X.-H. Xia, *ACS Nano*, 2011, **5**, 4350–4358.
62. I. Kruusenberg, S. Ratso, M. Vikkisk, P. Kanninen, T. Kallio, A. M. Kannan and K. Tammeveski, *J. Power Sources*, 2015, **281**, 94–102.
63. H. Ba, Y. Liu, L. Truong-Phuoc, C. Duong-Viet, X. Mu, W. H. Doh, T. Tran-Thanh, W. Baaziz, L. Nguyen-Dinh, J.-M. Nhut, I. Janowska, D. Begin, S. Zafeiratos, P. Granger, G. Tuci, G. Giambastiani, F. Banhart, M. J. Ledoux and C. Pham-Huu, *Chem. Commun.*, 2015, **51**, 14393–14396.
64. H. Ba, J. Luo, Y. Liu, C. Duong-Viet, G. Tuci, G. Giambastiani, J.-M. Nhut, L. Nguyen-Dinh, O. Ersen, D. S. Su and C. Pham-Huu, *Appl. Catal., B*, 2017, **200**, 343–350.
65. H. Ba, Y. Liu, L. Truong-Phuoc, C. Duong-Viet, J.-M. Nhut, D. L. Nguyen, O. Ersen, G. Tuci, G. Giambastiani and C. Pham-Huu, *ACS Catal.*, 2016, **6**, 1408–1419.
66. H. Jiang, Y. Zhu, Y. Su, Y. Yao, Y. Liu, X. Yang and C. Li, *J. Mater. Chem. A*, 2015, **3**, 12642–12645.
67. D. J. Scott, T. R. Simmons, E. J. Lawrence, G. G. Wildgoose, M. J. Fuchter and A. E. Ashley, *ACS Catal.*, 2015, **5**, 5540–5544.
68. D. W. Stephan, *J. Am. Chem. Soc.*, 2015, **137**, 10018–10032.
69. D. W. Stephan, *Acc. Chem. Res.*, 2015, **48**, 306–316.
70. A. Primo, F. Neatu, M. Florea, V. Parvulescu and H. Garcia, *Nat. Commun.*, 2014, **5**, 5291.

71. M. M. Trandafir, M. Florea, F. Neaţu, A. Primo, V. I. Parvulescu and H. García, *ChemSusChem*, 2016, **9**, 1565–1569.
72. X. Sun, B. Li, T. Liu, J. Song and D. S. Su, *Phys. Chem. Chem. Phys.*, 2016, **18**, 11120–11124.
73. M. Melchionna, S. Marchesan, M. Prato and P. Fornasiero, *Catal. Sci. Technol.*, 2015, **5**, 3859.
74. J. L. Figueiredo and M. F. R. Pereira, in *Carbon Materials for Catalysis*, ed. P. Serp and J. L. Figueiredo, John Wiley & Sons, 2008.
75. *Carbon Nanotubes and Related Structures, Synthesis, Characterization, Functionalization, and Applications*, ed. M. Guldi and N. Martin, Wiley-VCH Velag GmbH & Co. KGaA, Germany, 2010.
76. V. K. Thakur and M. K. Thakur, *Chemical Functionalization of Carbon Nanomaterials: Chemistry and Applications*, CRC Press, Boca Raton (FL, USA), 2015.
77. G. Giambastiani, S. Cicchi, A. Giannasi, L. Luconi, A. Rossin, F. Mercuri, C. Bianchini, A. Brandi, M. Melucci, G. Ghini, P. Stagnaro, L. Conzatti, E. Passaglia, M. Zoppi, T. Montini and P. Fornasiero, *Chem. Mater.*, 2011, **23**, 1923.
78. P. Cãnete-Rosales, V. Ortega, A. Álvarez-Lueje, S. Bollo, M. González, A. Ansón and M. T. Martínez, *Electrochim. Acta*, 2012, **62**, 163.
79. D. R. Dreyer, H.-P. Jia and C. W. Bielawski, *Angew. Chem., Int. Ed.*, 2010, **49**, 6813.
80. B. H. R. Suryanto and C. Zhao, *Chem. Commun.*, 2016, **52**, 6439.
81. B. H. R. Suryanto and C. Zhao, *RSC Adv.*, 2016, **6**, 78403.
82. J.-H. Yang, G. Sun, Y. Gao, H. Zhao, P. Tang, J. Tan, A.-H. Lu and D. Ma, *Energy Environ. Sci.*, 2013, **6**, 793.
83. C. L. Su, M. Acik, K. Takai, J. Lu, S. J. Hao, Y. Zheng, P. P. Wu, Q. L. Bao, T. Enoki, Y. J. Chabal and K. P. Loh, *Nat. Commun.*, 2012, **3**, 1298.
84. Q. Gu, G. Wen, Y. Ding, K.-H. Wu, C. Chenc and D. S. Su, *Green Chem.*, 2017, **19**, 1175.
85. J. Zhang, X. Liu, R. Blume, A. H. Zhang, R. Schlogl and D. S. Su, *Science*, 2008, **322**, 73.
86. X. Guo, W. Qi, W. Liu, P. Yan, F. Li, C. Liang and D. S. Su, *ACS Catal.*, 2017, **7**, 1424.
87. S. Yang, Y. Cui, Y. Sun and H. Yang, *J. Hazard. Mater.*, 2014, **280**, 55.
88. J. Wang, W. Fu, X. He, S. Yang and W. Zhu, *J. Environ. Sci.*, 2014, **26**, 1741.
89. J. Li and G. Zheng, *Adv. Sci.*, 2017, **4**, 1600380.
90. X. Lu, W.-L. Yim, B. H. R. Suryanto and C. Zhao, *J. Am. Chem. Soc.*, 2015, **137**, 2901.
91. Y. Gao, D. Ma, C. Wang, J. Guan and X. Bao, *Chem. Commun.*, 2011, **47**, 2432.
92. S. Wu, G. Wen, R. Schlőgl and D. S. Su, *Phys. Chem. Chem. Phys.*, 2015, **17**, 1567.
93. S. Wu, G. Wen, X. Liu, B. Zhong and D. S. Su, *ChemCatChem*, 2014, **6**, 1558.
94. T.-F. Yeh, J.-M. Syu, C. Cheng, T.-H. Chang and H. Teng, *Adv. Funct. Mater.*, 2010, **20**, 2255.

95. W. Cui, Q. Liu, N. Cheng, A. M. Asiri and X. Sun, *Chem. Commun.*, 2014, **50**, 9340.
96. X. Wang, C. Ouyang, S. Dou, D. Liu and S. Wang, *RSC Adv.*, 2015, **5**, 41901.
97. Y. Zhang, C. Chen, L. Peng, Z. Ma, Y. Zhang, H. Xia, A. Yang, L. Wang, D. S. Su and J. Zhang, *Nano Res.*, 2015, **8**, 502.
98. A. V. Kumar and K. R. Rao, *Tetrahedron Lett.*, 2011, **52**, 5188.
99. F. Hu, M. Patel, F. Luo, C. Flach, R. Mendelsohn, E. Garfunkel, H. He and M. Szostak, *J. Am. Chem. Soc.*, 2015, **137**, 14473.
100. F. Niu, J. Wu, L. Zhang, P. Li, J. Zhu, Z. Wu, C. Wang and W. Song, *ACS Catal.*, 2011, **1**, 1158.
101. Y. Kim, S. Somez and H. Lee, *Chem. Commun.*, 2013, **49**, 5702.
102. S. Verma, H. P. Mungse, N. Kumar, S. Choudhary, S. L. Jain, B. Sain and O. P. Khatri, *Chem. Commun.*, 2011, **47**, 12673.
103. Y. Gao, P. Tang, H. Zhou, W. Zhang, H. Yang, N. Yan, G. Hu, D. Mei, J. Wang and D. Ma, *Angew. Chem., Int. Ed.*, 2016, **55**, 3124.
104. D.-H. Lan, F.-M. Yang, S.-L. Luo, C.-T. Au and S.-F. Yin, *Carbon*, 2014, **73**, 351.
105. D.-H. Lan, L. Chen, C.-T. Au and S.-F. Yin, *Carbon*, 2015, **93**, 22.
106. Y.-B. Sun, C.-Y. Cao, S.-L. Yang, P.-P. Huang, C.-R. Wang and W.-G. Song, *Chem. Commun.*, 2014, **50**, 10307.
107. D. R. Dreyer, K. A. Jarvis, P. J. Ferreira and C. W. Bielawski, *Polym. Chem.*, 2012, **3**, 757.
108. D. R. Dreyer and C. W. Bielawski, *Adv. Funct. Mater.*, 2012, **22**, 3247.
109. L. Tan, B. Wang and H. Feng, *RSC Adv.*, 2013, **3**, 2561.
110. A. Dhakshinamoorthy, M. Alvaro, P. Concepción, V. Fornés and H. Garcia, *Chem. Commun.*, 2012, **48**, 5443.
111. S. Stankovich, D. A. Dikin, R. D. Piner, K. A. Kohlhaas, A. Kleinhammes, Y. Jia, Y. Wu, S. T. Nguyen and R. S. Ruoff, *Carbon*, 2007, **45**, 1558.
112. D. Li, M. B. Muller, S. Gilje, R. B. Kaner and G. G. Wallace, *Nat. Nanotechnol.*, 2008, **3**, 101–105.
113. W. Qi and D. S. Su, *ACS Catal.*, 2014, **4**, 3212.
114. B. H. R. Suryanto, S. Chen, J. Duan and C. Zhao, *ACS Appl. Mater. Interfaces*, 2016, **8**, 35513.
115. Y. Xu, M. Kraft and R. Xu, *Chem. Soc. Rev.*, 2016, **45**, 3039.
116. C. Hu and L. Dai, *Angew. Chem., Int. Ed.*, 2016, **55**, 11736.
117. K. Kokubo, K. Matsubayashi, H. Tategaki, H. Takada and T. Oshima, *ACS Nano*, 2008, **2**, 327.
118. X. Wang, R. Liu, M. M. Waje, Z. Chen, Y. Yan, K. N. Bozhilov and P. Feng, *Chem. Mater.*, 2007, **19**, 2395.
119. R. Liu, X. Wang, X. Zhao and P. Feng, *Carbon*, 2008, **46**, 1664.
120. J. Ji, G. Zhang, H. Chen, S. Wang, G. Zhang, F. Zhang and X. Fan, *Chem. Sci.*, 2011, **2**, 484.
121. D. R. Stellwagen, F. van der Klis, D. S. van Es, K. P. de Jong and J. H. Bitter, *ChemSusChem*, 2013, **6**, 1668.
122. K. Liu, C. Li, X. Zhang, W. Hua, D. Yang, J. Hu, Y. Yue and Z. Gao, *Catal. Commun.*, 2010, **12**, 217.

123. J.-P. Tessonnier, A. Villa, O. Majoulet, D. S. Su and R. Schlőgl, *Angew. Chem., Int. Ed.*, 2009, **48**, 6543.
124. A. Villa, J.-P. Tessonnier, O. Majoulet, D. S. Su and R. Schlőgl, *ChemSusChem*, 2010, **3**, 241.
125. A. Villa, J.-P. Tessonnier, O. Majoulet, D. S. Su and R. Schlőgl, *Chem. Commun.*, 2009, 4405.
126. Y. Sun, C. Cao, P. Huang, S. Yang and W. Song, *RSC Adv.*, 2015, **5**, 86082.
127. G. Tuci, L. Luconi, A. Rossin, E. Berretti, H. Ba, M. Innocenti, D. Yakhvarov, S. Caporali, C. Pham-Huu and G. Giambastiani, *ACS Appl. Mater. Interfaces*, 2016, **8**, 30099.
128. S. Wang, D. Yu and L. Dai, *J. Am. Chem. Soc.*, 2011, **133**, 5182.
129. G. Tuci, C. Zafferoni, P. D'Ambrosio, S. Caporali, M. Ceppatelli, A. Rossin, T. Tsoufis, M. Innocenti and G. Giambastiani, *ACS Catal.*, 2013, **3**, 2108.
130. G. Tuci, C. Zafferoni, A. Rossin, L. Luconi, A. Milella, M. Ceppatelli, M. Innocenti, Y. Liu, C. Pham-Huu and G. Giambastiani, *Catal. Sci. Technol.*, 2016, **6**, 6226.
131. G. Tuci, C. Zafferoni, A. Rossin, A. Milella, L. Luconi, M. Innocenti, L. Truong Phuoc, C. Duong-Viet, C. Pham-Huu and G. Giambastiani, *Chem. Mater.*, 2014, **26**, 3460.
132. L. Dai, Y. Xue, L. Qu, H.-J. Choi and J.-B. Baek, *Chem. Rev.*, 2015, **115**, 4823.
133. M. Zhou, H.-L. Wang and S. Guo, *Chem. Soc. Rev.*, 2016, **45**, 1273.
134. K.-H. Wu, D.-W. Wang, D.-S. Su and I. R. Gentle, *ChemSusChem*, 2015, **8**, 2772.
135. G. Tuci, L. Luconi, A. Rossin and G. Giambastiani, *Chimia*, 2017, **71**, 568–572.

CHAPTER 2

Non-covalent Methodologies for the Preparation of Metal-free Nanocarbons for Catalysis

ANASTASIOS STERGIOU[a], NIKOLAOS KAROUSIS[a] AND DIMITRIOS TASIS*[b]

[a]Theoretical and Physical Chemistry Institute, National Hellenic Research Foundation, 48 Vasileos Constantinou Avenue, 11635, Athens, Greece; [b]Department of Chemistry, University of Ioannina, 45110, Ioannina, Greece
*E-mail: dtassis@cc.uoi.gr

2.1 Introduction

Carbon nanotubes (CNTs), graphenes, carbon dots, nanodiamonds and fullerene buckyballs are dimensionally confined sp^2/sp^3 hybridized carbon allotropes which constitute the nanocarbon family. The exotic and often unprecedented intrinsic properties of nanocarbons (structural, conductive, electrocatalytic) have garnered tremendous interest for both fundamental and applied research. These properties are further coupled with a high surface area, high optical transmittance, curvature-dependent chemical reactivity and thermal stability. All properties are dependent upon a variety of factors such as C–C bond hybridization, nanocarbon size, dimensionality, layer number, density/type of defect sites and synthesis strategy. For example, multiwalled carbon nanotubes (MWCNTs) produced by arc-discharge

Catalysis Series No. 31
Metal-free Functionalized Carbons in Catalysis: Synthesis, Characterization and Applications
Edited by Alberto Villa and Nikolaos Dimitratos

Published by the Royal Society of Chemistry, www.rsc.org

have a more graphitic structure than the corresponding material prepared by a chemical vapor deposition process. Concerning the effect of defect sites, graphene oxide (GO) (a single layer of highly oxidized graphene) possesses a high population of sp^3-hybridized carbon atoms and a corrugated morphology. The presence of such defect sites significantly increases electrical resistance, making the nanomaterial an actual insulator. This behavior is in sharp contrast with the conductive properties of pristine graphene, in which the basal plane consists solely of sp^2-hybridized carbon atoms, thus permitting the electron cloud to be delocalized onto the graphitic lattice. Depending on the targeted application, it is desirable that some intrinsic properties of the nanocarbon are kept as in the parent form, whereas another family of properties should be altered in a beneficial optimization. This is a key strategy to enhance the performance and the multifunctional character of nanocarbons. For example, in energy conversion/storage applications, the carbon nanostructures should possess high electrical conductivity combined with optimized electrocatalytic properties for enhanced charge transfer interaction with their chemical environment. The performance, as well as the number of potential applications, of nanocarbons can be further increased by combining them with an additional component, thus forming a composite material. Many nanocarbon hybrid materials display enhanced and even new properties in relation to their individual components; these are commonly attributed to synergistic effects between the two components. Fabrication of nanocarbon hybrid structures can be achieved by surface modification in two general ways: (i) covalent attachment of functionalities through oxidation/addition reactions onto a graphitic network or (ii) physical adsorption of substances onto a nanostructured carbon surface through hydrophobic interactions. The latter approach of non-covalent modification mostly relies on stacking interactions between the adsorbates and carbon surfaces. Common chemical structures that exhibit stacking interactions with nanocarbon surfaces include pyrenes, macrocyclic complexes (for example, porphyrins), surfactants, polymers and metal-based nanoparticles among others. This chapter is dedicated to the synthesis of metal-free carbon-based hybrid nanostructures, obtained by non-covalent functionalization schemes. We will first present an overview of the various synthesis strategies of nanocarbon hybrids for each class of carbon allotropes. This includes physical adsorption of substances in a liquid environment through hydrophobic interactions, *in situ* processes, utilization of sacrificial templates, heteroatom doping and others. The hybrid nanomaterials, which will be discussed in this chapter, are utilized as multifunctional components in (electro)catalytic processes. Such nanocarbon hybrids show promise in a range of key applications, with the most promising ones highlighted in this chapter, including energy conversion, (photo)catalysis, energy storage, and sensing. For each application, we will discuss current challenges in the respective field and their potential solution by using nanocarbon hybrids and present some of the most intriguing examples.

2.2 Graphene

The functionalization of two-dimensional carbon nanostructures is considered as an important step for improving their dispersibility, self-assembly properties and integration in devices. At least, by comparison with one-dimensional nanomaterials such as single-walled CNTs, graphene has a larger specific surface area. This comes from the morphology of graphene-based nanostructures, which interact with the chemical environment with both sides of the graphitic surface. This leads to enhanced aggregation phenomena, due to π–π interactions of electron clouds between adjacent graphene sheets. Partial alteration of such weak interactions, followed by exfoliation of graphene in liquid media, may be succeeded through the combination of shear forces (sonication) and non-covalent functionalization of the graphene surface. Alternatively, chemical oxidation of graphite flakes towards graphene oxide (GO) or covalent functionalization through addition reactions have been widely studied.[1] As in the case of other carbon nanomaterials, the advantage of a physical adsorption strategy is clear, due to the preservation of the structural integrity of the graphitic lattice. Below, several synthetic protocols will be discussed, which involve the development of metal-free graphene-based (electro)catalytic systems by a physical adsorption approach. The studies discussed have been separated into four different families of synthetic approach. These include the non-covalent functionalization of graphene sheets by either low molecular weight substances, polymers, *in situ* processes or utilization of sacrificial templates.

2.2.1 Functionalization *via* Physisorption

2.2.1.1 Small Molecules

In the seminal work of Shi and co-workers, non-covalent functionalization of graphene sheets was addressed by π–π stacking interactions. Stable aqueous dispersions of graphene sheets were prepared by using a water-soluble pyrene derivative, 1-pyrenebutyrate (PB), as a stabilizer.[2] The mixing of the starting graphitic nanostructure, graphene oxide, with the oligonuclear derivative took place in an alkaline aqueous environment, in order for the carboxylic group to be deprotonated. The negatively charged stabilizer molecules adsorbed onto the graphitic lattice greatly improved the exfoliation efficiency of graphene sheets in an aqueous suspension, due to electrostatic repulsion forces. Partial graphitization of the defected graphene oxide sheets was carried out by chemical reduction with hydrazine, yielding reduced graphene oxide (RGO) sheets stabilized by adsorbed pyrene butyrate molecules. Such hybrid nanostructures were evaluated as functional counter electrodes (CEs) in dye-sensitized solar cells (DSSCs). The photovoltaic efficiency (2.2%) was almost half of that of a reference Pt electrode.

Non-covalently functionalized graphene sheets have been used as host components for the immobilization of enzymes towards the fabrication

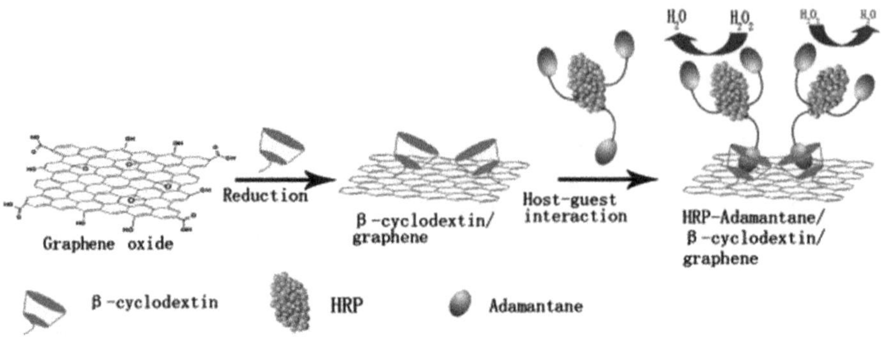

Figure 2.1 Schematic representation of a HRP-adamantane/β-cyclodextrin/graphene hybrid. Reprinted from *Biosensors and Bioelectronics*, **45**, L.-M. Lu, X.-L. Qiu, X.-B. Zhang, G.-L. Shen, W. Tan and R.-Q. Yu, Supramolecular assembly of enzyme on functionalized graphene for electrochemical biosensing, 102–107, copyright 2013, with permission from Elsevier.

of electrochemical biosensors.[3] Physical adsorption of β-cyclodextrin onto graphene sheets took place during the reduction of graphene oxide flakes in an alkaline hydrazine solution (Figure 2.1). The cavities of the physically adsorbed cyclodextrin substances were hosts for adamantane moieties during the self-assembly of a cyclodextrin-functionalized graphene and adamantane-modified horseradish peroxidase (ADA-HRP) hybrid. Such a host–guest supramolecular interaction yielded novel nanostructures in an aqueous solution. This concept of developing electrochemical platforms towards the reduction of hydrogen peroxide was adopted in a previous study by Song and co-workers in which the adsorbed cyclodextrin spacer functionalities were replaced by single-stranded DNA chains.[4] Both studies clearly revealed that such nanocomposites may act as efficient platforms for the immobilization of redox substances, leading to the development of electrochemical sensors.

By using an analogous protocol of graphene oxide-compatibilizer mixing and subsequent reduction by $NaBH_4$, tridodecylmethylammonium chloride-functionalized reduced graphene oxide (TDMAC/RGO) nanostructures were synthesized.[5] Adsorption of TDMAC molecules made the surface of RGO positively charged, leading to efficient dispersion *via* electrostatic repulsion among individual graphene sheets. The resultant functionalized graphene sheets have been shown to impart electrocatalytic activity as metal-free catalysts for the oxygen reduction reaction (ORR) in fuel cells (FCs). The mechanism of the ORR was confirmed to follow a four-electron pathway, leading to the formation of water. In addition, the hybrid exhibited excellent tolerance towards methanol and ethanol.

Using a slightly different synthesis protocol than the previous cases, Mou *et al.* have firstly reduced graphene oxide sheets by hydrazine in an alkaline environment.[6] In a subsequent step, chemically reduced graphene oxide was non-covalently functionalized by a xanthene-based dye, Eosin Y (EY). The formation of a stable aqueous EY-functionalized graphene (EY-RGO) suspension was suggested to be generated due to the non-covalent interaction between EY and the RGO surface *via* hydrogen bonding and π–π stacking interactions. The hybrid nanostructure was considered as a functional photocatalyst for hydrogen generation from water, under irradiation at either ultraviolet or visible wavelengths. The total amount of H_2 evolved from the EY-RGO system was higher than the one produced from the RGO, GO and EY-GO systems. The higher hydrogen evolution could be attributed to the broad absorption spectrum of the EY-RGO hybrid and the efficient electron transfer from photoexcited EY to RGO.

In an analogous concept, oligo(*p*-phenylenevinylene)s (OPVs) have been used as efficient dispersants of reduced graphene oxide in organic media, by non-covalent π–π interactions.[7] Each benzene ring is decorated with two hexadecyloxy functionalities in order to increase the solubility in organic media. Upon interacting with such a π-gelator, the dispersing ability of the RGO increased significantly in most non-polar and polar aprotic solvents, when compared to the unmodified nanostructure. The content of RGO suspensions was about 1.8 mg ml^{-1} in solvents such as tetrahydrofuran, 1,2-dichlorobenzene and toluene. Enhanced electrochemical performance was obtained by the hybrid containing 30 wt% RGO, which demonstrated high potential as an electrode material in supercapacitors (specific capacitance of 181 F g^{-1}).

2.2.1.2 Polymers

Besides the utilization of low molecular weight substances as physically adsorbed dopants of graphene, macromolecular systems have been widely used in energy-related applications. The main role of the macromolecular component within the graphene-based hybrid structure is to induce charge transfer phenomena with the graphitic nanostructure. Furthermore, photovoltaic efficiencies and/or electrocatalytic performances are greatly enhanced in cases where graphene and a conjugated polymer are integrated within the electrode material. The enhancement of the electrocatalytic performance of hybrids in the form of thin membranes is due to the high conductivity and the optimum film-forming ability of the polymer component, which far exceed that of neat graphene nanostructures. An additional reason for the integration of polymeric systems within the membrane-like electrode is that the macromolecular chains may be thermally decomposed and generate porous structures with high specific surface areas. The latter parameter is very important for electrocatalytic interactions between components. In most of the cases in the literature, two scenarios are met regarding the synthesis protocols in the liquid phase. Nevertheless, there are a few sophisticated processing approaches, which will be discussed in parallel.

The first scenario involves the stabilization of graphene oxide flakes in the presence of a macromolecular stabilizer and the subsequent chemical reduction process of the graphitic nanostructure. It is noted that the macromolecular system should possess a dual role, specifically to be both stabilizer and electrocatalytic component. Typical polycationic polymers, such as poly(diallyldimethylammonium chloride) (PDDA), have electron with-drawing properties inducing electron transfer from graphene to the polymer, leading to the formation of delocalized positive charges onto graphene's surface. Such hybrids have been used as potential electrocatalysts in the ORR, which proceeds by a four-electron pathway.[8] An analogous example involved the solution mixing of graphene oxide flakes with a conducting graft copolymer [poly(styrenesulfonic acid-*g*-pyrrole) (PSSA-g-PPY)] and the subsequent reduction by hydrazine in the solution phase.[9] The resulting nanocomposite could be well dispersed in water for at least two months with a solubility of 3 mg ml^{-1}. The functional nanocomposite showed high electrocatalytic activity toward the oxidation of hydrogen peroxide and uric acid and also as an electrocatalytic biosensor of hypoxanthine with a detection limit of 10 nM.

In a slightly different synthesis protocol, Wang and co-workers have firstly compounded graphene oxide and polypyrrole (PPY) in ethanol, drop cast the suspension onto FTO glass and then immersed the loaded electrode in a hydrazine solution for a reduction process.[10] The RGO/PPy film was used as a counter electrode for DSSCs and was found to exhibit a power conversion efficiency (8.14%) comparable with that for the cell based on a Pt cathode.

He and co-workers have prepared porous RGO/polypyrrole composites with superior catalytic performance for triiodide reduction in DSSCs, in a two-step electrochemical synthesis.[11] A GO/pyrrole suspension was subjected to static voltage conditions for the electropolymerization of pyrrole and subsequently to repeated voltage *via* cyclic voltammetry for the electroreduction of GO. The starting GO/PPy and the final rGO/PPy composite films had a porous character, in contrast to the dense and flat neat PPy films. The catalytic performance for triiodide reduction was measured in the order of rGO/PPy > GO/PPy > PPy. The energy conversion efficiency for the rGO/PPy counter electrodes was as high as 6.5%. In the case of rGO/PPy deposited onto a plastic ITO/PEN substrate, the energy conversion efficiency was 4.3%, a value comparable to that of the counterpart using a sputtered-Pt counter electrode (4.8%) on a plastic substrate.

By using the sophisticated process of layer-by-layer (LbL) assembly, Xu *et al.* have reported the facile synthesis of nanocomposite thin films consisting of reduced graphene oxide and poly(diallyldimethylammonium chloride).[12] The starting aqueous suspensions were those of negatively charged graphene oxide and positively charged poly(diallyldimethylammonium chloride). Alternate cycles of substrate dipping and rinsing gave rise to the formation of multi-layered films, in which the components remain adhered by electrostatic interactions. In a subsequent step, the graphene oxide sheets were electrochemically reduced by performing a cyclic voltammetry process. The resulting hybrid thin films were studied as potential counter electrodes for

DSSCs. The prepared devices reached absolute power conversion efficiencies of 9.5%, by using low volatility electrolytes. This value was higher than the one achieved by a platinized electrode (9.1%).

The second scenario is about preparing a colloidal suspension of graphene oxide sheets (either in parent, reduced or chemically modified form), followed by solution mixing with the macromolecular component. By using as a starting material reduced graphene oxide stabilized by pyrene butyrate, the group of Shi have prepared suspensions of chemically modified graphene and polystyrenesulfonate-doped poly(3,4-ethylenedioxythiophene) (graphene/PEDOT–PSS).[13] The graphene content was controlled by the volume ratios of each suspension. Composite films of graphene/PEDOT–PSS were prepared by spin coating onto ITO and applied as counter electrodes in DSSCs. A 60 nm-thick composite film, containing 1 wt% graphene, exhibited high transmittance (>80%) at visible wavelengths and high electrocatalytic activity. The energy conversion efficiency of the cell with this film as a counter electrode reached 4.5%, which was comparable to 6.3% of the cell with a reference platinum electrode.

Polymer composite films of high graphene content were prepared by filtration of the mixed components.[14] Hydrazine-reduced graphene oxide flakes were electrostatically stabilized in an aqueous environment by dialysis against ammonia solution (pH 10) and subsequently mixed with an aqueous solution of polyaniline (PANI) nanofibers under sonication. A LbL composite film was prepared by vacuum filtration of the mixed suspension, whereas the polymer nanofibers seemed to be intercalated between graphene layers. The conductivity of the composite film containing 44% graphene (550 S m^{-1}) was about one order higher than that of a neat polymer film. Supercapacitor devices based on this conductive flexible composite film showed large electrochemical capacitance (210 F g^{-1}) at a discharge rate of 0.3 A g^{-1}.

Following the described LbL strategy, Lu *et al.* have reported the facile synthesis of nanocomposite thin films, consisting of chemically modified reduced graphene oxide (S-RGO) and polyaniline.[15] GO was treated with taurine at a moderate temperature, followed by a $NaBH_4$-based reduction step, resulting in negatively charged sulfonic-acid-grafted RGO. Polyaniline was diluted in an acidic aqueous solution, resulting in its cationic form. The LbL process was carried out by automated dip coating equipment and the hybrid was deposited onto ITO. The multilayer films of S-RGO/polyaniline could be used as flexible transparent electrodes. The enhancement of electrocatalytic properties could be attributed to the drastically increased electrical and ionic conductivities of the S-RGO/polyaniline films brought by the graphitic structure of the S-RGO sheets and the sulfonic acid groups attached to S-RGO, which lead to non-diffusion-controlled redox processes of the polymer component.

Analogous LbL assembled graphene oxide/polyaniline multilayered films were used as counter electrodes in DSSCs.[16] Promising power conversion efficiencies were obtained, which were found to be highly dependent on the layer number of the superstructure. Specifically, the hybrid film PANI-(GO/PANI)$_4$-based counter electrode had the highest efficiency of 7.4%, arising from the enhanced interfacial area in the PANI-(GO/PANI)$_4$ multilayer.

As an extension to the previous work, the same group have chemically attached polyaniline chains onto single-walled carbon nanotubes by a combination of reflux at 184 °C and an *in-situ* polymerization process. Multilayered films, obtained by alternate LbL electrostatic assembly of CNT-PANI and GO flakes, were fabricated with the aim of accelerating charge transport and increasing the interface area for triiodide electroreduction reaction.[17] Covalent bonding between PANI and CNTs was considered a critical factor for the decreased charge-transfer resistance within the complex, resulting to counter electrodes for DSSCs with a decent power conversion efficiency of 6.9% for the (PANI - 4 wt‰ SWCNT/GO)$_5$ assembly.

2.2.2 Graphene/Polymer Assemblies by *In situ* Polymerization

Due to the bulky character of macromolecular systems, a major drawback for the synthesis of nanocomposites is the lack of homogeneity in the formed hybrid nanostructures. A simple and versatile approach towards the synthesis of metal-free (electro)catalytic systems is the *in-situ* polymerization of conjugated monomers in the presence of carbon-based nanostructures. Due to the much smaller size of the monomer itself, the homogeneity of the final hybrid would be improved appreciably. It is noted here that *in situ* polymerization processes in the presence of graphitic nanostructures may give rise to the formation of both chemically attached and physically adsorbed polymer chains in the vicinity of a graphitic surface. In the seminal independent works of the groups of Wang[18] and Wu,[19] a flexible graphene/polyaniline hybrid material was synthesized by an *in-situ* polymerization–reduction/de-doping/re-doping process. A suspension of GO was mixed with aniline and was polymerized in an acidic environment. The adduct was further treated with a base to de-dope the polymer and reduce the GO. After re-doping of polyaniline in an acidic solution, a flexible conducting graphene/polyaniline hybrid was obtained. The composite material showed better electrochemical performances than the individual components. As a supercapacitor electrode, a high specific capacitance up to 1126 F g^{-1} with a retention life of 84% after 1000 cycles was obtained.

Along the same lines, graphene oxide/polyaniline nanostructures were obtained by dilute polymerization of aniline monomer in an aqueous suspension of GO.[20] The morphologies of polymer nanowires onto the graphene surface could be controlled by adjusting the amount of aniline. In addition, the initial mass ratio of GO and monomer was a critical factor in optimizing the electrochemical capacitance of the hybrids, which were studied as supercapacitor electrode materials. Using a slightly different synthesis protocol, graphene oxide was firstly reduced within an autoclave at 180 °C. Subsequently, *in situ* polymerization of aniline took place in the presence of RGO flakes in an aqueous environment.[21] The as-prepared polyaniline/RGO composites were studied as CEs in DSSCs, which achieved a conversion efficiency of 6.1%, comparable to that of the cell with a Pt counter electrode (6.9%).

Wang *et al.* have synthesized graphene/polyaniline hybrids by a two-step hydrothermally-assisted chemical oxidation polymerization process, aiming for the preparation of uniform and ultrathin layers of conjugated polymers onto a graphene surface as supercapacitors.[22] In the first step, aniline was polymerized by ammonium persulfate in the presence of graphene oxide sheets and the suspension was transferred into an autoclave reactor at a temperature range between 120 and 180 °C. During the hydrothermal treatment, GO flakes were reduced and the nanostructured PANI decorated both sides of each graphene resulting in a porous ultrathin film of 10–20 nm. The latter two-dimensional nanocomposites reached specific capacitances of 532.3 to 304.9 F g^{-1} at scan rates of 2 to 50 mV s^{-1} and capacitance retention as high as 99.6% due to the porous structure of the ultrathin polymer layer.

Instead of performing electrodeposition with both the starting components in a liquid environment, researchers have pre-patterned graphene deposits onto a specific substrate and subsequently electrodeposited polyaniline architectures upon it.[23] Following this strategy, an all-solid-state flexible micro-supercapacitor was fabricated and the morphology of PANI nanorods could be controlled by the concentration of aniline and the growth time during the electrodeposition process, resulting in a micro-supercapacitor possessing an electrochemical capacitance up to 970 F g^{-1} at a discharge current density of 2.5 A g^{-1}.

In order to substitute costly FTO/ITO-based CEs in DSSCs, researchers have turned their attention towards cheaper alternative substrates such as flexible polymer films. Poly(ethylene terephthalate) (PET) membranes are considered potential candidates for the preparation of such transparent conductive substrates. In the work of Park and co-workers, a graphene monolayer, synthesized by a chemical vapor deposition (CVD) method, was transferred onto a PET substrate.[24] On top of this, a prepolymerized PEDOT solution was deposited by a spin coating technique. The graphene/PEDOT film was obtained by post-polymerization at 70 °C. These metal-free graphene/PEDOT films on a PET substrate were used as counter electrodes in DSSCs resulting in a conversion efficiency of 6.3%, whereas the efficiencies of DSSCs with Pt/ITO and PEDOT counter electrodes were 6.7% and 5.6%, respectively.

The aforementioned *in situ* chemical polymerization processes can also be realized through an electrodeposition process.[25] In such conditions and depending on the scan range, GO is reduced, and the monomer is electropolymerized simultaneously.[26] By adopting the electrodeposition strategy, thermally reduced graphene oxide flakes were mixed with aniline and were refluxed at 210 °C, yielding aqueous dispersions of a graphene/aniline complex. The latter was subjected to electropolymerization/deposition onto FTO substrates by applying a constant potential of 0.8 V (*vs.* Ag/AgCl/saturated KCl).[27] The electropolymerization was stopped when the desired amount of charge was passed. The composite films were used as CEs in DSSCs. Under optimized conditions, the power conversion efficiency of the device with the hybrid film was measured to be 7.2%, similar to the one of the reference Pt electrode.

Similarly, EDOT monomer was electropolymerized onto a flexible sheet of exfoliated graphite flakes.[28] The Pt-free and TCO-free counter electrodes made from the conductive exfoliated graphite/PEDOT composite afforded enhanced photovoltaic efficiency in DSSCs, ascribed to the high catalytic activity and electron conductivity of the complex structure. Utilization of such counter electrodes showed an energy conversion efficiency of 5.7% with a solid polymer electrolyte, which was significantly higher than conventional Pt electrodes (4.4%) under similar device architecture conditions.

2.2.3 Preparation of Porous Supramolecular Architectures

2.2.3.1 Soft Templates

The development of high surface area catalytic nanomaterials for energy-related applications has been a high priority in related research. Although heteroatom-doped graphenes have shown appreciable performance as metal-free catalysts, further improvement is still feasible. A basic limitation for enhancing the catalytic performance of graphene-based nanostructures—at least in the cases discussed previously—is the low surface population of catalytic sites. During recent years, there have been some efforts towards the development of high surface area porous carbons through the utilization of sacrificial templates. Porous networks of RGO/single-walled CNT hybrids were shown to behave as excellent CEs in DSSC devices.[29] Graphene oxide flakes were reduced by zinc powder towards the formation of RGO material. Both components (RGO and CNTs) were separately exfoliated in liquid polyethylene glycol and were mixed in various mass ratios. After deposition of the resulting gel onto FTO-coated glass, the polymer was thermally decomposed at 430 °C. A porous network consisting of 20 wt% CNT material had afforded the optimum photovoltaic efficiency exceeding the ones of either neat rGO, CNTs or reference Pt electrodes by a relative percentage of 8–10%.

Using a dynamic template-assisted approach, porous reduced graphene oxide networks were fabricated during electrodeposition of graphene oxide suspensions.[30] The role of the dynamic template was played by hydrogen bubbles, which were generated by experimental conditions. In a subsequent step, pyrrole was electropolymerized onto deposited graphene networks, followed by annealing at high temperature. Thus, nitrogen-doped graphene assemblies were obtained *via* the three-step protocol and the resultant electrode was considered as an efficient electrocatalyst for the ORR with a current density value of 5.56 mA cm^{-2}.

A sophisticated strategy was adopted by the group of Feng in order to pattern functional moieties with well-defined architectures onto two-dimensional nanostructures.[31] The principle of the patterning involved self-organization of block copolymer chains into micelles in a selective medium. Specifically, a polystyrene-poly(ethylene oxide) block copolymer was self-assembled in a THF–water mixture, with the latter solvent being the selective one. The resulting polymeric micelles were mixed with a graphene oxide suspension, followed by the addition of pyrrole monomer and initiator. The spherical PS-b-PEO micelles were intimately organized on both sides of graphene oxide flakes *via* hydrogen bonding and, as sacrificial templates, guided the oriented polymerization of conjugated monomer. Upon removal of block copolymer micelles by rinsing, patterned polypyrrole nanostructures on the graphene surface were achieved. The mesoporous architecture of such polymeric assemblies gave rise to an enhanced specific surface area (85 $m^2\ g^{-1}$) in the prepared hybrid nanostructures, providing excellent specific capacitance and rate performance for supercapacitors.

Polymers have been widely used as multifunctional cohesive components for the fabrication of stable graphene/polymer films.[32] To this end, Aksay and co-workers have prepared functionalized graphene sheets through thermal exfoliation of graphite oxide sheets at 1000 °C. The thermally exfoliated graphene material was then dispersed in a poly(ethylene oxide) aqueous solution by the aid of an amphiphilic triblock copolymer (PEO-PPO-PEO, pluronic type). The resulting suspension was spin-coated onto FTO and annealed to 350 °C, where decomposition of the polymer occurs, forming a porous network of graphene. Similarly, ethyl cellulose was used as a sacrificial binder, resulting in the lowest charge transfer resistance value at a decomposition temperature of 370 °C (20% weight loss).[33] The latter porous material was employed as a CE in DSSCs with comparable efficiencies to CEs based on thermally decomposed chloroplatinic acid.

Using the concept of sacrificial soft templates, Zhang and co-workers have utilized a macroporous nickel foam as a substrate for the deposition of precursor components.[34] The first step in the synthesis involved the impregnation of an aqueous suspension containing graphene oxide, polyvinylpyrrolidone (PVP) and sulfonated polystyrene microspheres into the metal foam. PVP acted as a graphene stabilizer and N-dopant, whereas sulfonated polystyrene microspheres acted as sacrificial templates and an S-dopant. Due to the size difference between graphene oxide sheets and polymer microspheres, it was suggested that graphitic nanostructures were adsorbed onto the polystyrene spheres. Elimination of water through freeze drying followed by calcination at 800 °C in an inert atmosphere gave rise to the formation of porous doped graphene nanostructures, decorating the metal foam. Thus, by an *in-situ* approach, a hierarchically porous structure, highly conductive network, and heteroatom dual doping were combined. Such porous assemblies were suggested as potential electrode materials in energy storage applications.

2.2.3.2 *Hard Templates*

In some of the previous studies discussed, utilization of soft templates, such as polymer micelles, has been adopted in order to develop porous nanostructures. Alternatively, inorganic nanomaterials have been widely used as sacrificial hard templates. To this end, the group of Antonietti synthesized mesoporous nitrogen-doped carbon materials with high surface areas of up to 1500 $m^2 g^{-1}$ by the carbonization of nucleobases dissolved in an all-organic ionic liquid (1-ethyl-3-methylimidazolium dicyanamide).[35] The sacrificial templates were silica nanoparticles of 12 nm size. After carbonization at 1000 °C and subsequent dissolution of the silica nanoparticles, high surface area mesoporous nitrogen-doped carbons were obtained, with a local graphitic structure. Such systems exhibited appreciable catalytic activity in the ORR, leading to a four-electron-mediated mechanism.

Using a similar approach, the ionic liquid *N*-butyl-3-methylpyridinium dicyanamide (BMP-dca) was employed as a direct precursor for the synthesis of mesoporous carbon, through the utilization of silica nanoparticles as a hard template.[36] The adduct obtained after carbonization at 800 °C and silica dissolution had a mean pore size of about 13 nm and a nitrogen content of 16 wt%. Data extracted from the rotating disk electrode technique revealed that the ORR was a two-electron process, implying that oxygen was converted to hydrogen peroxide and not water.

A slight variation of the synthesis protocol adopted by Antonietti's group gave rise to the fabrication of nitrogen-doped porous carbon nanosheets.[37] In detail, a mesoporous silica shell was initially grown on GO sheets to generate graphene-based silica nanosheets (G-silica). The latter were then functionalized with poly(diallyl dimethylammonium chloride) and subsequently assembled with negatively charged colloidal silica nanoparticles by electrostatic interactions. The graphene/polymer/silica hybrid was dispersed in an aqueous solution of dopamine, which self-polymerized into polydopamine (PDA), yielding a thin organic layer on top of silica nanoparticles. The subsequent thermal annealing of the PDA-coated graphene/silica assembly followed by dissolution of silica generated nitrogen-doped carbon nanosheets with defined mesopores. Such hybrids have shown electrocatalytic performance in the ORR comparable to commercially available Pt–C catalysts.

2.2.4 Other Approaches

Although GO-based assemblies are preferably studied as catalytic components, they have a major drawback: the rather high electrical resistance, which is greatly correlated with its defective structural integrity. Moving away from graphene oxide, Kim and co-workers have synthesized edge-carboxylated graphene nanoplatelets (ECGnPs) by a simple, efficient and eco-friendly reaction, involving ball milling of graphite in the presence of dry ice (solid carbon dioxide).[38] Mechanochemically driven graphitic C–C bond scissions in a ball mill apparatus generate transient carbon species (mainly carbon radicals and carbanions), which react with carbon dioxide yielding graphene nanoplatelets (GnPs) with edge-carboxylate groups. ECGnPs were used as oxygen-rich metal-free counter electrodes (CEs) in organic DSSCs and their photovoltaic activity was about 8% higher than the one of reference Pt. Such behavior was justified by the lower charge transfer resistance value of the graphene-based counter electrode, a parameter which is related to electron interactions between CEs and electrolyte ions.

Decoration of a graphitic lattice with various N-containing species was accomplished by pyrolysis of graphene oxide/polymer nanostructures, prepared by *in situ* polymerization of pyrrole or aniline in the presence of GO, at 850 °C in an inert atmosphere.[39] The bonding state of the N atom (pyridinic and pyrrolic) was found to have a significant effect on the selectivity and catalytic activity for the ORR. In contrast, the effect of the total amount of nitrogen was negligible for the aforementioned reaction.

Nitrogen-doped graphene nanostructures were prepared by a two-step synthesis protocol. Firstly, graphene oxide and aniline monomer were mixed, and the hybrids were prepared by an *in situ* chemical polymerization process. Decoration of the graphitic lattice with various N-containing species was accomplished by pyrolysis of the resulting graphene oxide/polymer nanostructures at 1000 °C in an inert atmosphere (Figure 2.2).[40] The resulting N-doped graphene contained 2.4 at.% N, whereas electrochemical characterizations revealed that the doped nanomaterial possessed excellent catalytic activity toward the oxygen reduction reaction (ORR) in an alkaline electrolyte, including a desirable four-electron pathway for the formation of water and long-term stability. In addition, the authors demonstrated that the heteroatom-doped graphene also had high catalytic activity toward the oxygen evolution reaction (OER), rendering its potential application as a bi-functional catalyst for both the ORR and OER.

By extending the scope of previous work, Li *et al.* have developed graphene-based nanostructures co-doped with nitrogen and phosphorus in a two-step procedure.[41] In the first step, GO, aniline and phytic acid (phosphorus-containing precursor) were mixed and the *in situ* chemical polymerization of aniline monomer was performed. The obtained GO/polyaniline/phytic acid composite was pyrolyzed at 850 °C to promote the (dual)doping process. The B,N co-doped graphene showed outstanding catalytic activity towards both the ORR and OER. The enhanced bifunctional catalytic performance was attributed to the synergistic effects between the boron and nitrogen active sites within the graphitic surface, the high conductivity of the doped graphitic nanostructure and the large surface area.

For further enhancement of the charge transfer interactions between a graphene-based layer and electrochemically deposited PEDOT film, Ho and co-workers have utilized a commercially available nitrogen-doped graphene (NGr) dispersed in an ethanolic solution of Nafion polymer and drop cast onto an FTO substrate.[42] In a subsequent step, EDOT monomer was electropolymerized onto N-doped graphene sheets and the prepared NGr/PEDOT

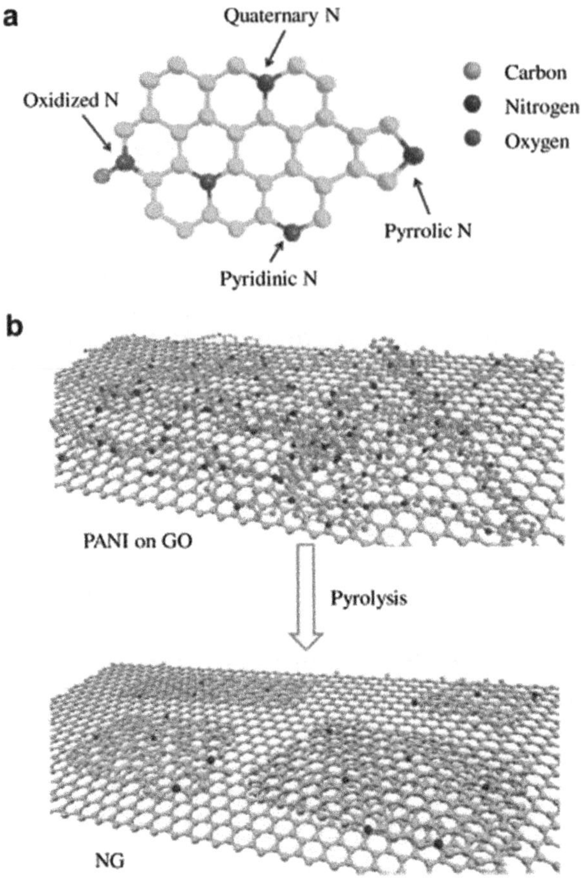

Figure 2.2 Schematic illustration of (a) the structure of nitrogen-doped graphene and (b) the preparation of N-doped graphene through the pyrolysis of PANI/GO supramolecular assemblies synthesized by *in situ* polymerization. Reprinted from *Carbon*, **45**, Z. Lin, G. H. Waller, Y. Liu, M. Liu and C.-P. Wong, Simple preparation of nanoporous few layer nitrogen-doped graphene for use as an efficient electrocatalyst for oxygen reduction and oxygen evolution reactions, 130–136, copyright 2013, with permission from Elsevier.

composite was studied as a potential CE in DSSCs. The composite film-based device exhibited a power conversion efficiency of 8.3%, which was slightly higher than the one of reference Pt. The efficiency of the NGr/PEDOT-based cell was attributed to the chemical binding between the nitrogen moieties of graphene and the sulfur atoms of the polymer; nevertheless, in a similar study of Paterakis *et al.*,[43] the latter conclusion was not verified.

Jang and co-workers have developed an alternative method for the preparation of CEs for DSSCs, based on the synthesis of CVD-grown three-dimensional graphene nanonetworks (3D-GNs).[44] These types of assemblies possessed a diminutive degree of defects (oxygen atoms), retaining appreciable conductivity (5 S cm^{-1}). DSSCs based on the pristine single component 3D-GNs without any treatment recorded comparable photovoltaic efficiency to reference Pt (~8% as absolute value). Superior performance to Pt was accomplished through immersion of 3D graphene assemblies into nitric acid solutions at specific dilutions, resulting in heteroatom doping. It was found that p-doped graphitic nanostructures with an optimized nitrogen content of 0.5% exhibited a maximum photoconversion efficiency of about 8.5%, which was 6% greater than that exhibited by Pt-based DSSCs.

2.3 Fullerenes

Fullerenes, especially the C_{60} cage, are considered as efficient light harvesting molecules with extraordinary photocatalytic properties. A summary for photocatalytic transformations employing fullerene derivatives is presented in Table 2.1.

Fullerenes are also effective catalysts in the absence of light. The C_{60} dianion (C_{60}^{2-}), produced by the $NaBH_4$-assisted reduction of the supramolecular assembly of C_{60} with two γ-cyclodextrins as capping agents, was employed for the reduction of C=O, C=C and N–N bonds, and specifically for the reduction of triphenyl tetrazolium chloride, styrene, acrylonitrile, methyl acrylate and acetophenone, without irradiation, under ambient conditions to afford the corresponding dihydro-substituted products.[53] Functionalized fullerenes are capable of participating in supramolecular architectures for energy conversion applications. C_{60} functionalized with a barbiturate additive proceeds to the formation of multiple intermolecular hydrogen bonds forming highly stable supramolecular dimers, trimers, *etc.*[54] C_{60}-barbiturates have been successfully exploited by Pagona *et al.* for the preparation of supramolecular

Table 2.1 Summarized data for photocatalytic transformations employing fullerene derivatives.

Fullerene	Application	Comments	Ref.
C_{60}	Photocatalytic reduction of nitrobenzene	1 atm H_2, r.t. (In dark 4–5 MPa H_2, 150 °C)	45
C_{60}	Photooxidation of sec-benzylamines imines	$^1O_2$, UV irradiation	46
C_{60}	Photoreduction of –N=N–	Photoinduced electron transfer to N=N bond	47
C_{60}	Hajos–Parrish–Eder–Sauer–Wiechert reaction	Activation of proline-N, conversion 40%	48
C_{60} or C_{60}/C_{60}^-	Photocatalytic reduction of 4-substituted nitrobenzenes	4 wt% C_{60} At ambient conditions Conversion, selectivity >99%	49
C_{60}, C_{70}	Photooxidation of olefins	$^1O_2$, UV irradiation	50
$(C_{59}N)_2$, $HC_{59}N$	Oxidation of unsaturated compounds	$^1O_2$, Visible light	51, 52

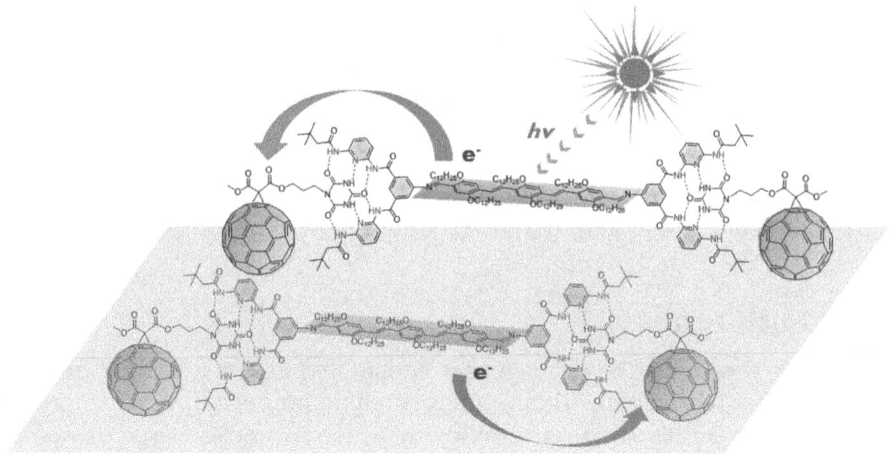

Figure 2.3 Illustration of the electron transfer path within the supramolecular C_{60}-Hamilton-C_{60} triad. Adapted from ref. 55 with permission from the PCCP Owner Societies.

architectures with a dimer Hamilton receptor based on an oligo-phenylvinylene bridge, *via* a host–guest strategy.[55] The formation of multiple hydrogen bonds between the barbiturate moiety and the Hamilton functionality affords stable C_{60}-Hamilton-C_{60} ensembles with enhanced intra-hybrid electron transfer assigned to the strong H-bonding between the individual components, which could afford promising candidates for the preparation of thin large surface area photovoltaics (Figure 2.3).

In the field of energy conversion schemes, a C_{60}-based dyad was prepared by non-covalent attachment of a non-metal phthalocyanine (Pc), a π-extended chromophore with strong absorption in the visible light spectrum. The hybrid was isolated by simple mixing of the components, which were assembled through weak van der Waals forces between the fullerene cage and the Pc ring. Nevertheless, the electron transfer from the Pc to the fullerene was more efficient for the case of the corresponding covalent ensembles rather than the supramolecular Pc/C_{60}.[56]

Beyond experiments, theory is also contributing to the design of effective materials based on the supramolecular functionalization of C_{60} and C_{59}N-heterofullerene cages interacting with extended tetrathiafulvalene units (exTTF).[57] Theoretical approximations for C_{60}/exTTF and C_{59}/exTTF nanohybrids revealed enhanced electron communication due to the p-extended backbone of exTTF and charge transfer phenomena from the excited exTTF to the fullerenes.

The electron accepting properties of C_{60} can be further improved by chemical functionalization. The most efficient C_{60} derivative is 6,6-phenyl-C_{61}-butyric acid methyl ester (PCBM). PCBM has been widely used in bulk heterojunction solar cells as an electron acceptor within the active layer in conjunction with a series of chromophores (single molecules and macromolecules). Recently,

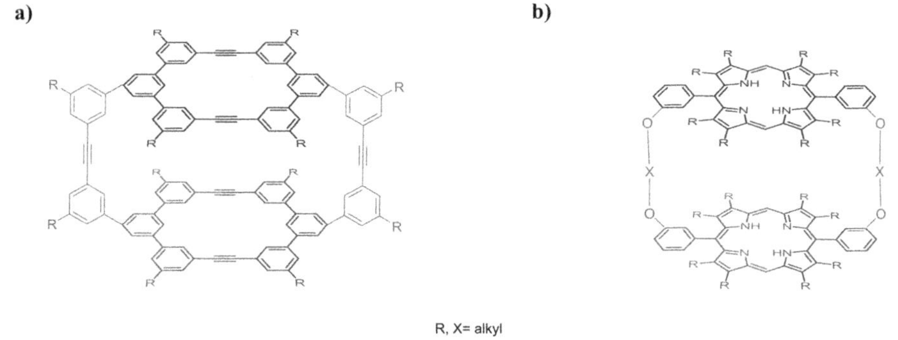

Figure 2.4 Representative structures of (a) arylene-ethynylene and (b) porphyrin-based fullerene "catchers".

PCBM was employed in supramolecular assemblies with N-doped carbon quantum dots (N-CQDs), which were functionalized with thiophene derivatives.[58] PCBM/CQDs were prepared *via* simple mixing of the individual components and processed in the form of thin films. It was observed that photoinduced charge separation occurred, namely electron transfer from the CQDs to the PCBM.

More advanced energy systems, based on arylene ethynylenes (Figure 2.4), can be employed as hosts for the fullerene cage affording hybrids where the C_{60} stands between π-extended organic rings. Yu *et al.* have recently reviewed thoroughly these remarkable supramolecular assemblies with fullerenes and metal-free fullerene "catchers" with high potential for energy conversion applications and optoelectronics.[59]

2.4 Carbon Nanotubes

2.4.1 Solubilization of CNTs and Physisorption of Small Molecules

CNTs are considered as rolled-up graphene sheets and present extraordinary electronic and mechanical properties.[60] However, they suffer from the major drawback of low solubility, which limits greatly their potential in most applications. In this context, a series of protocols have been developed focusing on CNT-based assemblies derived from either π–π stacking or van der Waals interactions with numerous substances. These include small aromatic molecules, like benzonitrile,[61] 1-pyrene acetic acid,[62] anionic perylenes,[63] or macromolecular systems like epoxy-polymers,[64] poly(4-vinylpyridine),[65] poly(4-vinylpyridine)-*block*-polystyrene,[66] polybenzoxazine,[67] or ionic liquids.[68] The aforementioned supramolecular hybrids, which in most cases are derived by simple blending of CNTs with the different surfactants, present enhanced colloidal stability in solution. Moreover, LbL self-assembly is also a powerful technique for the non-covalent modification of CNTs based on the same principles.[69] As discussed earlier when referring to graphene and fullerene composites, the exploitation of counter charged individuals or hydrogen bonding strategies allows a variation of self-assembly motifs, which can be efficiently applied in many catalytic processes like photocatalysis, energy and sensing applications, or to enhance the mechanical and thermal properties of the target materials.

Designing the surfactant moiety properly is a convenient way to access functional CNT-based materials. Employing a pyrene-terminated C_{60}-derivative, Guldi *et al.* have prepared a novel supramolecular donor acceptor dyad based on the π–π stacking interactions between the fullerene derivative and the sidewalls of SWCNTs.[70] Following the same concept, Sandanayaka *et al.* have studied the dependence of the nanotube diameter on the electron transfer rate between the two components of the assembly.[71]

Aqueous dispersions of supramolecular assemblies of a soluble polymer carrying pyrene units *via* disulfide bridges have shown self-reduction activity in the presence of dithiols.[72] The disulfide bridges are reduced, releasing the water-soluble polymer from the pyrene units assembled onto the CNTs and subsequently inducing the oxidation of the dithiols to disulfides.

Towards the preparation of electrocatalytic materials, α-cyclodextrin/CNTs ensembles were prepared by simple blending and were deposited onto high ordered pyrolytic graphite (HOPG) electrodes. The latter were found to be efficient for the oxidation of thymine in alkaline buffered electrolyte (pH 10.5) with detection limit up to 5×10^{-6} mol L^{-1}.[73] In an analogous strategy, β-cyclodextrin/CNT modified electrodes have been tested for the oxidation of adenine and guanine. Using the hybrid electrocatalyst, well-defined oxidation peaks for the nucleosides have been produced and detection limits of 0.1 and 0.2 μg ml^{-1} for adenine and guanine, respectively, were obtained.[74]

Non-covalent functionalization of CNTs with ketone-containing molecules exhibited enhanced reactivity in the heterogeneous catalysis of nitrobenzene. Phenanthraquinone derivatives assembled onto CNTs *via* π–π stacking in the presence of hydrazine hydrate catalyzed the formation of aniline with high conversion (97.5%) and retained high activity (86.5%) for up to four catalytic cycles.[75]

2.4.2 Functionalization with Polymers

Studies of CNT hybrids with glucose oxidase (GOx) in aqueous solutions demonstrated that the activity of the enzyme was preserved after its successful immobilization onto the nanotubes' sidewalls. This finding has revealed that the assembly of GOx onto the CNTs' backbone does not disturb the secondary structure of the enzyme.[76,77] In another approach, flexible counter electrodes were constructed using enzyme-dispersed MWCNTs deposited on a polyester fabric.[78] Lipase from *Candida rugosa*, glucose oxidase from *Aspergillus niger*, and laccase from *Trametes versicolor* were used as the enzyme dispersants of MWCNTs. The highly concentrated colloidal suspension of enzyme-MWCNTs was prepared by a simple mixing of pristine MWCNTs in an aqueous solution of the enzyme for 12 hours at room temperature. After filtration and washing, the enzyme-MWCNT suspension was deposited onto polyester fabric following a simple tape casting technique under drying air at 50 °C. This method allowed the fabrication of hybrid materials from MWCTNs and enzymes with different structures, thus, the fine-tuning of their electrocatalytic properties.

Poly(ethylenenimine) (PEI)/N-doped CNTs have been proposed for the electrocatalytic reduction of CO_2 to a formate anion. In this context, PEI participates as a co-catalyst by attracting the CO_2 molecules onto the surface of the electrocatalyst, stabilizing the $CO_2^{\bullet-}$ intermediate radicals and thus significantly reducing the overpotential of the reaction. The hybrid electrocatalyst was prepared in a three-step method: (i) deposition of CNTs onto the surface of a glassy carbon electrode, (ii) doping of CNTs with ammonia plasma and (iii) deposition of the polymer.[79] Similarly, PEI/CNTs have been employed as electrocatalysts for the oxidation of glucose.[80]

More advanced polylysine (Plys)/CNT ensembles were used as templates for the immobilization of GOx, affording modified electrodes coated with Plys/CNT/GOx hybrid electrocatalysts for the oxidation of glucose in aqueous media. The Plys/CNT blend was formed by mixing the individuals in solution. Polylysine acted as a polycationic electrolyte allowing the self-assembly of the enzyme. Hybrid electrodes with two layers of GOx have shown the maximum activity due to the optimized distribution of the enzyme onto the surface of the Plys/CNT substrate.[81]

A carbon-based counter electrode for DSSCs based on electropolymerized poly(3,4-ethylenedioxythiophene) (PEDOT) on a SWCNT film on a plain plastic substrate was proposed, presenting an efficiency similar to that of conventional platinum-on-ITO based electrodes. The PEDOT-SWCNTs hybrid was synthesized through electropolymerization of 3,4-ethylenedioxythiophene in an aqueous solution followed by electrochemical deposition of PEDOT on an acid-purified SWNT film placed on a plastic substrate.[82]

2.4.3 Encapsulation of Molecules (Peapods)

Taking advantage of their hollow empty space, CNTs can effectively act as a host material for different kinds of metal-free moieties, from small organic molecules like perylenes, coronenes[83] and tetrathiafulvalenes,[84] to other carbon-based derivatives like fullerenes,[85] alkyl- and aryl-fullerenes (Figure 2.5),[86] aza-fullerenes[87] and graphene nanoribbons,[88] or even polymerized materials. A typical encapsulation procedure first consists of the acid-aided cap-opening of CNTs, followed by mild annealing and encapsulation of the desired molecules following vapor-phase doping in a sealed vessel, under a vacuum and high temperature.

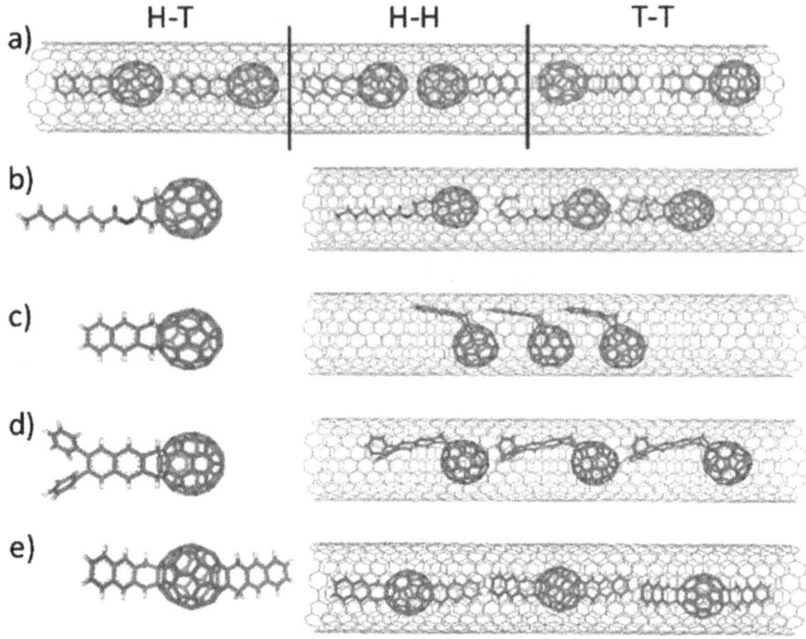

Figure 2.5 (a) Different configurations of mono-functionalized fullerenes inside SWNTs. (b)–(d) Packing of mono- and (e) bis-functionalized fullerenes, depending on rigid and/or bulky character of functional groups. Adapted from ref. 86 with permission from the Royal Society of Chemistry.

Furthermore, the inert internal cavity of CNTs facilitates numerous reactions to take place in a protected and controlled environment. In this context, encapsulated coronene was condensed inside SWCNTs, forming oligomers (trimers, tetramers and pentamers).[89]

In a similar approach, the synthesis of quaterrylene inside SWCNTs from the thermal treatment of encapsulated perylene was reported.[90] Photophysical studies indicated that excitation energy transfer occurred from the quaterrylene to the SWCNTs and was strongly dependent on the diameter of the SWCNTs.

Electronic interactions between the encapsulated moieties and the CNT backbone were also thoroughly investigated in terms of spectroscopic characterization and showed an intense electronic communication between the encapsulated moieties and the sidewalls of the nanotubes.[91–94]

2.4.4 Functionalization with Other Graphitic Carbons

Depending on their preparation method, CNTs may carry a significant amount of nanographitic impurities *i.e.* amorphous carbon, which is quite difficult to remove. Recent studies showed that CNTs (single- or double-walled) carrying such nanographite impurities present enhanced activity towards the electroreduction of azo groups, namely the –N=N– bond of methyl orange, compared to extra pure CNTs.[95]

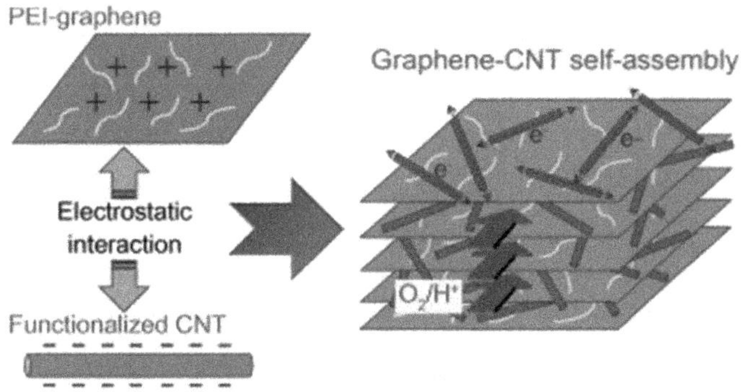

Figure 2.6 Preparation of a graphene/CNT supramolecular electrostatic self-assembly for the ORR. Reprinted from *Applied Catalysis B: Environmental*, **144**, C. H. Choi, M. W. Chung, H. C. Kwon, J. H. Chung and S. I. Woo, Nitrogen-doped graphene/carbon nanotube self-assembly for efficient oxygen reduction reaction in acid media, 760–776, copyright 2014, with permission from Elsevier.

In a more advanced approach, a graphene/CNT self-assembly (GCA) was synthesized through the electrostatic interaction between graphene and CNTs, and was applied as a catalyst for the ORR in acidic media after modification with N-doping.[96] As shown in Figure 2.6, self-assembly was performed by an electrostatic interaction between PEI-modified graphene having a positive charge and acid-treated CNT having a negative charge. For application as ORR catalysts, nitrogen was doped into the prepared carbon materials through pyrolysis with dicyandiamide and small amounts of transition metal chlorides (Co and Fe), followed by acid treatment to eliminate the metals. These hybrids are ideal candidates as cathode electrodes for fuel cell applications.

In an analogous approach, a suspension of graphene quantum dots was mixed with acid-purified MWCNTs in water to form a metal-free hybrid nanocomposite, which showed enhanced electrocatalytic activity for the ORR.[97] In an earlier work, graphene oxide-CNT hybrids with an optimal mass ratio of GO to CNTs of 1 : 4 were synthesized by sonication-assisted mixing in ethanol. This oxygen-enriched graphene-CNT nanocomposite was tested in ORR studies and as a metal-free cathode electrode in a hybrid electrolyte Li–air battery. Both experimental data and theoretical calculations indicated the critical role of the oxygen networks of these metal-free hybrids towards the electrocatalysis of O_2.[98]

2.4.5 Doping of CNTs

Doping with heteroatoms, involving nitrogen, sulfur, phosphorus or other non-metallic elements, results in the electron enrichment of a CNT surface, thus affecting not only their structural but also their electronic properties. The most commonly used element for doping of CNTs is nitrogen, which is found to enhance greatly their electronic properties due to the interactions of the lone pair of electrons of nitrogen with the extended π-system of the outer surface of CNTs. N-doping of CNTs was achieved either by an *in-situ* process during the growth of the carbon nanomaterial or by post-treatment of CNTs with appropriate precursors. The *in situ* direct synthesis follows one of the common methods of CNT preparation, such as CVD, arc discharge or laser ablation, in the presence of an appropriate N-containing compound. Typical examples include polypyrrole, polyvinylpyrrolidone, polyaniline and/or polyacrylonitrile, which may be thermally decomposed thus providing the nitrogen atoms on the graphitic lattice.[99] The major drawback of this approach is that the as-prepared material must be purified from residual metal catalysts thus increasing the risk of destroying the N-doped sites of CNTs. The above-mentioned N-doped carbon materials (amorphous carbon and/or CNTs) are effectively used as metal-free electrocatalysts for the ORR under alkaline conditions. Among the aforementioned precursors, the polypyrrole-derived N-doped narrow CNTs exhibited the best performance in terms of electrochemical stability.

An alternative pathway for the synthesis of N-doped CNTs consists of the thermal treatment of already prepared CNTs with nitrogen-containing species, preferably NH_3 or polyaniline. In addition, similarly synthesized N-doped graphene/CNTs hybrid materials were also reported.[100] In a similar way, a simple method based on heat treatment of nitrogen-containing precursors 1*H*-benzotriazole, 1*H*-1,2,3-triazolo[4,5-*b*]pyridine, and 5-(2-pyridyl)-1*H*-tetrazole and MWCNTs at 700 °C under argon proposed novel carbon structures enriched in nitrogen for the reduction of oxygen in an alkaline medium.[101]

N-doped CNTs have shown enhanced catalytic activity for the selective oxidation of the harmful H_2S to elemental sulfur[102] and the selective hydrochlorination of acetylene.[103] The material demonstrated high selectivity and high turnover frequency values, while it is also non-toxic compared to conventional $HgCl_2$ catalysts. Furthermore, N-doped CNT hybrids with heteropolyacid (HPA) have shown solid acid catalytic activity towards the hydrolysis of ethyl acetate, the Beckmann rearrangement of cyclohexanone oxime and alkylation of toluene.[104]

A direct comparison between the catalytic activity of CNTs and N-doped CNTs in the oxidation of cyclohexanone in liquid phase was also proposed. In this regard, neat CNTs were produced by chemical vapor deposition (CVD) using liquefied petroleum gas as a carbon source over a $FeMo/Al_2O_3$ catalyst in a hydrogen and nitrogen atmosphere. The catalytic activity of these metal-free CNTs was found to be 2–10 times higher than the corresponding metal-based (Au or Fe) catalysts. In the same context, N-doped CNTs were synthesized by the aforementioned CVD method using aniline as a carbon source in an NH_3 atmosphere, presenting a 4.5% N atomic ratio as evidenced by XPS and a similar BET surface area to undoped CNTs. The as-prepared N-doped CNTs showed high selectivity and high conversion rates in the production of adipic acid from cyclohexane.[105]

In another novel procedure, S,N co-doped CNTs derived by *in situ* polymerization of polydopamine were deposited on a uniform hard template of CdS nanowires, with subsequent thermal annealing. In this case, polydopamine played a dual role as a source of both nitrogen and carbon while CdS nanowires provided the sulfur and played the role of a sacrificial template. These S,N co-doped CNTs showed enhanced catalytic activity in the reduction reaction of *p*-aminophenol in the presence of $NaBH_4$.[106] In a different strategy, a sequential doping process was developed for the preparation of a N,S co-doped graphene/CNT nanocomposite, which showed increased catalytic activity in the ORR in alkaline media.[107] In this context, a blend of oxidized CNTs and graphene oxide was first treated with NH_3 at 500 °C and then with phenyldisulfide at 900 °C. The main drawback of this procedure was the low yield of the final nanocomposite due to extended ammonia etching at high temperatures.

2.5 Carbon Nanohorns

Carbon nanohorns (CNHs) are a synthetic conical-shaped nanoform consisting of a single wall of sp^2 hybridized carbon atoms and usually form aggregates with an average diameter of 100 nm.[108] Sonication of CNH dispersions in the presence of β-cyclodextrin (β-CD) afforded a β-CD/CNH hybrid material, capable of the detection of naringin, hesperidin, and naringin dihydrochalcone.[109] Chitosan polysaccharides wrapped around pristine CNHs and crosslinked *via* acid catalyzed addition of glutaraldehyde were found to be an efficient CNH-based template *i.e.* for the immobilization of nanoparticles towards bioimaging applications.[110] Polyelectrolytes (*i.e.* polystyrene sulfonate) can also effectively solubilize CNHs by wrapping the nanocone and were used as templates for the immobilization of molecules or other particles.[111,112] According to this methodology, CNHs wrapped with poly(maleic anhydride-*alt*-1-octadecene-poly(ethylene glycol)) (C18PMH-PEG) formed stable dispersions of CNH/C18PMH-PEG in water. CNH/C18PMH-PEG ensembles were highly stable and biocompatible.[113] Moreover, their strong absorption in the NIR and excellent photothermal properties have been exploited for the photothermal therapy of tumors in mice.

CNHs can be easily decorated with π-extended molecules *via* π–π stacking interactions yielding non-covalent hybrids that are stable in water. Water-soluble tetra-cationic non-metal porphyrin was mixed with CNHs and sonicated to access stable dispersions of CNH/Porph ensembles.[114] The latter approach had a two-fold scope: (i) to solubilize the CNHs and (ii) to afford functional donor–acceptor systems. A non-metal tetra-benzoic acid porphyrin (Porph-COOH), was successfully attached onto the surface of amine-modified CNHs (CNH–NH_2) *via* π–π stacking interactions and complementary H-bonding between the carboxyl groups of the porphyrin and the amine groups on the CNH–NH_2.[115] The prepared ensembles were efficient materials for photovoltaics, since—after photoexcitation of the porphyrin—electron transfer to the CNHs was observed and the large lifetime

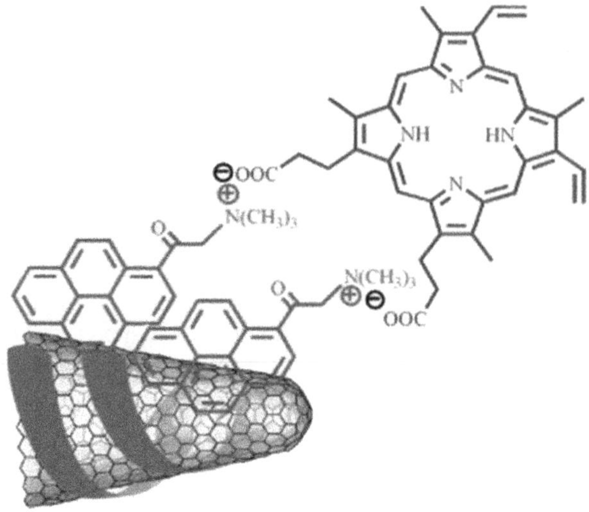

Figure 2.7 Electrostatic assembly of CNHs/ammonium-terminated pyrene coupled to a negatively-charged non-metal porphyrin forming a ternary CNH/pyrene+/porphyrin− assembly. Reprinted from *Diamonds and Related Materials*, **16**, G. Pagona, A. S. D. Sandanayaka, Alan Maignè, J. Fan, C. Papavassiliou, I. D. Petsalakis, B. R. Steele, M. Yudasaka, S. Iijima and N. Tagmatarchis, Aqueous carbon nanohorn–pyrene–porphyrin nanoensembles: Controlling charge-transfer interactions, 1150–1153, copyright 2007, with permission from Elsevier.

of the charge separated state allowed the subsequent transfer of the electron to methyl viologen (MV^{2+}). Concerning energy applications, CNHs are known to have excellent charge transport properties as demonstrated by the preparation and study of CNH-based counter electrodes in DSSCs.[116] CNHs mixed with ammonium-terminated pyrene derivatives in water yielded water-stable CNH/pyrene ensembles, which further coupled to negatively-charged non-metal porphyrin forming a ternary CNH/pyrene$^+$/porphyrin$^-$ assembly (Figure 2.7).[117] Photoexcitation of the porphyrin induces transfer of electrons to the CNHs mediated by pyrene, demonstrating the excellent electron accepting properties of CNHs. Following the same strategy, a positively-charged derivative of an extended tetrathiafulvalene (exTTF) has been electrostatically attached to a CNH/pyrene$^+$ hybrid forming a novel ternary CNH/pyrene$^+$/exTTF$^-$ system managing photoinduced electron transfer from the exTTF to the CNHs.[118]

2.6 Nanodiamonds

Nanodiamonds (NDs) are small carbon particles where most of the carbon atoms are sp^3 hybridized, although sp^2 sites can also be identified. They have attracted scientific interest for bio-applications since they are biocompatible and not cytotoxic.[119] Moreover, NDs are excellent candidates towards antibacterial applications and handling of oxidative stress in cells.[120] NDs have also been employed in oxidative catalytic systems for reactions with industrial interest *i.e.* the oxidative hydrogenation of ethylbenzene,[121] due to their remarkable stability under the harsh conditions required for the synthesis. Their properties can be significantly engineered by the introduction of other atoms *via* elemental doping. For example, thermal treatment of NDs with nitrogen-containing compounds afforded N-doped nanostructures, which were found to be capable of photocatalyzing the oxidation of benzyl alcohol to benzaldehyde.[122]

Currently, the chemistry and functionalization of NDs are still in expansion. Regarding their non-covalent functionalization, only a few works have been presented; nevertheless, the results are promising in terms of applying these unique carbon particles in real life. NDs suffer from aggregation and usually have to be treated in order to dismantle the bulk material into individualized nanodiamonds with diameters usually around 5 nm. An effective one-step procedure for the individualization of aggregated NDs in water has been developed by Skaltsas *et al.*, where polymers such as quaternized poly(2-vinylpyridine), poly(*N*-isopropylacrylamide), poly(*n*-butylacrylate-*b*-*N*-isopropylacrylamide), poly(2-vinylpyridine), poly[styrene-*b*-(ethylene oxide)], poly[styrene-*b*-(2-vinylpyridine)] and poly(methacrylic acid) were employed.[123] Pristine NDs were simply mixed with the polymers in water and sonicated. After removal of any aggregated residues, the content of ND/polymer ensembles in diamond was as high as 83%. The ND/PNIPAM hybrids have shown good stability in water allowing further hybridization with bovine serum albumin (BSA) through electrostatic interactions. The strategy of polymer-assisted de-aggregation of NDs could create ND-based template materials for the immobilization of a series of molecules and macromolecules with biological interest.

Treating pristine NDs with strong acids induces charges on their surfaces allowing their functionalization with counter-charged molecules or macromolecules. Lysozyme, an anti-bacterial enzyme, was effectively immobilized onto NDs and tested for its catalytic activity, following such a supramolecular electrostatic approach. NDs with diameters of 100 nm, enriched in –COOH groups after acidic treatment, easily stabilized lysozyme onto their surface and, interestingly, the antibacterial activity of the enzyme was retained, indicating that the electrostatic attachment didn't affect or interfere with the enzyme's active site.[124] The addition of supplementary biomolecules (*i.e.* cytochrome b) is able to regulate, enhance or block, the enzyme's hydrolytic activity. Along the same lines, these ensembles have also been tested for their activity in a sample containing *E. coli* bacteria, demonstrating their advantage as nanoprobes for bio-applications.[125]

Following the same immobilization protocol, lysozyme was immobilized onto a series of different size NDs (5, 50, 100, 200, 300, 400 and 500 nm) and the corresponding ND/lysozyme ensembles have been evaluated for the size-dependent hydrolytic activity of the supported enzyme.[126] Lysozyme on NDs with a diameter of 100 nm has shown the maximum activity, which is as high as 75% of the activity of free lysozyme. The smaller NDs (5 and 50 nm) have only 15% of the free enzyme's activity and the bigger NDs have slightly less activity (~60% of the free lysozyme). The size-dependent activity is related to the surface of the NDs and the possible presence of extended non-diamond fractions in the smaller NDs. These systems are biocompatible and capable of the fluorescent labeling of single bacteria and mapping cultures.[127]

Another approach, developed by Qi *et al.*, employed non-covalent hybrids of boron-doped NDs (BNDs) and RGO, prepared by simple blending, as electrocatalysts for the degradation of *E. coli* bacteria in the absence of any hydrolytic enzymes (*i.e.* lysozyme).[128] Under voltage conditions, charges are transferred from the bacteria's membrane to the conducting RGO and the BNDs convert water molecules to hydroxyl radicals inducing oxidative stress to *E. coli* resulting in the corrosion of the cells' membrane and finally their death.

2.7 Carbon Dots

Fluorescent carbon nanoparticles with crystalline 2D (sp^2 carbon network) domains have been widely adopted in catalysis, optoelectronics and energy conversion applications. In general, such particles can be freestanding or surface passivated by polymers and small molecules for enhanced stability in solvents or engineering of their electronic and optical properties. These carbon particles will be referred to as carbon quantum dots (CQDs) and, besides their structure, their chemical composition can also vary depending on the precursors used for their synthesis affording doped CQDs with heteroatoms such as sulfur, boron, nitrogen and phosphorus.[129] CQDs have been effectively employed in a series of catalytic applications starting from classical chemical reactions (aldol, aza-Michael, and polymerization) to advanced energy applications (oxygen reduction or evolution and hydrogen evolution reactions).

2.7.1 Functionalization *via* Physisorption

CQDs, due to their partial 2D crystalline nature, can be effectively functionalized *via* π–π electrostatic interactions with π-extended aromatic molecules. N-doped CQDs were prepared by the hydrothermal treatment of citric acid with urea in a 3 : 1 molar ratio. Two types of perylene-diimide derivatives terminated either to a quaternized ammonium group or carboxylic acids were employed as candidates for the electrostatic functionalization of N-doped CQDs. Titration essays in phosphate-buffered D_2O (pH 7.2) at room temperature, followed by recording the UV-Vis absorption and fluorescence spectra, have shown that the positively charged PDI-ammonium derivative was successively attached onto the surface of the carbon dots in contrast to the carboxyl-terminated PDI. The light-harvesting properties of N-doped CQDs and the accepting properties of the PDI-derivatives could be effectively combined in photovoltaic applications.[143]

Carbon nanoforms are usually modified with, or are imbedded into, a polymer matrix, forming composites, which enhance their mechanical properties, stability *etc*. The same stands for CQDs, since their electronic and photophysical properties can be exploited in optoelectronic applications as flexible transparent photoresponsive thin films and photodiodes. Interestingly, the broadband emission of CQDs could be realized by selecting appropriate precursors and synthesis methods in order to afford white-light-emitting materials.[144] A novel composite derived by mixing two different CQD/polymer nanoensembles, synthesized by different carbon precursors [6-*O*-(*O*-*O*′-di-lauroyl-tartaryl)-D-glucose and 6-*O*-(*O*-*O*′-dilauroyl-tartaryl)-L-ascorbic acid] mixed with an elastomer and annealed in a two-step procedure (at 75 and 127 °C) in a 3 : 1 weight ratio was successfully tested as a warm-light-emitting hybrid material. Under these processes was realized the formation and mixing of green-light- and yellow-light-emitting (under UV light irradiation) CQDs embedded in a polymer matrix, which can be further engineered by substituting the precursors or the polymer matrix according to the target application.

Moving a step forward, CQDs produced *via* pyrolysis of green tea leaves can both contribute to chain elongation and catalyze polymerization *via* the functional groups on their surface. A representative work by Zhu *et al.* demonstrated the effective catalysis of the *in situ* free-radical polymerization of poly(4-styrenesulfonate) and poly(methyl acrylate) and the formation of valuable CQDs/polymer composites for optoelectronic applications due to a combination of the electronic properties and flexibility of the polymer matrix and the optical properties of the CQDs. The as-prepared CQDs increase the molecular weight and decrease the polydispersity of the polymer and, consequently, the CQD/polymer composite, indicating a smoother synthesis and a more homogeneous material.[145]

Along the same lines, studies on the ring-opening polymerization of ε-caprolactone catalyzed by tartaric acid with CQDs derived from the pyrolysis of a lotus leaf as co-catalysts under visible light irradiation have revealed that the photoexcited CQDs activate the monomer, accelerate the reaction, and the final product is a biocompatible and biodegradable light-responsive CQD/polycaprolactone composite.[146] The composite was easily synthesized in a one-step procedure at 110 °C under argon for 18 h and isolated by precipitation from a 10-fold excess of *n*-hexane. The hydrophobic nature, biocompatibility and biodegradability of the latter supramolecular CQD/polymer hybrid meet the requirements for applications in biomedicine, photocatalysis, bioimaging and environmental analysis.

2.7.2 Functionalization with Graphitic Nanostructures

Instead of elemental doping (discussed earlier and also summarized in Table 2.2 for individual CQDs), CQDs could also be hybridized with other materials to enhance their properties and long-term stability. Graphene derivatives

Table 2.2 A summary of reactions catalyzed by individual undoped, doped and co-doped CQDs.

Doping	Precursors	Size	Synthesis method	Application	Comments	Ref.
–	Graphite	1–4 nm	Electrochemical ablation of graphite	Photoinduced aldol reaction	Activation of carbonyl group *via* strong H-bonding	130
–	Graphite	5–10 nm	Electrochemical ablation of graphite	Photoinduced esterification, Beckman rearrangement	Activation of carbonyl group *via* strong H-bonding	131
–	β-carotene	2–5 nm	Hydrothermal treatment	Aza–Michael reaction, *O*-arylglycoside synthesis	Ambient conditions, 80% conversion, reaction time: 15–50 min	132
–	Graphite	1–4 nm	Alkali-assisted electrochemical method	NIR-light induced arylalcohol oxidation to acids	Conversion: NIR light >85%, no light <54%	133
–	Shoot from wood	1–6 nm	Thermal treatment with HNO_3	Photodegradation of methyl blue	O-content and pH dependent activity	134
–	Glucose	3 nm	Ultrasonic treatment with HCl	Photodegradation of rhodamine B	pH dependent activity	135
–	Glucose	8 nm	Ultrasonic treatment with HCl	Photodegradation of rhodamine B, methylene B, photoreduction of 4-nitrophenol	NIR-light dependent activity	136
N	Glucose, ammonia	10 nm	Ultrasonic treatment	Photodegradation of methyl orange with H_2O_2	Two-fold efficiency increment under visible light	137
N, Cl	Ethylene glycol (+NH_3 or $SOCl_2$)	2–4 nm	Pyrolysis in autoclave	Photodegradation of methyl blue	Doping-dependent activity	138
N	Soya milk	20–30 nm	Hydrothermal treatment	Oxygen reduction reaction (ORR)	$3e^-$ mechanism, high stability	139
N	TBAP/MeCN	2–5 nm	Electrochemical	ORR	24 h stability, methanol tolerance	140
N	Willow leaves	1–2 nm	Pyrolysis	ORR	$4e^-$ mechanism, 2000 cycles stability	141
N, S, P, B	Salts	2–4 nm	Oil candle flame	ORR	B,N co-doped $4e^-$ mechanism, high current density close to Pt–C P,N co-doped $4e^-$ mechanism	142

(GO, RGO, N-doped graphene *etc*) offer a series of advantages since graphene derivatives have remarkable electronic, structural and mechanical properties. CQDs/graphene nanoribbon (GNR) supramolecular nanohybrids have been synthesized by thermal treatment of a methylbenzene/hexabromobenzene mixture with sodium.[147] Under these conditions, CQDs with diameters of 2 nm have been successfully introduced onto a 1 μm-long GNR and were found to be active towards the ORR reaction with a more positive onset potential and a diffusion current density 40% higher than Pt–C. The electrocatalytic activity of the CQD/GNR ensembles is attributed to charge transfer between CQDs and GNRs.

Hybrid materials consisting of CQDs and CNTs have also been presented.[148] SWCNTs of different diameters treated with poly(4-vinylbenzyl trimethylamine) (PVBTA) resulted in positively charged CNT derivatives, which were subsequently used as templates for the stabilization of negatively charged N-CQDs observed by the pyrolysis of a citric acid–urea mixture. The electrostatic attraction and stabilization of the N-CQD/SWCNT material is located between the negatively charged carboxylates on the N-CQDs surfaces and the ammonium cations on the SWNT-PVBTA assemblies. Upon photoexcitation, charge injection into the SWNTs was observed, indicating the photoinduced electron transfer from the N-CQDs to the nanotubes.

Treating acetic acid with phosphorus pentoxide (P_2O_5) results in the formation of fluorescent CQDs in a single-step procedure without heating.[149] *Via* simple injection of CQDs into a GO suspension, CQD/GO ensembles have been formed. The ensembles have shown ultrafast electron transfer from the nanodots to the GO, indicating useful candidates for the fabrication of efficient photoanodes.

Following a two-step pyrolysis of a GO/dopamine mixture and a two-step calcination process, hybrid ensembles consisting of 5 nm (average diameter) N-CQDs deposited onto thermally reduced graphene have been prepared.[150] In detail, dopamine was thermally polymerized in the presence of GO, resulting in a GO/PDA, which was subsequently calcined in a two-step process to obtain the final N-CQDs/rGO ensembles. The observed material has shown ORR activity and good methanol tolerance.

In a procedure employing citric acid, urea and GO, CQDs/GO hybrids doped with nitrogen were efficiently prepared in one step.[151] The preparation of undoped CQDs/GO hybrids is also possible; nevertheless, the undoped material has significantly different activity towards the ORR. CQD/GO ensembles, where only the CQDs are N-doped, have also been successfully prepared and tested for their ORR activity, which proceeds *via* a four-electron mechanism.[152]

CQDs synthesized from petroleum coke and GO sheets (Hummer's method) have been assembled and post-treated with ammonia at 180 °C, resulting in nitrogen doping of the ensembles.[153] The as-prepared N-doped (CQD/GO) material has shown ORR activity close to commercial 20% Pt–C catalysts and the electron transfer number was calculated to be about 3.9, close to the four-electron mechanism for water formation over peroxide.

Moreover, they possessed superior durability and methanol tolerance in contrast to the commercial platinum catalyst.

S,N co-doped CQDs prepared by thermal carbonization of L-cysteine were mixed with graphite oxide, derived from Hummer's method, and sonicated well until a stable suspension was formed. CQDs with an average diameter of 4 nm were distributed well onto the GO flakes (~500 nm) as indicated by TEM, AFM and SEM imaging.[154] The observed S,N co-doped CQD/GO hybrids were tested as photoelectrodes based on electron transfer from the doped CQDs to the GO sheets.

An interesting method to fabricate an all B,N co-doped CQDs/GO assembly was developed by Fei *et al.* and was based on the synthesis of individual CQDs from coal pyrolysis blended with CQDs, followed by a post B,N co-doping/annealing process (Figure 2.8). The resulting B,N co-doped hybrid was found to be active towards the electrocatalysis of ORR and followed the $4e^-$ mechanism with a higher kinetic current than commercial Pt–C.[155]

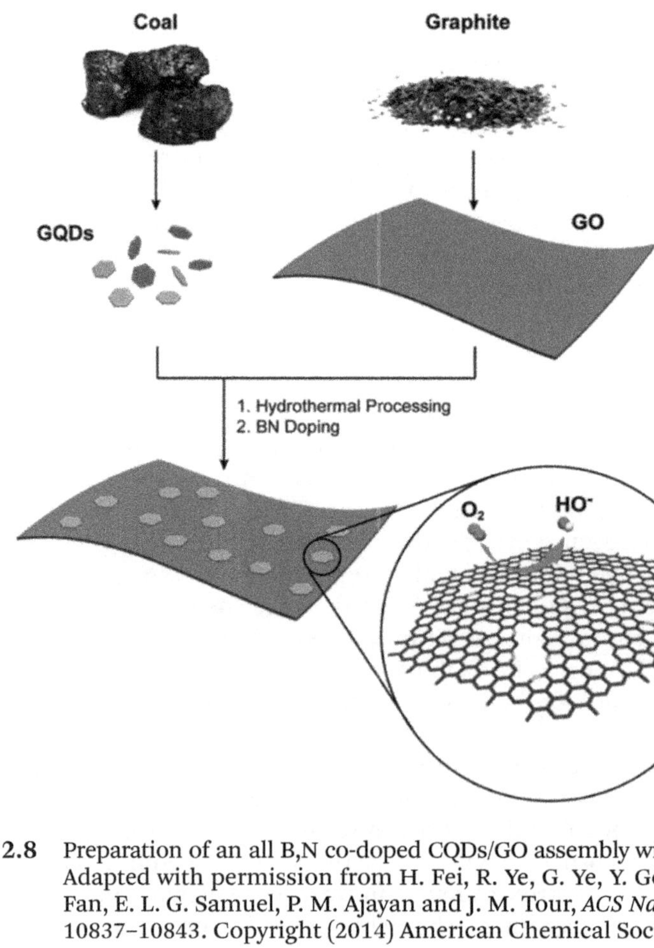

Figure 2.8 Preparation of an all B,N co-doped CQDs/GO assembly with ORR activity. Adapted with permission from H. Fei, R. Ye, G. Ye, Y. Gong, Z. Peng, X. Fan, E. L. G. Samuel, P. M. Ajayan and J. M. Tour, *ACS Nano*, 2014, **8**(10), 10837–10843. Copyright (2014) American Chemical Society.

A more sophisticated microbial-engaged synthesis of CQDs/rGO carried out by *S. oneidensis* proceeded to the formation of S,N,P co-doped CQDs/rGO ensembles.[156] The microbe initially reduced the oxygen functionalities of GO and the observed mixture (microbe and rGO) was subjected to a hydrothermal treatment inducing the carbonization of the microbe. Between the nanodots and the rGO, electron transfer phenomena occurred, which justified the ORR activity of the ensembles with an acceptable methanol tolerance and a current density close to commercial Pt–C.

In a case study, different amounts of CQDs, derived by graphite ablation, were mixed with graphitic carbon nitride (g-C_3N_4) and heated in an autoclave reactor affording g-C_3N_4 nanosheets decorated with CQDs (3–6 nm diameter).[157] This simple procedure allows the preparation of a variety of CQD/g-C_3N_4 nanoensembles with application in photocatalysis (*i.e.* methylene blue degradation) *via* the photoinduced electron transfer from g-C_3N_4 to CQDs.

The electron–hole separation–recombination rate can be further controlled by hybridization with g-C_3N_4. Trials with CQDs/g-C_3N_4 hybrids, with CQDs content varying from 0.1–1%, degrade effectively the phenol molecules used as target substrates *via* the generation of oxidative hydroxyl and peroxide radicals.[158] The composite was active under visible light, due to the absorption properties of g-C_3N_4, which is the electron donating moiety.

CQDs, prepared by pyrolysis of EDTA, supported on g-C_3N_4 have been demonstrated as efficient photoelectrocatalysts for the hydrogen evolution reaction (HER).[159] This work demonstrated the first CQD-based NIR photoelectrocatalyst for the HER. For the preparation of the CQD/g-C_3N_4 hybrids, a hydrothermal method was followed, and a mixture of the individuals was heated at 180 °C for 4 h and collected through centrifugation in a powdered form. The CQD loading can be varied, although the optimum value for the application is 10% w/w and the hydrogen production is 7 μmol h^{-1} under 808 nm laser irradiation.

CQD/g-C_3N_4 hybrids have also been tested for the HER under UV-Vis irradiation.[160] In this case, L-ascorbic acid was used as a precursor for the CQDs. The formation of CQD/g-C_3N_4 nanohybrids prepared *via* the hydrothermal procedure has been described previously; the deposited carbon dots have an average diameter of 5 nm. The use of CQDs enhances the photocurrent produced by individual g-C_3N_4 by 30% with excellent stability and the production of H_2 is 2.2 μmol h^{-1}.

2.8 Conclusions

Metal-free, non-covalent functionalization of carbon nanoforms represents a facile and cost-effective strategy for the synthesis of effective catalysts for a variety of applications including electrocatalysis, photocatalysis, bio-sensing, energy storage and others. The major advantage of non-covalent modification of carbon nanostructured materials relates to the maintenance of all their exceptional mechanical and structural properties, due to non-disruption of their basic backbone. At the same time, the absence of any metallic adjuvants renders these materials as environmentally friendly, low-cost nanocatalysts with great potential for bio-applications. However, their drawbacks—lower stability and weaker immobilization—compared to the corresponding metal-based and covalently-functionalized carbon nanocatalysts, limit their potential in nanoapplications, and, in this context, the exploration of their chemistry needs to be further developed.

References

1. A. Ciesielski and P. Samori, *Chem. Soc. Rev.*, 2014, **43**, 381–398.
2. Y. Xu, H. Bai, G. Lu, C. Li and G. Shi, *J. Am. Chem. Soc.*, 2008, **130**, 5856–5857.
3. L.-M. Lu, X.-L. Qiu, X.-B. Zhang, G.-L. Shen, W. Tan and R.-Q. Yu, *Biosens. Bioelectron.*, 2013, **45**, 102–107.
4. Q. Zhang, Y. Qiao, F. Hao, L. Zhang, S. Wu, Y. Li, J. Li and X.-M. Song, *Chem.-Eur. J.*, 2010, **16**, 8133–8139.
5. M. S. Ahmed and S. Jeon, *J. Power Sources*, 2012, **218**, 168–173.
6. Z. Mou, Y. Dong, S. Li, Y. Du, X. Wang, P. Yang and S. Wang, *Int. J. Hydrogen Energy*, 2011, **36**, 8885–8893.
7. B. Vedhanarayanan, B. Babu, M. M. Shaijumon and A. Ajayaghosh, *ACS Appl. Mater. Interfaces*, 2017, **9**, 19417–19426.
8. S. Wang, D. Yu, L. Dai, D. W. Chang and J.-B. Baek, *ACS Nano*, 2011, **5**, 6202–6209.
9. J. Zhang, J. Lei, R. Pan, Y. Xue and H. Ju, *Biosens. Bioelectron.*, 2010, **26**, 371–376.
10. F. Gong, X. Xu, G. Zhou and Z.-S. Wang, *Phys. Chem. Chem. Phys.*, 2013, **15**, 546–552.
11. W. Liu, Y. Fang, P. Xu, Y. Lin, X. Yin, G. Tang and M. He, *ACS Appl. Mater. Interfaces*, 2014, **6**, 16249–16256.
12. X. Xu, D. Huang, K. Cao, M. Wang, S. M. Zakeeruddin and M. Grätzel, *Sci. Rep.*, 2013, **3**, 1489.
13. W. Hong, Y. Xu, G. Lu, C. Li and G. Shi, *Electrochem. Commun.*, 2008, **10**, 1555–1558.
14. Q. Wu, Y. Xu, Z. Yao, A. Liu and G. Shi, *ACS Nano*, 2010, **4**, 1963–1970.
15. J. Lu, W. Liu, H. Ling, J. Kong, G. Ding, D. Zhou and X. Lu, *RSC Adv.*, 2012, **2**, 10537–10543.
16. H. Cai, Q. Tang, B. He, M. Wang, S. Yuan and H. Chen, *Electrochim. Acta*, 2014, **121**, 136–142.
17. M. Wang, Q. Tang, H. Chen and B. He, *Electrochim. Acta*, 2014, **125**, 510–515.
18. H. Wang, Q. Hao, X. Yang, L. Lu and X. Wang, *Nanoscale*, 2010, **2**, 2164–2170.
19. K. Zhang, L. L. Zhang, X. S. Zhao and J. Wu, *Chem. Mater.*, 2010, **22**, 1392–1401.
20. J. Xu, K. Wang, S.-Z. Zu, B.-H. Han and Z. Wei, *ACS Nano*, 2010, **4**, 5019–5026.

21. G. Wang, W. Xing and S. Zhuo, *Electrochim. Acta*, 2012, **66**, 151–157.
22. R. Wang, M. Han, Q. Zhao, Z. Ren, X. Guo, C. Xu, N. Hu and L. Lu, *Sci. Rep.*, 2017, **7**, 44562.
23. M. Xue, F. Li, J. Zhu, H. Song, M. Zhang and T. Cao, *Adv. Funct. Mater.*, 2012, **22**, 1284–1290.
24. K. S. Lee, Y. Lee, J. Y. Lee, J.-H. Ahn and J. H. Park, *ChemSusChem*, 2012, **5**, 379–382.
25. Y. Tang, N. Wu, S. Luo, C. Liu, K. Wang and L. Chen, *Macromol. Rapid Commun.*, 2012, **33**, 1780–1786.
26. Y. S. Lim, N. M. Huang and H. N. Lim, *Sci. Rep.*, 2014, **4**, 5305.
27. C.-Y. Liu, K.-C. Huang, P.-H. Chung, C.-C. Wang, C.-Y. Chen, R. Vittal, C.-G. Wu, W.-Y. Chiu and K.-C. Ho, *J. Power Sources*, 2012, **217**, 152–157.
28. S. Nagarajan, P. Sudhagar, V. Raman, W. Cho, K. S. Dhathathreyanc and Y. S. Kang, *J. Mater. Chem. A*, 2013, **1**, 1048–1054.
29. H. Zheng, C. Y. Neo and J. Ouyang, *ACS Appl. Mater. Interfaces*, 2013, **5**, 6657–6664.
30. X. Yang, A. Liu, Y. Zhao, H. Lu, Y. Zhang, W. Wei, Y. Li and S. Liu, *ACS Appl. Mater. Interfaces*, 2015, **7**, 23731–23740.
31. S. Liu, P. Gordiichuk, Z.-S. Wu, Z. Liu, W. Wei, M. Wagner, N. Mohamed-Noriega, D. Wu, Yi. Mai, A. Herrmann, K. Müllen and X. Feng, *Nat. Commun.*, 2015, **6**, 8817–8825.
32. D. J. Bozym, C. Punckt and I. A. Aksay, *ACS Nano*, 2010, **4**, 6203–6211.
33. J. D. Roy-Mayhew, G. Boschloo, A. Hagfeldt and I. A. Aksay, *ACS Appl. Mater. Interfaces*, 2012, **4**, 2794–2800.
34. Z.-L. Wang, D. Xu, H.-G. Wang, Z. Wu and X.-B. Zhang, *ACS Nano*, 2013, **7**, 2422–2430.
35. W. Yang, T.-P. Fellinger and M. Antonietti, *J. Am. Chem. Soc.*, 2011, **133**, 206–209.
36. T.-P. Fellinger, F. Hasché, P. Strasser and M. Antonietti, *J. Am. Chem. Soc.*, 2012, **134**, 4072–4075.
37. W. Wei, H. Liang, K. Parvez, X. Zhuang, X. Feng and K. Müllen, *Angew. Chem., Int. Ed.*, 2014, **53**, 1570–1574.
38. M. J. Ju, I.-Y. Jeon, K. Lim, J. C. Kim, H.-J. Choi, I. T. Choi, Y. K. Eom, Y. J. Kwon, J. Ko, J.-J. Lee, J.-B. Baek and H. K. Kim, *Energy Environ. Sci.*, 2014, 7, 1044–1052.
39. L. Lai, J. R. Potts, D. Zhan, L. Wang, C. K. Poh, C. Tang, H. Gong, Z. Shen, J. Lin and R. S. Ruoff, *Energy Environ. Sci.*, 2012, **5**, 7936–7942.
40. Z. Lin, G. H. Waller, Y. Liu, M. Liu and C.-P. Wong, *Carbon*, 2013, **53**, 130–136.
41. R. Li, Z. Wei and X. Gou, *ACS Catal.*, 2015, **5**, 4133–4142.
42. P.-Y. Chen, C.-T. Li, C.-P. Lee, R. Vittal and K.-C. Ho, *Nano Energy*, 2015, **12**, 374–385.
43. G. Paterakis, D. Raptis, A. Ploumistos, M. Belekoukia, L. Sygellou, M. S. Ramasamy, P. Lianos and D. Tasis, *Appl. Surf. Sci.*, 2017, **423**, 443–450.
44. H.-J. Ahn, I.-H. Kim, J.-C. Yoon, S.-I. Kim and J.-H. Jang, *Chem. Commun.*, 2014, **50**, 2412–2415.

45. B. Li and Z. Xu, *J. Am. Chem. Soc.*, 2009, **131**, 16380–16382.
46. R. Kumar, E. H. Gleisser, E. G. V. Tiu and Y. Yamakoshi, *Org. Lett.*, 2016, **18**, 184–187.
47. Y. Guo, W. Li, J. Yan, B. Moosa, M. Amad, C. J. Werth and N. M. Khashab, *Chem.-Asian J.*, 2012, **7**, 2842–2847.
48. G. A. Rance and A. N. Khlobystov, *Nanoscale*, 2014, **6**, 11141–11146.
49. J. Niemeyer and D. Erker, *ChemCatChem*, 2010, **2**, 499–500.
50. M. Orfanopoulos and S. Kambourakis, *Tetrahedron Lett.*, 1994, **35**, 1945–1948.
51. N. Tagmatarchis and H. Shinohara, *Org. Lett.*, 2000, **2**, 3551–3554.
52. N. Tagmatarchis, H. Shinohara, M. Fujitsuka and O. Ito, *J. Org. Chem.*, 2001, **66**, 8026–8029.
53. S. Takekuma, H. Takekuma and Z. Yoshida, *Chem. Commun.*, 2005, 1628–1630.
54. N. D. McClenaghan, C. Absalon and D. M. Bassani, *J. Am. Chem. Soc.*, 2003, **125**, 13004–13005.
55. G. Pagona, A. Stergiou, H. B. Gobeze, F. D'Souza and N. Tagmatarchis, *Phys. Chem. Chem. Phys.*, 2016, **18**, 811–817.
56. A. Gouloumis, S.-G. Liu, Á. Sastre, P. Vázquez, L. Echegoyen and T. Torres, *Chem.–Eur. J.*, 2000, **6**, 3600–3607.
57. I. Petsalakis, D. Tzeli, I. Kerkines and G. Theodorakopoulos, *Comput. Theor. Chem.*, 2011, **965**, 168–175.
58. A. Privitera, M. Righetto, D. Mosconi, F. Lorandi, A. A. Isse, A. Moretto, R. Bozio, C. Ferrante and L. Franco, *Phys. Chem. Chem. Phys.*, 2016, **18**, 31286–31295.
59. C. Yu, Y. Jin and W. Zhang, *Chem. Rec.*, 2015, **15**, 97–106.
60. D. Tasis, N. Tagmatarchis, A. Bianco and M. Prato, *Chem. Rev.*, 2006, **106**, 1105–1136.
61. O. P. Ferreira, L. Otubo, A. L. Aguiar, J. J. A. Silva, J. M. Filho, A. G. Souza Filho, S. B. Fagan and O. L. Alves, *J. Nanopart. Res.*, 2009, **11**, 2163–2170.
62. F. Lu, X. Wang, M. J. Meziani, L. Cao, L. Tian, M. A. Bloodgood, J. Robinson and Y.-P. Sun, *Langmuir*, 2010, **26**, 7561–7564.
63. C. Backes, U. Mundloch, C. D. Schmidt, J. N. Coleman, W. Wohlleben, F. Hauke and A. Hirsch, *Chem.–Eur. J.*, 2010, **16**, 13185–13192.
64. C. M. Damian, G. Sorina, E. Vasile and H. Iovu, *Composites, Part B*, 2012, **43**, 3507–3515.
65. N. Su, Y. N. Wei, L. Y. Qi, Y. C. Pang, Y. Z. Guo, P. Li, J. Li and H. P. Li, *eXPRESS Polym. Lett.*, 2011, **12**, 1102–1112.
66. W. Xue, L. Qi, X. Li, S. Huang, H. Li, X. Guan, G. Bai and L.-E. Liu, *Chem. Eng. J.*, 2012, **209**, 118–125.
67. C.-C. Yang, Y.-C. Lin, P.-I. Wang, D.-J. Liaw and S.-W. Kuo, *Polymer*, 2014, **55**, 2044–2050.
68. T. Fukushima and T. Aida, *Chem.–Eur. J.*, 2007, **13**, 5048–5058.
69. Y. Xiang, S. Lu and S. P. Jiang, *Chem. Soc. Rev.*, 2012, **7**, 7291–7321.
70. D. M. Guldi, E. Menna, M. Maggini, M. Marcaccio, D. Paolucci, F. Paolucci, S. Campidelli, M. Prato, G. M. A. Rahman and S. Schergna, *Chem.–Eur. J.*, 2006, **12**, 3975–3983.

71. A. S. D. Sandanayaka, E. Maligaspe, T. Hasobe, O. Ito and F. D'Souza, *Chem. Commun.*, 2010, **46**, 8749–8751.
72. S. Y. An, S. Sun and J. K. Oh, *Macromol. Rapid Commun.*, 2016, **37**, 705–710.
73. Z. Wang, Y. Wang and G. Luo, *Electroanalysis*, 2003, **15**, 1129–1133.
74. Z. Wang, S. Xiao and Y. Chen, *J. Electroanal. Chem.*, 2006, **589**, 237–242.
75. X. Gu, W. Qi, S. Wu, Z. Sun, X. Xu and D. Su, *Catal. Sci. Technol.*, 2014, **4**, 1730–1733.
76. A. Guiseppi-Elie, S.-H. Choi, K. E. Geckeler, B. Sivaraman and R. A. Latour, *NanoBiotechnology*, 2008, **4**, 9–17.
77. F. Gutiérrez, M. D. Rubianes and G. A. Rivas, *Sens. Actuators, B*, 2012, **161**, 191–197.
78. A. A. Arbab, K. C. Sun, I. A. Sahito, M. B. Qadi and S. H. Jeong, *Phys. Chem. Chem. Phys.*, 2015, **17**, 12957–12969.
79. S. Zhang, P. Kang, S. Ubnoske, M. K. Brennaman, N. Song, R. L. House, J. T. Glass and T. J. Meyer, *J. Am. Chem. Soc.*, 2014, **136**, 7845–7848.
80. M. D. Rubianes and G. A. Rivas, *Electrochem. Commun.*, 2007, **9**, 480–484.
81. Y. Jalit, M. C. Rodríguez, M. D. Rubianes, S. Bollo and G. A. Rivas, *Electroanalysis*, 2008, **20**, 1623–1631.
82. K. Aitola, M. Borghei, A. Kaskela, E. Kemppainen, A. G. Nasibulin, E. I. Kauppinen, P. D. Lund, V. Ruìz and J. Halme, *J. Electroanal. Chem.*, 2012, **683**, 70–74.
83. T. Okazaki, Y. Iizumi, S. Okubo, H. Kataura, Z. Liu, K. Suenaga, Y. Tahara, M. Yudasaka, S. Okada and S. Iijima, *Angew. Chem., Int. Ed.*, 2011, **50**, 4853–4857.
84. A. Chuvilin, E. Bichoutskaia, M. C. Gimenez-Lopez, T. W. Chemberlain, G. A. Rance, N. Kuganathan, J. Biskupek, U. Kaiser and A. N. Khlobystov, *Nat. Mater.*, 2011, **10**, 687–692.
85. O. Vostrowsky and A. Hirsch, *Angew. Chem., Int. Ed.*, 2004, **43**, 2326–2329.
86. T. W. Chamberlain, R. Pfeiffer, J. Howells, H. Peterlik, H. Kuzmany, B. Kräutler and T. Da Ros, *Nanoscale*, 2012, **4**, 7540–7548.
87. Y. Iizumi, T. Okazaki, Z. Liu, K. Suenaga, T. Nakanishi, S. Iijima, G. Rotas and N. Tagmatarchis, *Chem. Commun.*, 2010, **46**, 1293–1295.
88. F. Banhart, *Nat. Mater.*, 2011, **10**, 651–652.
89. M. Fujihara, Y. Miyata, R. Kitaura, Y. Nishimura, C. Camacho, S. Irle, Y. Iizumi, T. Okazaki and H. Shinohara, *J. Phys. Chem. C*, 2012, **116**, 15141–15145.
90. T. Koyama, T. Tsunekawa, T. Saito, K. Asaka, Y. Saito, H. Kishida and A. Nakamura, *J. Phys. Chem. C*, 2014, **118**, 21671–21681.
91. H. Kuzmany, W. Plank, Ch. Schaman, R. Pfeiffer, F. Hasi, F. Simon, G. Rotas, G. Pagona and N. Tagmatarchis, *J. Raman Spectrosc.*, 2007, **38**, 704–713.
92. S.-K. Joung, T. Okazaki, S. Okadaa and S. Iijima, *Phys. Chem. Chem. Phys.*, 2010, **12**, 8118–8122.
93. A. I. Chernov, P. V. Fedotov, A. V. Talyzin, I. S. Lopez, I. V. Anoshkin, A. G. Nasibulin, E. I. Kauppinen and E. D. Obraztsova, *ACS Nano*, 2013, **7**, 6346–6353.

94. I. V. Anoshkin, A. V. Talyzin, A. G. Nasibulin, A. V. Krasheninnikov, H. Jiang, R. M. Nieminen and E. I. Kauppinen, *ChemPhysChem*, 2014, **15**, 1660–1665.
95. E. J. E. Stuart and M. Pumera, *Chem.–Asian J.*, 2011, **6**, 804–807.
96. C. H. Choi, M. W. Chung, H. C. Kwon, J. H. Chung and S. I. Woo, *Appl. Catal., B*, 2014, **144**, 760–776.
97. X. Zhou, Z. Tian, J. Li, H. Ruan, Y. Ma, Z. Yang and Y. Qu, *Nanoscale*, 2014, **6**, 2603–2607.
98. S. Wang, S. Dong, J. Wang, L. Zhang, P. Han, C. Zhang, X. Wang, K. Zhang, Z. Lan and G. Cui, *J. Mater. Chem.*, 2012, **22**, 21051–21056.
99. A. Zhao, J. Masa, M. Muhler, W. Schuhmann and W. Xia, *Electrochim. Acta*, 2013, **98**, 139–145.
100. P. Chen, T.-Y. Xiao, Y.-H. Qian, S.-S. Li and S.-H. Yu, *Adv. Mater.*, 2013, **25**, 3192–3196.
101. A. Morozan, P. Jégou, M. Pinault, S. Campidelli, B. Jousselme and S. Palacin, *ChemSusChem*, 2012, **5**, 647–651.
102. K. Chizari, A. Deneuve, O. Ersen, I. Florea, Y. Liu, D. Edouard, I. Janowska, D. Begin and C. Pham-Huu, *ChemSusChem*, 2012, **5**, 102–108.
103. K. Zhou, B. Li, Q. Zhang, J.-Q. Huang, G.-L. Tian, J.-C. Jia, M.-Q. Zhao, G.-H. Luo, D. S. Su and F. Wei, *ChemSusChem*, 2014, **7**, 723–728.
104. W. Qi, W. Liu, S. Liu, B. Zhang, X. Gu, X. Guo and D. Su, *ChemCatChem*, 2014, **6**, 2613–2620.
105. H. Yu, F. Peng, J. Tan, X. Hu, H. Wang, J. Yang and W. Zheng, *Angew. Chem., Int. Ed.*, 2011, **50**, 3978–3982.
106. F. Wang, S. Song, K. Li, J. Li, J. Pan, S. Yao, X. Ge, J. Feng, X. Wang and H. Zhang, *Adv. Mater.*, 2016, **28**, 10679–10683.
107. D. C. Higgins, M. A. Hoque, F. Hassan, J.-Y. Choi, B. Kim and Z. Chen, *ACS Catal.*, 2014, **4**, 2734–2740.
108. N. Karousis, I. Suarez-Martinez, C. P. Ewels and N. Tagmatarchis, *Chem. Rev.*, 2016, **116**, 4850–4883.
109. H. Dai, C. Yang, X. Ma, Y. Lina and G. Chen, *Chem. Commun.*, 2011, **47**, 11915–11917.
110. J. Li, Z. He, C. Guo, L. Wang and S. Xu, *J. Lumin.*, 2014, **145**, 74–80.
111. G. Mountrichas, T. Ichihashi, S. Pispas, M. Yudasaka, S. Iijima and N. Tagmatarchis, *J. Phys. Chem. C*, 2009, **113**, 5444–5449.
112. X. Liua, H. Li, F. Wanga, S. Zhua, Y. Wanga and G. Xu, *Biosens. Bioelectron.*, 2010, **25**, 2194–2199.
113. D. Chen, C. Wang, X. Nie, S. Li, R. Li, M. Guan, Z. Liu, C. Chen, C. Wang, C. Shu and L. Wan, *Adv. Funct. Mater.*, 2014, **24**, 6621–6628.
114. G. Pagona, A. S. D. Sandanayaka, Y. Araki, J. Fan, N. Tagmatarchis, M. Yudasaka, S. Iijima and O. Ito, *J. Phys. Chem. B*, 2006, **110**, 20729–20732.
115. A. S. D. Sandanayaka, O. Ito, T. Tanaka, H. Isobe, E. Nakamura, M. Yudasaka and S. Iijima, *New J. Chem.*, 2009, **33**, 2261–2266.
116. F. Lodermeyer, M. Prato, R. D. Costa and D. M. Guldi, *Nanoscale*, 2016, **8**, 7556–7561.

117. G. Pagona, J. Fan, A. Maignè, M. Yudasaka, S. Iijima and N. Tagmatarchis, *Diamond Relat. Mater.*, 2007, **16**, 1150–1153.
118. G. Pagona, A. S. D. Sandanayaka, A. Maignè, J. Fan, C. Papavassiliou, I. D. Petsalakis, B. R. Steele, M. Yudasaka, S. Iijima, N. Tagmatarchis and O. Ito, *Chem.–Eur. J.*, 2007, **13**, 7600–7607.
119. A. M. Schrand, H. Huang, C. Carlson, J. J. Schlager, E. Ōsawa, S. M. Hussain and L. Dai, *J. Phys. Chem. B*, 2007, **111**, 2–7.
120. S. Liu, T. H. Zeng, M. Hofmann, E. Burcombe, J. Wei, R. Jiang, J. Kong and Y. Chen, *ACS Nano*, 2011, **5**, 6971–6980.
121. J. Zhang, D. Su, A. Zhang, D. Wang, R. Schlögl and C. Hébert, *Angew. Chem., Int. Ed.*, 2007, **46**, 7319–7323.
122. Y. Lin and D. Su, *ACS Nano*, 2014, **8**, 7823–7833.
123. T. Skaltsas, S. Pispas and N. Tagmatarchis, *Mater. Res. Express*, 2015, **2**, 115005.
124. T.-T.-B. Nguyen, H.-C. Chang and V. W.-K. Wu, *Diamond Relat. Mater.*, 2007, **16**, 872–876.
125. E. Perevedentseva, C.-Y. Cheng, P.-H. Chung, J.-S. Tu, Y.-H. Hsieh and C.-L. Cheng, *Nanotechnology*, 2007, **18**, 315102.
126. E. Perevedentseva, P.-J. Cai, Y.-C. Chiu and C.-L. Cheng, *Langmuir*, 2011, **27**, 1085–1091.
127. A. Chatterjee, E. Perevedentseva, C.-Y. Cheng, Y.-S. Ye, P.-H. Chung and C.-L. Cheng, *J. Biomed. Opt.*, 2015, **20**, 051014.
128. X. Qi, T. Wang, Y. Long and J. Ni, *Sci. Rep.*, 2015, **5**, 10388.
129. S. Y. Lim, W. Shen and Z. Gao, *Chem. Rev.*, 2015, **44**, 362–381.
130. Y. Han, H. Huang, H. Zhang, Y. Liu, X. Han, R. Liu, H. Li and Z. Kang, *ACS Catal.*, 2014, **4**, 781–787.
131. H. Li, R. Liu, W. Kong, J. Liu, Y. Liu, L. Zhou, X. Zhang, S.-T. Lee and Z. Kang, *Nanoscale*, 2014, **6**, 867–873.
132. B. Majumdar, S. Mandani, T. Bhattacharya, D. Sarma and T. K. Sarma, *J. Org. Chem.*, 2017, **82**, 2097–2106.
133. H. Li, R. Liu, S. Lian, Y. Liu, H. Huang and Z. Kang, *Nanoscale*, 2013, **5**, 3289–3297.
134. S. Hu, M. Tian, L. Wu, Q. Zhao, J. Yang, J. Liu and S. Cao, *Chem.–Asian J.*, 2013, **8**, 1035–1041.
135. P. Yang, J. Zhao, L. Zhang, L. Li and Z. Zhu, *Chem.–Eur. J.*, 2015, **21**, 8561–8568.
136. H. Wang, J. Zhuang, D. Velado, Z. Wei, H. Matsui and S. Zhou, *ACS Appl. Mater. Interfaces*, 2015, **7**, 27703–27712.
137. Z. Ma, H. Ming, H. Huang, Y. Liu and Z. Kang, *New J. Chem.*, 2012, **36**, 861–864.
138. S. Hu, R. Tian, Y. Dong, J. Yang, J. Liu and Q. Chang, *Nanoscale*, 2013, **5**, 11665–11671.
139. C. Zhu, J. Zhai and S. Dong, *Chem. Commun.*, 2012, **48**, 9367–9369.
140. Y. Li, Y. Zhao, H. Cheng, Y. Hu, G. Shi, L. Dai and L. Qu, *J. Am. Chem. Soc.*, 2012, **134**, 15–18.

141. S. Gao, Y. Chen, H. Fan, X. Wei, C. Hu, L. Wangb and L. Qu, *J. Mater. Chem. A*, 2014, **2**, 6320–6325.
142. Y. Han, D. Tang, Y. Yang, C. Li, W. Kong, H. Huang, Y. Liu and Z. Kang, *Nanoscale*, 2015, **7**, 5955–5962.
143. V. Strauss, J. T. Magraf, K. Dirian, Z. Syrgiannis, M. Prato, C. Wessendorf, A. Hirsch, T. Clark and D. M. Guldi, *Angew. Chem., Int. Ed.*, 2015, **54**, 8292–8297.
144. S. K. Bhunia, S. Nandi, R. Shikler and R. Jelinek, *Nanoscale*, 2016, **8**, 3400–3406.
145. C. Zhu, M. Yan, X. Shi, J. Fan and H. Bi, *RSC Adv.*, 2016, **6**, 38470–38474.
146. M. Yan, M. Zhou, J. Chen, T. Zhao, L. Tang and H. Bi, *Mater. Sci. Eng., C*, 2017, **79**, 76–83.
147. H. Jin, H. Huang, Y. He, X. Feng, S. Wang, L. Dai and J. Wang, *J. Am. Chem. Soc.*, 2015, **137**, 7588–7591.
148. V. Strauss, J. T. Magraf, T. Clark and D. M. Guldi, *Chem. Sci.*, 2015, **6**, 6878–6885.
149. P. Yu, X. Wen, Y.-R. Toh, Y.-C. Lee, K.-Y. Huang, S. Huang, S. Shrestha, G. Conibeer and J. Tang, *J. Mater. Chem. C*, 2014, **2**, 2894–2901.
150. B. Zhang, C. Xiao, Y. Xiang, B. Dong, S. Ding and Y. Tang, *ChemElectroChem*, 2016, **3**, 864–870.
151. W.-J. Niu, R.-H. Zhu, Yan-Hua, H.-B. Zeng, S. Cosnier, X.-J. Zhang and D. Shan, *Carbon*, 2016, **109**, 402–410.
152. M. Fan, C. Zhu, J. Yang and D. Sun, *Electrochim. Acta*, 2016, **216**, 102–109.
153. A. Privitera, M. Righetto, D. Mosconi, F. Lorandi, A. A. Isse, A. Moretto, R. Bozio, C. Ferrantea and L. Franco, *Phys. Chem. Chem. Phys.*, 2016, **18**, 31286–31295.
154. Z. Yang, J. Xiao, J.-Y. Wan, Z.-G. Liu, T.-T. Cao, W.-J. Zhanga and H.-X. Wang, *Phys. Chem. Chem. Phys.*, 2015, **17**, 32283–32288.
155. H. Fei, R. Ye, G. Ye, Y. Gong, Z. Peng, X. Fan, E. L. G. Samuel, P. M. Ajayan and J. M. Tour, *ACS Nano*, 2014, **8**, 10837–10843.
156. L. Zhou, P. Fu, Y. Wang, L. Suna and Y. Yuan, *J. Mater. Chem. A*, 2016, **4**, 7222–7229.
157. T. Qin, Z. You, H. Wang, Q. Shen, F. Zhang and H. Yang, *J. Electrochem. Soc.*, 2017, **164**, H211–H214.
158. H. Zhang, L. Zhao, F. Geng, L.-H. Guo, B. Wan and Y. Yang, *Appl. Catal., B*, 2016, **180**, 656–662.
159. X. Xia, N. Deng, G. Cui, J. Xie, X. Shi, Y. Zhao, Q. Wang, W. Wanga and B. Tang, *Chem. Commun.*, 2015, **51**, 10899–10902.
160. X. Wang, J. Cheng, H. Yu and J. Yu, *Dalton Trans.*, 2017, **46**, 6417–6424.

CHAPTER 3

Carbon Nitrides (g-C_3N_4) and Covalent Triazine Frameworks (CTFs)

DEVADUTTA NEPAK, VIJAY K. TOMER AND KAMALAKANNAN KAILASAM*

Institute of Nano Science and Technology (INST), Habitat Centre, Phase X, Sector 64, Mohali-160062, Punjab, India
*E-mail: kamal@inst.ac.in, kkamal17@gmail.com

3.1 Introduction

Triazine- and heptazine-based polymeric systems hold great promise for applications in a range of fields, namely optoelectronics, photocatalytic water splitting, gas adsorption, and sensors, and, most importantly, as multifunctional heterogeneous metal-free catalysts and catalyst support materials for several organic transformations.[1–4] Despite being described as the oldest materials in the chemical literature, they have attracted much attention after pioneering work done by Schnick and co-workers and Antonietti and co-workers. Carbon nitride (g-CN, commonly termed as g-C_3N_4)[†] is one of the

[†]g-C_3N_4 is most commonly used to denote polymeric graphitic carbon nitrides, which is incorrect as there is always around 1–3% of hydrogen left in the carbon nitride structure depending on the temperature used during the thermal condensation process; for example, at 550 °C, it is just polymeric melon. Hence, it should be termed g-CN but as it is referred to as g-C_3N_4 in most of the literature, we will refer to it as g-C_3N_4 to avoid confusion.

Catalysis Series No. 31
Metal-free Functionalized Carbons in Catalysis: Synthesis, Characterization and Applications
Edited by Alberto Villa and Nikolaos Dimitratos

Published by the Royal Society of Chemistry, www.rsc.org

familiar materials that belongs to such a class (triazine- and heptazine-based polymers).[5] The initial investigations on carbon nitrides have been known in the literature since the 1830s. The material is a linear polymer of interconnected tri-*s*-triazines through a secondary nitrogen with the general formula $(C_3N_3H)_n$, first reported by Berzelius; the name "melon" was coined by Liebig.[6,7] Later on, seven different phases of carbon nitrides, namely α-C_3N_4, β-C_3N_4, cubic C_3N_4, pseudocubic C_3N_4, g-h-triazine, g-o-triazine and g-h-heptazine, were identified.[8] All these allotropes were constructed with triazine (C_3N_3) and tri-*s*-triazine/heptazine (C_6N_7) rings as basic tectonic building blocks. However, among all the allotropes, tri-*s*-triazine-based graphitic carbon nitride (g-C_3N_4) was reported as a highly stable phase, which was due to its poly(tri-*s*-triazine) structure and the high degree of condensation (Figure 3.1).[9] This was also supported by many theoretical studies.[10,11] g-C_3N_4 material is structurally similar to that of graphite bearing strong covalent C–N bonds in conjugation connected through tertiary nitrogen atoms, while the layers are held by van der Waals forces. The defect-rich, N-bridged polymer with a medium-band gap, an in-plane quantum confinement effect and indirect semiconductor features predominantly favour its usage in many applications, in particular, as a multifunctional heterogeneous metal-free catalyst.[2] Apart from the usage of g-C_3N_4 materials in hydrogen evolution and photocatalytic dye degradation, these materials find interesting applications in Friedel–Crafts type reactions,[12] O_2 activation for the selective oxidation of alcohols,[13] oxidative coupling of amines[14] and CO_2 activation/reduction[15–17] reactions.

Like g-C_3N_4, another class of dynamic polymerization systems with porous crystalline or amorphous triazine-based frameworks is obtained through the trimerization of aromatic polynitriles and such materials were named as covalent triazine frameworks (CTFs), introduced by Antonietti and Thomas and their co-workers.[18,19] The first report on the dynamic condensation reaction of boronic acids for the successful synthesis of covalent organic frameworks (boroxine COFs) by Yaghi and co-workers[20,21] has spurred many researchers to work in this field and in particular on covalent triazine frameworks (CTFs), where the triazines in the latter are thermally and chemically much more stable than the boroxine rings in the former. The tailored construction of such extended porous frameworks from versatile molecular building blocks paves the way to set up a connection between molecular and solid properties in a well-defined, pre-ordered chemical environment. These porous solid materials have significant applications in sorption, gas storage, catalysis (as active catalysts or supports), and also as electrodes in batteries and supercapacitors.[22,23] In particular, covalent triazine frameworks (CTFs) are a special and emerging class of COFs[24,25] because they are completely inert in water, unlike many metal organic frameworks (MOFs) known in the literature. Additionally, they are cheap compared to a number of functional materials and they offer modularity to functionalization of the surface and tunable pore sizes. Hence, the rational synthesis and design of extended arrays of organic matter in bulk, solution, crystals and as thin films has always been desirable.

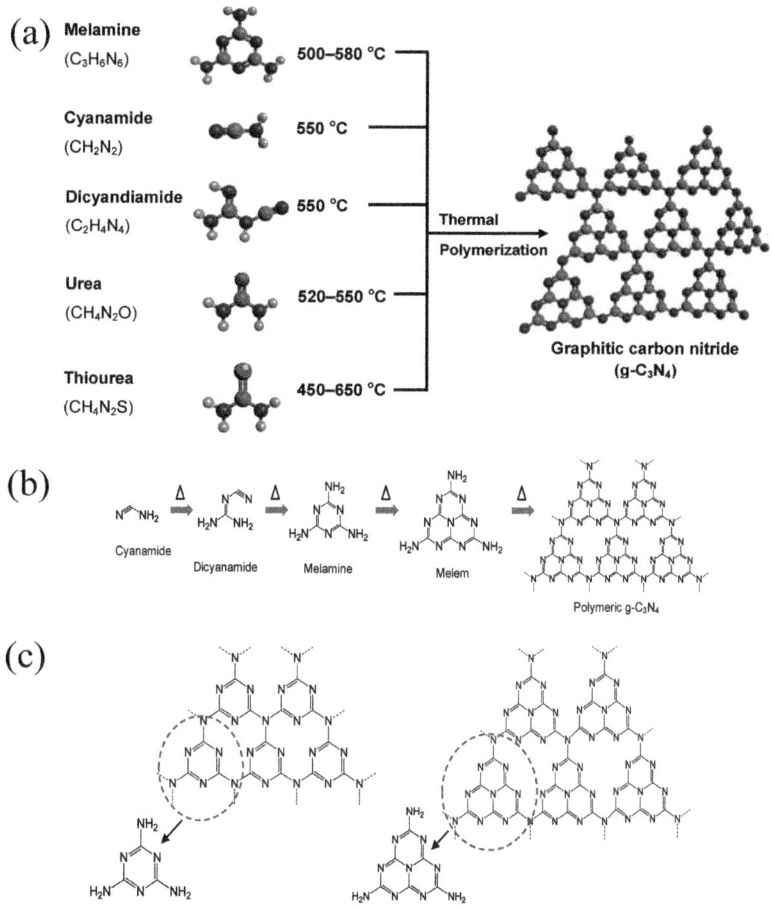

Figure 3.1 (a) Schematic illustration of g-C_3N_4 synthesis from various precursors, (b) reaction pathway for the preparation of g-C_3N_4 from cyanamide, and (c) presence of *s*-heptazine and *s*-triazine units. Reprinted with permission from W.-J. Ong, L.-L. Tan, Y. H. Ng, S.-T. Yong and S.-P. Chai, *Chem. Rev.*, 2016, **116**(12), 7159. Copyright 2016 American Chemical Society.[3]

The most challenging hurdle in classical synthetic approaches is to obtain long-range regularity in random-natured covalent polymerization reactions[26,27] and the requirement of linking individual organic constituents, which should be reversible. This allows the scaffold to arrange in such a way that it is a thermodynamically well-ordered product (2D and 3D structures) rather than a kinetic, amorphous structure.[28,29]

Intrinsically, specific physicochemical properties of both C_3N_4 and CTFs, such as an appealing electronic band structure due to a natural earth-abundance, good thermal and chemical stabilities, π-conjugated structures, and structural manipulation makes them suitable for diversified applications, particularly in heterogeneous catalysis. However, applications of these materials as metal-free heterogeneous catalysts were not explored much until 2006, when a mesoporous g-C_3N_4 was employed in a Friedel–Crafts reaction.[12,30] Further, the implication of g-C_3N_4 as a visible-light-driven photocatalyst for hydrogen evolution was disclosed in 2009 by Wang *et al.*[31] Since then, numerous studies on the catalytic,[32] electrochemical,[33] and photocatalytic performance[18,19] of g-C_3N_4 and CTFs have increased tremendously and they find a special place in material science, specifically in heterogeneous catalysis. However, reports on the performance of CTFs in catalysis are limited.[34]

Interestingly, g-C_3N_4 and CTFs are novel all-organic, metal-free polymeric heterogeneous (and photo-) catalysts different from inorganic materials. They have high tunable surface functionality, compositions, sizes, thicknesses, pore structures, pore size distributions and morphologies. From the point-of-view of quantitative structure–property relationships, these properties can prosper to form various highly tailorable hybrid functional materials and photocatalysts. This chapter briefly discusses various synthetic protocols for obtaining high performance triazine- and heptazine-based conjugated polymeric materials with myriad applications, with the focus on catalysis. Furthermore, these functional covalent organic frameworks are also effectively used as support materials in metal-/metal oxide-based catalysis, and are coupled with graphene, semiconductors and metals *etc.* to form composite materials, which could boost their physicochemical properties and, therefore, their activity; this is not within the scope of this chapter and readers are referred to the literature.[2,3,35–38]

3.2 Methods of Synthesis of g-C_3N_4

The introduction of a tailored nanostructure and porosity in a g-C_3N_4 matrix by following various synthetic protocols is a facile and effective pathway to enhance its performance and multifunctional activity in different types of applications (Figure 3.2). Recently, many design and synthesis methodologies, namely the template-assisted method, exfoliation, hydro- and solvo-thermal methods, a top-down strategy and combined methods have been reported for the preparation of nanostructured g-C_3N_4 materials. The synthesis methods and their advantages in tuning the physicochemical properties and hence the catalytic performances of g-C_3N_4 will be discussed briefly. We hope that this chapter will provide researchers with an immense input for extending the generation of the g-C_3N_4 family by a novel state-of-the-art design to synthesize these materials and to bring out abundant micro-/nanostructures and morphologies with various applications.

3.2.1 Bulk g-C_3N_4

Initially, g-C_3N_4 is prepared in bulk form and it commonly stands as a benchmark material for comparisons. Bulk g-C_3N_4 can be obtained by a simple synthesis route through the thermal condensation of any kind of low-cost nitrogen-rich precursors, which serves as the main advantage. Generally,

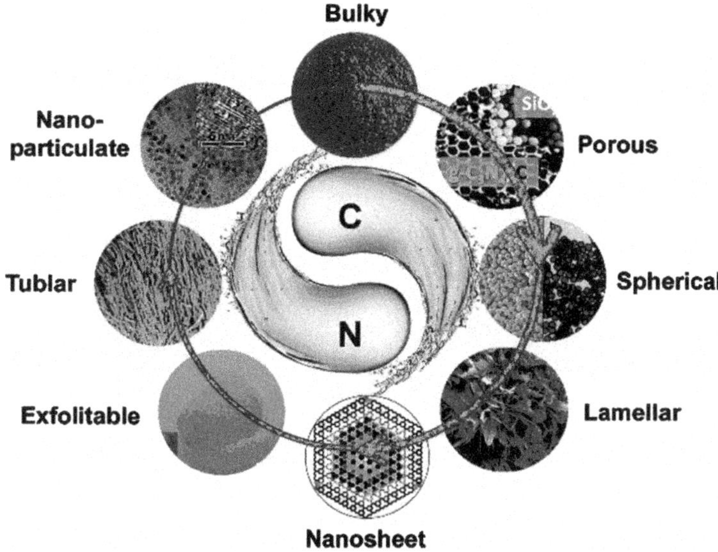

Figure 3.2 Image showing various morphologies of g-C_3N_4 with the dimensions ranging from bulk to quantum dots. Reprinted from ref. 2 with permission from The Royal Society of Chemistry.

cyanamide, dicyandiamide, melamine, thiourea, urea, or mixtures of these precursors are used. However, the materials obtained possess a very low surface area, usually lower than 10 m^2 g^{-1}. Moreover, following topotactic transformations developed by Jun[39] and Shalom[40] independently, using a pre-organized supramolecular complex of melamine and cyanuric acid, specific surface areas up to 60–80 m^2 g^{-1} can be achieved.

For instance, Yan *et al.*[41] used melamine as a precursor to prepare this material and observed an increase in the C–N ratio from 0.721 to 0.742 when the precursor was heated in a semi-closed system at different temperatures (500 to 580 °C). Achieving stoichiometric fabrication of carbon nitride is very difficult due to incomplete condensation *i.e.* because of the presence of 1° and 2° amino groups. The types of precursor and synthesis conditions (such as time and temperature of the thermal treatment) have a great influence on the optical, electronic and textural properties; for instance, a surface area (of *ca.* 8 m^2 g^{-1}) was reported for melamine-derived g-C_3N_4.[41] However, a very high temperature leads to decomposition and a reduction in surface area. Among the various precursors used for preparing g-C_3N_4, urea is also a superior precursor and has been found to attribute a high specific surface area (~100 m^2 g^{-1}) due to the formation of sheet-like g-C_3N_4 of a much smaller thickness.[42–46] Martin *et al.*[47] reported higher activity in hydrogen evolution over urea-derived g-C_3N_4 compared to DDCA and thiourea-derived g-C_3N_4. Furthermore, various synthetic approaches and exfoliation methods explored to derive porous structures with high surface areas are described in following sections.[48–55]

3.2.2 Template-assisted Method

The template-assisted method is one of the most effective methods to synthesize porous or nano-architectured materials, where the inorganic or organic nanostructures are used as templates.[56] Hard-templating or soft-templating approaches are exploited to obtain different size, morphology, and pore structure tailored g-C_3N_4 materials by the selection of different templates.

3.2.2.1 Hard Template Approach

The hard template approach is a most practicable strategy for obtaining a controllable and precise g-C_3N_4 nanostructure. This method is a highly feasible strategy for the design of different geometries that cover various structured and textured g-C_3N_4 in length scales from several nanometers over the micro- to the macro-scale.[57] Further, this method enables the development of high performance, functional, porous materials, in particular, heterogeneous (photo)catalysts and catalyst supports by incorporating hierarchical pore architectures, large surface areas, accessible open pore walls, a high light-harvesting ability, accelerated mass transfer, and unique semiconductor properties.

3.2.2.1.1 Mesoporous g-C_3N_4. Initial reports on mesoporous graphitic carbon nitride (mpg-C_3N_4) showed that it was synthesized by nanocasting/replication of mesoporous silica matrices as sacrificial templates followed by the thermal condensation of precursors.

In a typical synthesis, a known amount of cyanamide and commercially available silica sol as a template are mixed in an aqueous solution. The suspension is stirred in an oil bath until the complete removal of water and the composite obtained is transferred to a semi-closed crucible with a lid. Then the contents are thermally condensed by heating at 550 °C for 4 h to transform into polymeric carbon nitride. The silica template is leached out by desolution using an aqueous solution of 4M ammonia hydrogen difluoride (NH_4HF_2) to obtain the mpg-C_3N_4. Thermal condensation of cyanamide with varying silica–cyanamide mass ratios from 0 to 1.5 results in a mesoporous three-dimensionally interconnected framework with variation in surface areas from 8 to 373 $m^2\ g^{-1}$.[58,59] However, the resulting g-C_3N_4 does not exhibit a perfectly ordered porous structure due to the competitive absorption of moisture/H_2O or the solvent with precursor molecules on the templates. To rectify this issue, Park *et al.* developed a solvent-free synthetic approach.[60] A further development for inclusion of g-C_3N_4 precursors in silica templates has been achieved by the surface acidification of silica, sonication and vacuum treatment.[3] The post-acidified silica (SBA-15) surface can react efficiently with NH_2-rich precursor molecules based on the acid–base interaction.[61] Thus, prepared composite is treated with NH_4HF_2 solution to result in ordered mesoporous g-C_3N_4 (ompg-CN) (Figure 3.3(a)). ompg-CN has exhibited a better free-charge migration, well-assembled 2D

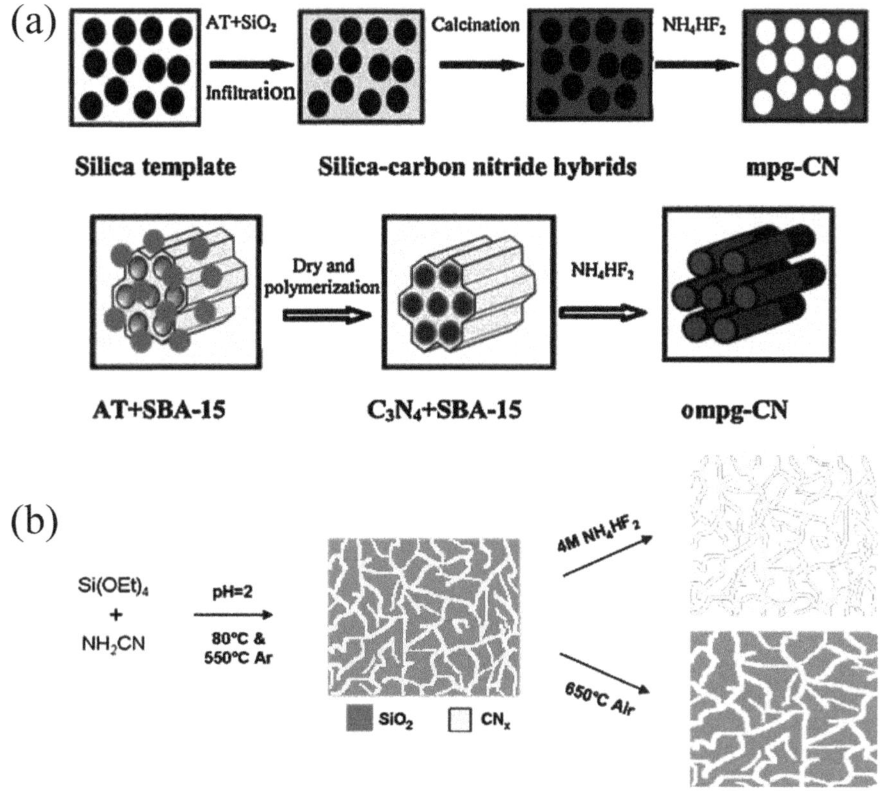

Figure 3.3 (a) Schematic diagram for the fabrication of mesoporous and ordered mesoporous graphitic carbon nitride (mpg-C_3N_4 and ompg-C_3N_4) using SiO_2 and SBA-15 as hard templates. Reprinted from *Journal of Photochemistry and Photobiology C: Photochemistry Reviews*, **20**, G. Dong, Y. Zhang, Q. Pan, J. Qiu, A fantastic graphitic carbon nitride (gC_3N_4) material: Electronic structure, photocatalytic and photoelectronic properties, 33–50, Copyright 2014, with permission from Elsevier. (b) The sol–gel route to obtain porous carbon nitride and silica by a hard template approach. Reprinted from ref. 68 with permission from The Royal Society of Chemistry.

hexagonal framework and, hence, showed higher photocatalytic reactivity that the g-C_3N_4 obtained by a conventional solvent-assisted method. Vinu and co-workers have contributed various studies regarding the synthesis of carbon nitride and N-doped carbon materials using various hard templates and carbon, nitrogen precursors, and their applications in various fields such as catalysis, sensing and CO_2 capture.[62–67] Talapaneni *et al.*[62] demonstrated a hard template synthesis approach using mesoporous silica KIT-6 with a 3D porous structure for obtaining a 3D mesoporous carbon nitride with Ia3d symmetry (MCN-6) using ethylenediamine and carbon tetrachloride as the sources for N and C, and C, respectively. Thus, the material obtained possesses bimodal pores that can be tunable. The C–N ratio of the

MCN-6 material is 4.3, which indicates the presence of a large number of free NH_2 groups originated from ethylenediamine, which can offer basic sites. They exhibited excellent activity in the Knoevenagel condensation of benzaldehyde and malononitrile and afforded a high yield of the corresponding α,β-unsaturated nitrile under ambient conditions.[62]

A sol–gel-assisted thermal condensation approach using cyanamide and TEOS (tetraethylorthosilicate) as a silica source has obtained a high surface area g-C_3N_4 (SG-CN) (Figure 3.3(b)). The photocatalytic activity of SG-CN in the reduction of H_2O was investigated and its catalytic superiority in comparison with bulk and even mesoporous carbon nitrides prepared using pre-formed silica templates is inferred by Kailasam *et al.*[68] The high surface area and well-connected continuous mesoporous structure are desirable for enhancing the activity and selectivity because of the easy accessibility of guest molecules, and catalytic cofactors can be allocated in the structure (such as charge separation and transfer). In addition, inspired by the biological process carried out by photosynthetic organisms in the design of thylakoid structures, the synthesis of hollow g-C_3N_4 nanospheres (HCNSs)[69–71] using hollow silica nanospheres as templates has also been reported.[70,72] These could harvest light efficiently for H_2 evolution with visible light.[70] This is a unique class of hollow nanostructures with a thermally and chemically stable polymeric framework, which can promote light capture and distribution, and separation of oxidative and reductive species in its nano-domain and, hence, leads to efficient light-induced redox catalysis in a cascading manner.[70,73]

3.2.2.1.2 g-C_3N_4 Nanosheets and Thin Films. A novel, size-controllable synthesis of graphitic carbon nitride nanosheets (g-CNNSs) using smectite clays [saponite (spt) and montmorillonite (mmt)] as templates has been carried out, where the interlayer spaces of these clays can act as confined nanoreactors for cyanamide condensation.[74]

In a typical synthesis of g-CNNSs, initially, an aqueous suspension of cyanamide and clay is stirred for a fixed time, and then centrifuged, washed with water and air dried. The dry powder samples are denoted CS (for cyanamide/spt) and CM (cyanamide/mmt), respectively. Further, the dry powder obtained is heated in a quartz ampule to 550 °C for 4 h. Thus, the resulting yellow-gray powder is acid treated, washed and dried to get the final samples from spt and mmt, which are denoted as g-CNNSs-S and g-CNNSs-M, respectively.[75] In addition to size and texture differences, the interlayer space confinement of each clay results in products with different C–N compositions.

To overcome limitations such as serious agglomeration of g-CNNSs and/or the deformation of nanopores in g-C_3N_4 hollow nanospheres (HCNS), interconnected nanosheets in nanospheres with highly open surface structures have been developed. Following this strategy, a three-dimensional (3D) hierarchical framework nanospherical g-C_3N_4 built-up with 2D nanosheets with sharp edge tips (NS-g-C_3N_4) has been engineered using spherical silica as the template.[76] This kind of hierarchical nanostructure can induce a fast charge separation and collection at the tips, called the "lightning rod effect",[77,78] which is well-known in metal materials.[77–79]

Liu *et al.* reported a novel nano-confinement strategy to synthesize graphitic carbon nitride films on conductive substrates.[80] The fabrication of microstructured carbon nitride films on the substrates was carried out by *in-situ* chemical vapor deposition (CVD) using an anodic aluminium oxide (AAO) stamp penetrated with cyanamide aqueous solution (ink). This method was modified by Bian *et al.* by simply putting a substrate at the mouth of the crucible with the film precursor placed in a "CVD" chamber. However, control over the film morphology was lost in the absence of the AAO template.[81] Further developments were carried out by Jia and coworkers in formulating a hierarchical macro-/mesoporous carbon nitride film on any substrate by utilizing dual templates (P123 and polystyrene spheres); their application in sensing has been investigated.[82]

3.2.2.1.3 One-dimensional g-C_3N_4 Nanowires/Nanorods/Nanotubes. One-dimensional nanostructured g-C_3N_4 has its own place in nanoscience and technology due to its promising optical and electronic properties, and photocatalytic applications arising from smooth carrier transport. Li *et al.* reported the fabrication of g-C_3N_4 nanorods *via* confined thermal condensation of cyanamide inside the nanochannels of an anodic aluminium oxide (AAO) membrane template. However, the inter-rod hexagonal packing of the template didn't allow the completion of the process.[83] Furthermore, Liu *et al.* reported the synthesis of morphologically diverse carbon nitride micro-/nanoclusters (nanowires, nanoribbons, and nanosheets) by polycondensation of cyanamide condensing inside diatom frustule channels as templates. The morphologies of g-C_3N_4 can be altered by varying the cyanamide to diatom substrate composition ratios and the condensation temperature in an inert atmosphere. The final hierarchical nanostructure arrays could be obtained after removing the diatom template.[84,85] Zhao *et al.* described a thermal evaporation method for the large-scale preparation of nitrogen-rich carbon nitride microfibers from as-synthesized graphitic carbon nitride.[86]

Bian *et al.* followed a polymerization reaction of ethylenediamine with carbon tetrachloride using porous anodic aluminium oxide membranes as templates to obtain carbon nitride nanotubes. However, the process leads to material with a carbon-to-nitrogen ratio of 4.9, where 0.75 is the ratio for an ideal carbon nitride, thus the resulting material can be named an N-doped carbon. Such materials are used as catalyst supports as reported for cyclohexene dehydrogenation.[87] Another peculiar procedure reported is the fabrication of helical g-C_3N_4 nanorods using chiral mesoporous SiO_2 templates.[88,89] These chiral g-C_3N_4 nanorods can have both left- and right-handed helical nanostructures and the semiconductor absorption edge shows distinctive optical activity in polarized light. The peculiar optical properties of these helical g-C_3N_4 nanorods promote charge separation and mass transfer comparable to pristine g-C_3N_4, which makes them more efficient catalysts for photocatalytic water splitting and CO_2 reduction.[88,89]

3.2.2.2 Soft Template Approach

This approach is greener and simpler than the hard template approach. In addition, tuning the nanostructure and architecture has been made easy through the choice of different soft templates. Generally, surfactants, amphiphilic block polymers, or ionic liquids are used as soft structure-directing agents and the desired porous structures and surface morphologies can be formulated by using specific types of soft templates. Wang *et al.* reported the synthesis of mesoporous g-C_3N_4 by polycondensation of dicyandiamide using various soft templates, such as non-ionic surfactants, amphiphilic block polymers (*e.g.*, Triton X-100, P123, F127, Brij30, Brij58, and Brij76) and some ionic liquid surfactants (*e.g.*, BmimPF6, BmimDCN).[90,91] However, the formation of useful pore systems was achieved with very few specific templates (*e.g.*, Triton X-100 and ionic liquids) and the inefficient process of removing the template residue during the synthesis led to side reactions resulting in a high carbon content in the carbon nitride samples. These drawbacks disrupt the structural features of g-C_3N_4 and lower the photocatalytic performance.[90,91] Yan *et al.* used a less reactive precursor melamine and Pluronic P123 as a template to form a worm-like mesoporous g-C_3N_4 with a small carbon content, a large surface area, and a red-shifted absorbance edge. This porous g-C_3N_4 showed an increased H_2 evolution rate of 148.2 μmol h^{-1} compared to that of bulk g-CN (60.5 μmol h^{-1}) and also extended the visible light absorption above 700 nm.[92]

Furthermore, modified g-C_3N_4 nanoarchitectures were obtained with dicyandiamide or urea and using ionic liquids as soft templates.[90,91] A simple one-pot process has been reported on the use of 1-butyl-3-methylimidazolium tetrafluoroborate ionic liquid as a multifunctional modifier to obtain g-C_3N_4 by self-polymerization of urea, with tunable texture, surface chemistry, and semiconducting properties.[90,91,93] Following a similar protocol B, F co-doped and P-doped g-C_3N_4 nanosheets, also obtained from dicyandiamide and 1-butyl-3-methylimidazolium hexafluorophosphate (Figure 3.4), showed improved photocatalytic activity in H_2 evolution and degradation of rhodamine B, methyl orange and oxidation of hydrocarbons compared with pure g-C_3N_4.[90,91,93,94] These soft template approaches can further advance synthetic protocols for obtaining engineered carbon nitride for various applications, but choice and cross-reactions should be taken into account.

3.2.2.3 Supramolecular Preorganization Method

The supramolecular preorganization method is a kind of self-templating method. Unlike template-assisted synthesis, nanostructured materials are prepared without an external template or a post-treatment procedure for removing the template. The ordered building blocks are designed by contorted molecular self-assembly by non-covalent interactions (*e.g.*, hydrogen bonding) to obtain the desired architecture.[95]

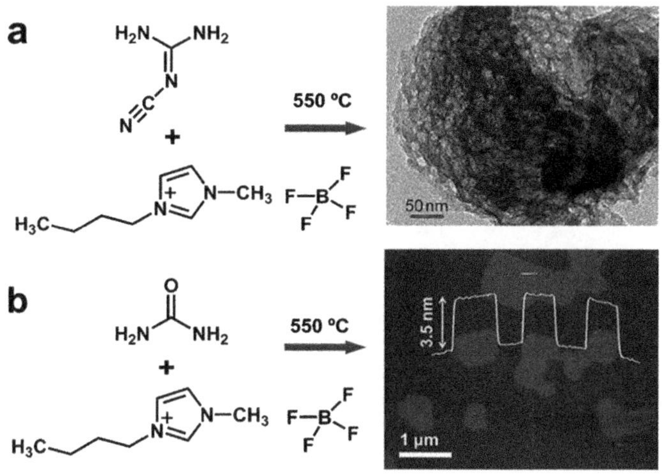

Figure 3.4 Ionic liquid (1-butyl-3-methylimidazolium tetrafluoroborate)-assisted synthesis of B- and F-containing g-C_3N_4 using (a) dicyanamide (reprinted with permission from ref. 91. Copyright © 2010 WILEY-VCH Verlag GmbH & Co. KGaA, Weinheim) and (b) urea (reprinted with permission from ref. 93. © 2014 WILEYVCH Verlag GmbH & Co. KGaA, Weinheim).

This method has an appealing approach for tuning the morphology, texture, and photophysical properties of g-C_3N_4.[96] Hydrogen-bonded molecular assemblies of mixtures of cyanuric acid/melamine,[40] melamine/cyanuric acid(MCA), or melamine/trithiocyanuric acid can lead to the formation of nanostructured g-C_3N_4 by this approach.[39,97,98] The ordered hollow g-C_3N_4 structures formed using a cyanuric acid/melamine complex in ethanol showed enrichment in charge transfer and photocatalytic performance in the degradation of rhodamine B.[40] Rationally modified 3D macroscopic assemblies of low-dimensional g-C_3N_4 can be obtained by integrating the supramolecular approach with a molecular design strategy and controlling the ratio of monomers, precipitation temperature, solvent, and hydrogen-bonding donor–acceptor pairs (Figure 3.5).[96,97] Ishida *et al.* followed a triple supramolecular complex approach using cyanuric acid, melamine, and 2,4-diamino-6-phenyl-1,3,5-triazine as precursors to obtain g-C_3N_4 with a surface area of 81 $m^2 g^{-1}$, which is comparatively higher than bulk g-C_3N_4 (12 $m^2 g^{-1}$), and with improved light absorption in the visible region, showing 16-fold higher photocatalytic activity in the degradation of rhodamine B compared to bulk g-C_3N_4.[96] An ion-doped roll-like g-C_3N_4 was also prepared by a one-pot supramolecular approach using cyanuric acid/melamine precursors[95] to modulate the electronic and chemical properties of the g-C_3N_4. Thus, the one-pot, self-templating synthetic pathway called the supramolecular preorganization method is an efficient method that opens up new directions for the optimization of the structure, texture, and optical properties of g-C_3N_4 for the creation of efficient catalytically/photoactive materials.[40] However, the method is not explored much due to the

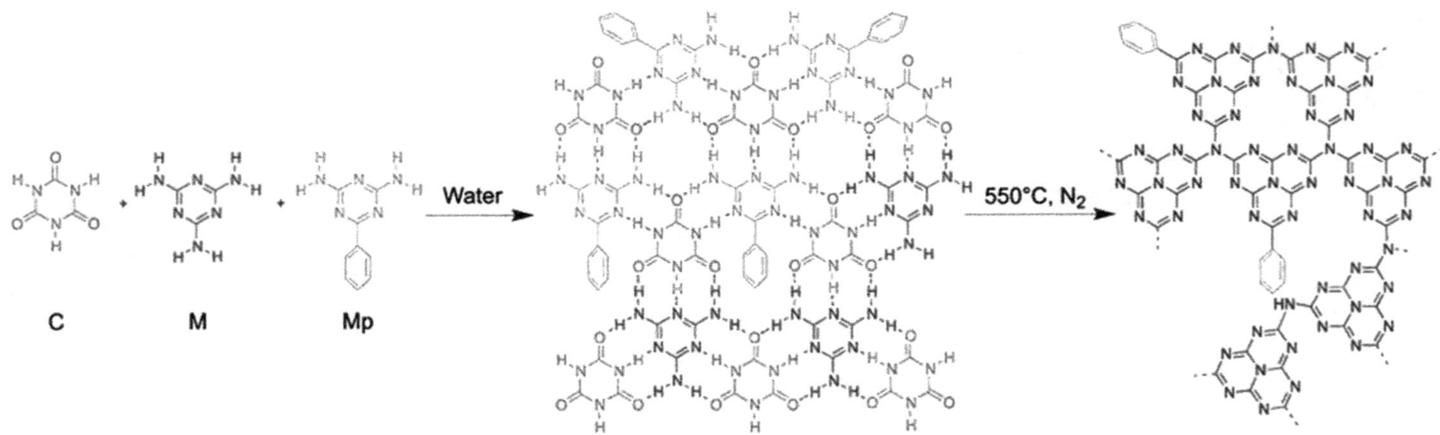

Figure 3.5 Hydrogen-bonded supramolecular C–M–Mp complex and the proposed carbon nitride structure after calcination [cyanuric acid (C), melamine (M), and 2,4-diamino-6-phenyl-1,3,5-triazine (Mp)]. Reprinted with permission from Y. Ishida, L. Chabanne, M. Antonietti and M. Shalom, *Langmuir*, 2014, **3**, 447. Copyright 2014 American Chemical Society.[96]

limited choice of precursors with an affinity to hydrogen-bonded molecular assemblies.

3.2.3 Ionothermal Approach

In general, a condensation process to obtain carbon nitride products leads to insufficient crystallinity; this could be overcome by following the ionothermal approach. That is, synthesis in molten salts leads to a great opportunity to form crystalline, condensed $g-C_3N_4$.[99–101] For the first time, triazine-based $g-C_3N_4$ (TGCN) was successfully synthesized using dicyandiamide/melamine by following the ionothermal reaction.[102,103] $g-C_3N_4$ films with thicknesses of a few to several hundreds of atomic layers, consisting of stacked, two-dimensional (2D) crystals have been obtained by intercalation with halides in a eutectic salt melt (*e.g.*, KCl/LiCl, KBr/LiBr, and $LiCl{\cdot}H_2O$/KCl/NaCl).[102–105] Bojdys *et al.* followed this approach using dicyandiamide as a precursor and a salt melt of lithium chloride and potassium chloride as a solvent to synthesize crystalline $g-C_3N_4$ based on sheets consisting of hexagonally arranged *s*-heptazine (C_6N_7) units.[106] Wirnhier *et al.* and Zou and co-workers reported the influence of alkali coordination on the basic structural unit (stabilization of *s*-triazine) and in inducing non-uniform spatial charge distribution on C–N planes, which promotes efficient separation of photogenerated carriers and leads to the formation of novel non-radiative paths.[101,104] Schwinghammer *et al.* reported the synthesis of crystalline poly(triazineimide) (PTI) nanosheets by a simple one-step, liquid-phase exfoliation in water.[107] In addition, facile exfoliation has resulted in highly crystalline, and chemically and colloidally stable nanosheets with thicknesses of 1–2 nm, which showed significantly high photocatalytic efficiency for visible-light-driven hydrogen evolution compared to bulk crystalline PTI.[107] Bhunia *et al.* adopted a two-step approach by combining supramolecular aggregation and polycondensation in the ionic melt, along with controlled variation in a chemically compatible dopant for the rational design of photoactive crystalline $g-C_3N_4$ from a diverse range of inexpensive precursors. This approach has improved the condensation process and offers high crystallinity.[108,109] Furthermore, a polycondensation process in an ionic melt with nanostructural engineering, copolymerization, and supramolecular assembly can provide a rational preparation pathway to design a broad range of triazine-based $g-C_3N_4$ structures with modified functions because of the availability of versatile organic/inorganic dopants.[110,111] The structural and electronic properties of $g-C_3N_4$ with adjustable crystallinity provide versatility for the conversion of solar energy.[110,111]

3.2.4 Bottom-up and Top-down Strategies

One-pot, bottom-up design, soft solution-processing occurs in the presence of an aqueous/non-aqueous solvent under sub-/supercritical conditions in a closed system by molecular design, solution assembly, crystallization, and covalent cross-linking chemistry. This process is specifically known as the solvothermal synthesis approach.[112,113] In addition, following this approach enables the assembly of large polymeric superstructures at relatively low temperatures, and it is beneficial for tuning the electronic and chemical structures of g-C_3N_4 by facile organic chemistry procedures. It also prevents the depletion of nitrogen from the material;[112,113] this is more advantageous as it avoids the destruction of the precursor that is usually employed to produce carbon nitrides through high temperature calcination processes.

The solvothermal syntheses of g-C_3N_4 through the thermal condensation of melamine and cyanuric acid to fabricate superstructures, such as nanorod networks, using various solvents such as isopropylethylamine and acetonitrile have been reported by Montigaud *et al.*[114] and Cui *et al.*[112,113] The as-obtained g-C_3N_4 shows a reduced band gap, improved light-harvesting properties, and photocatalytic reactivity for pollutant degradation and water splitting. Considering the simplicity of this process, new and modified approaches have been established by integrating the solvothermal method for the design of highly efficient g-C_3N_4-based semiconductors for various applications in catalysis, energy conversion, and environmental remediation.[112,113]

Appreciating the exfoliation strategy to obtain graphene from bulk graphite, the process was extended to layered g-C_3N_4. Thin layer g-C_3N_4 nanosheets were prepared from bulk g-C_3N_4 by following a top-down approach, which relies mainly on the exfoliation method by physical and chemical protocols (*e.g.*, ultrasonic, thermal oxidation, and the addition of strong oxidizers). Many researchers have prepared g-C_3N_4 nanosheets and explored their use in water splitting and pollution degradation.[115,116] A sonication-assisted liquid exfoliation method to form g-C_3N_4 nanosheets was developed by Zhang *et al.*[117–119] Niu *et al.*[120] reported a simple thermal oxidation etching process to exfoliate bulk g-C_3N_4 (derived from dicyandiamide) into nanosheets with a high surface area of 300 m^2 g^{-1} and thickness of about 2 nm. Xu *et al.*[48] followed a chemical exfoliation method to synthesize ultrathin g-C_3N_4 nanosheets with a single atomic thickness of 0.4 nm. These are easy processes to prepare nanostructured carbon nitrides with high porosity, which can be good candidates for catalytic reactions and for many other emerging applications.

Metal-free, polymeric g-C_3N_4 QDs can stand alongside traditional QDs because of their bright fluorescence, good stability, biocompatibility, and nontoxic properties. Groenewolt *et al.* reported the silica-host-matrices-assisted synthesis of carbon nitride QDs. They observed blue-shifted photoluminescence in 5 nm-sized QDs compared to their bulky counterparts.[58] Fabrication of zero-dimensional (0D) g-C_3N_4 quantum dots (QD) from the 2D-sheet is an interesting way to increase the number of catalytically active sites[121] due to enhancement in pyridinic N, graphitic N and edge amine groups.[122] In general, g-C_3N_4 QDs can be prepared efficiently using nanosheets *via* hydrothermal reactions.[123] In addition, they find many applications in photodetectors, sensors, bioimaging and also in solar cells because of their highly versatile absorption range covering the visible to near-infrared regions of the solar spectrum.[124,125] Many g-C_3N_4 QDs with nanocrystal composites have been reported for enhanced visible light photocatalytic and solar cell applications.[126–129] However, systematic studies to improve the performance of $g-C_3N_4$ QDs through various modifications have yet to be undertaken.

3.3 Methods of Synthesis of Covalent Triazine Frameworks (CTFs)

Recently, conjugated organic framework materials have attracted much attention due to their high surface areas and porous structures.[130] Numerous studies have been conducted on covalent organic frameworks (COFs)[20,131] and conjugated microporous polymers (CMPs).[132] A number of reports have focused on the development of new synthetic strategies, designing new building blocks and developing novel applications.[133–135] Thus far, these materials have found applications in adsorption and separation,[136] heterogeneous catalysis,[137] photocurrent generation,[138] photocatalytic water splitting,[139] sensing[140] and so on.

A unique class of nitrogen-rich organic porous materials, similar to $g-C_3N_4$,[30,102] called covalent triazine frameworks (CTFs), have also emerged as fascinating materials due to their good thermal stabilities, good electron mobilities in the 2D plane due to π-conjugated structures and the existence of both crystalline and amorphous structures[24] as described in the introduction. They occupy an exceptional place in many applications, such as gas separation and sensing,[141–146] organic dye removal,[147] catalysis[38,148] and many other emerging areas. This chapter discusses the various synthetic approaches that have been taken to form desired engineered CTFs with tailor-made functionalities for suitable applications.

3.3.1 Ionothermal Synthesis

CTFs are synthesized by the trimerization of aromatic nitriles using a Lewis or Brønsted acid catalyst, typically either $ZnCl_2$ or trifluoromethanesulfonic acid (TFMS). Depending on the precursor and synthesis conditions, the nitrogen content, and the amorphous/crystalline nature and nonporous/porous nature can be tailored.

In a typical synthesis of CTFs, a mixture of 1,4-dicyanobenzene and anhydrous $ZnCl_2$ of a known ratio is placed in a glove box. The contents are sealed under vacuum in a 20 mL quartz ampoule; the sealed tube is placed in a muffle furnace heated to a known temperature (400–700 °C) up to a fixed time. The resulting $ZnCl_2$ melt is used as a catalyst as well as a solvent, resulting in all the precursor monomers being polymerized into CTFs. The quartz tube is then allowed to cool to room temperature, giving a black monolith material. Then, the contents are ground into a fine powder and dispersed in diluted HCl solution to dissolve the $ZnCl_2$. After 24 h of treatment, the contents are washed thoroughly with water and then dried in air at 110 °C for 12 h to obtain the final material CTFs (Figure 3.6(a)).[24,134,149,150]

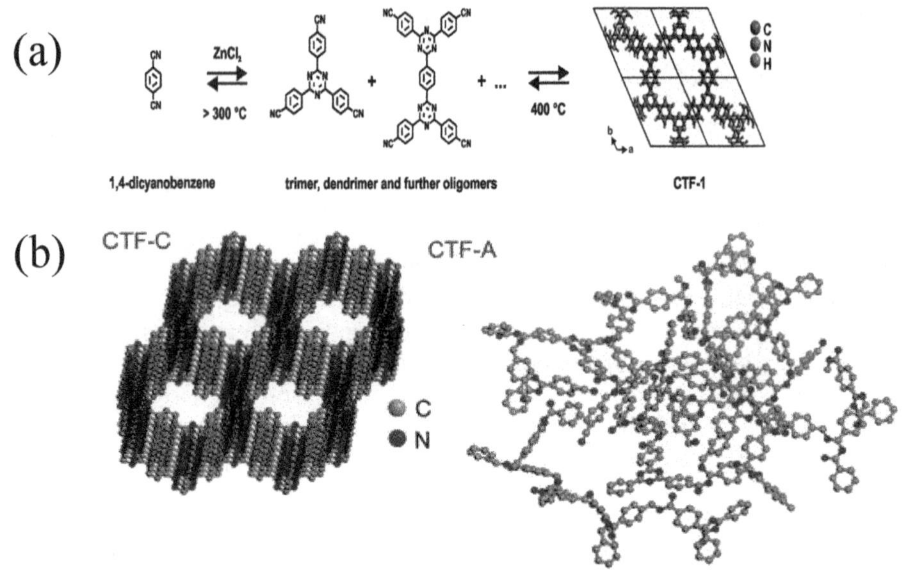

Figure 3.6 (a) Reaction mechanism of CTF-1 starting from 1,4-dicyanobenzene, which leads to the formation of oligomers *via* trimerization and, ultimately, to an extended 2D CTF (right). The idealized crystal structure of fully condensed, crystalline CTF-1 is shown in an AA-type eclipsed configuration (viewed along the c-direction). Reprinted from ref. 134 with permission from The Royal Society of Chemistry and (b) schematic illustration of the structures of CTF-C (crystalline) and CTF-A (amorphous). Reprinted from ref. 157 with permission from The Royal Society of Chemistry.

Kuhn *et al.*[24] reported that the amorphous and crystalline phase CTF structures (CTF-A and CTF-C) prepared by varying the molar ratio of 1,4-dicyanobenzene and anhydrous $ZnCl_2$, *i.e.*, 1 : 2 and 1 : 1, were used to obtain CTF-A and CTF-C, respectively (Figure 3.6(b)). CTF-C is formed by extension of the trimers of 1,4-dicyanobenzene along the two-dimensional (2D) plane where the layers eclipsed each other, whereas highly porous but amorphous CTF-A materials led to a three-dimensional (3D) cross-linking architecture of the trimers of 1,4-dicyanobenzene.[151] The triazine rings broke slightly and partially carbonized in CTF-A, which resulted in a higher electronic conductivity than that of CTF-C. The enhancement in conductivity would lead CTF-A to perform better in energy storage,[151–155] as a catalyst support,[38,156] and in gas sensing.[145] UV-Vis analysis and Mott–Schottky measurements confirmed that the CTFs have suitable conduction-band (CB) and valence-band (VB) edge positions and a band gap energy of 1.6 eV, which provides a better visible light photocatalytic degradation activity on the organic dye methylene blue (MB).[157] Thomas and co-workers observed that the crystalline or amorphous nature and porous structure formation of extended and periodic networks can be tuned by the adjustment of dicyano-compound–$ZnCl_2$ ratios and with a variation in polymerization temperature.[150,158,159] Amorphous, but nitrogen-rich, carbonaceous CTFs with a high surface area of 3300 m^2 g^{-1} and a porosity up to 2.4 cm^3 g^{-1} have been obtained by using a dicyano-compound–$ZnCl_2$ molar ratio of 1:5. These CTFs are employed as catalysts for the synthesis of cyclic and linear carbonates from carbon dioxide and epoxides[158] and as support materials for the synthesis of Co_3O_4/CTF catalysts used in the liquid-phase aerobic oxidation of alcohols.[159] In addition, the group reported a modified synthesis pathway, where this approach does not start from the cyano-monomers, but follows a two-step process: firstly, a pre-CTF is prepared using 1,4-dicyanobenzene in $CHCl_3$/TFMS at ambient temperatures. This is followed by a second step, *i.e.* the transformation of this amorphous pre-CTF into an ordered CTF by thermal treatment of pre-CTF together with $ZnCl_2$ salt melt in open crucibles, without evaporation of the organic precursor. A drastic shortening of the reaction time from 40 h to below 10 min reduced the carbonization significantly to obtain CTF-1 materials with colors ranging from yellow to brown.[150,160] Further, they showed appreciable photocatalytic activity in hydrogen evolution.[160]

A crystalline, porous polymeric framework consisting of a 2D honeycomb structure composed of benzene and triazine rings with an eclipsed AAA stacking and P6/mmm symmetry was synthesized by ionothermal synthesis using *p*-dicyanobenzene or tris(4-cyanophenylbenzene) as a precursor.[161] A high degree of ordered micropores, and triazine and bipyridine moieties were obtained at a reaction temperature of 400 °C with specific surface areas of up to 1100 m^2 g^{-1}, and hierarchical mesopores with exceptionally large surface areas of 3200 m^2 g^{-1} were formed at synthesis temperatures above 450 °C (Figure 3.7).[162] The ionothermal synthesis method is highly versatile due to its simple approach. However, it suffers from the requirement of quartz ampoules, a high temperature, longer reaction times and a high level of carbonization of the frameworks.

3.3.2 Low-temperature and Microwave-assisted Synthesis

A room temperature and/or microwave-assisted synthesis of a porous covalent triazine-based framework (PCTF) using a single component tetra(4-cyanophenyl)ethylene source in the presence of TFMS as the catalyst has been employed. This approach was extended to many precursors, in which the tetraphenylethylene (TPE) units were directly connected to strong electron-accepting triazine rings. Furthermore, this method leads to frameworks that have advantages over the PCTFs obtained using the ionothermal $ZnCl_2$ route.[24,163–165] The frameworks obtained by the use of TFMS as a catalyst[166] have significant advantages: (1) a low reaction temperature and time, (2) synthesis of a defect-free framework by circumventing decomposition and unwanted side reactions that lead to carbonization, and (3) elimination of $ZnCl_2$ contamination in the porous materials. These features add benefits to obtain PCTFs that can exhibit a broad variety of physical properties.

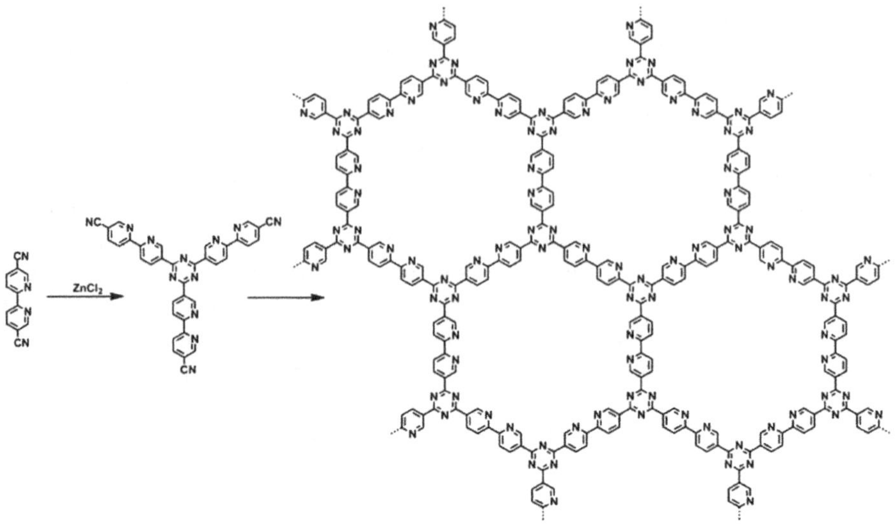

Figure 3.7 Schematic representation of the synthesis of *bipy*-CTF following an ionothermal approach. Reprinted from ref. 162 with permission from The Royal Society of Chemistry.

In the general procedure, a known amount of TFMS charged in a Schlenk flask containing dry $CHCl_3$ is stirred at 0 °C in a N_2 atmosphere. To this acid solution, a mixture of tetra(4-cyanophenyl)ethylene in dry $CHCl_3$ is added drop by drop over 30 min and stirred for a fixed time while maintaining the temperature at 0 °C. Further stirring is carried out overnight at room temperature and then aqueous ammonia solution is added to neutralize the medium. The resulting yellow product is washed and purified by Soxhlet extraction for 24 h with acetone and is vacuum dried to yield the final product $(C_{30}H_{16}N_4)_n$.[146]

Microwave-assisted synthesis is an emerging process, with a high scope for industrial applications in organic synthesis and materials preparation.[143,167] CTFs are prepared *via* this method in a similar way to the ionothermal method, but the trimerization of nitriles using TFMS or $ZnCl_2$ is carried out by microwave heating. This process is simple, operated at a range of temperatures and is a time-efficient technique for the synthesis of porous, fluorescent microporous CTF networks. CTF-type polymers named P1-P6 and P1M-P6M have been prepared using TFMS as an acid catalyst instead of $ZnCl_2$ with microwave heating (Figure 3.8). The polymers showed defect-free, preferred-orientation networks with some crystallinity, with a glassy solid-like powder in a range of colors from pale yellow to brown.[166] These CTFs are employed for the capture and storage of CO_2 and H_2.[162] However, only a few reports on microwave-assisted synthesis have been published; a lot of work is needed to optimise microwave-assisted synthesis using a variety of precursors to obtain CTFs in a more facile way.

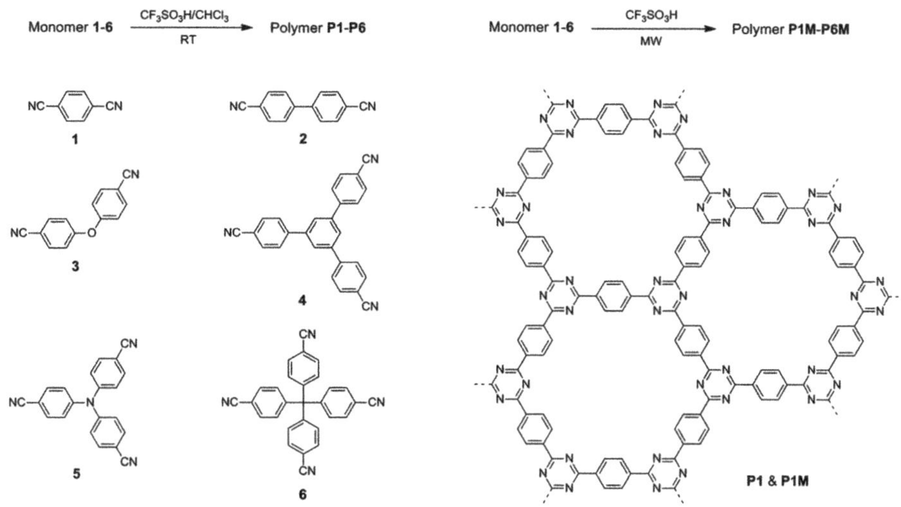

Figure 3.8 Representative CTF-type polymer structures obtained by room-temperature and microwave-assisted methods using various precursors. Reprinted with permission from ref. 166. Copyright © 2012 WILEYVCH Verlag GmbH & Co. KGaA, Weinheim.

3.3.3 Solid-state Synthetic Method

The utilization of CTFs in various applications, such as photocatalysis, is limited due to its harsh synthesis conditions, which could result in partial carbonization of the polymer leading to uncontrollable conduction-band (CB) and valence-band (VB) positions. In addition, morphology control has been considered as one of the significant tools in heterogeneous (photo)catalysis,[168–171] where hollow architectures are highly fascinating because they facilitate mass transfer and light absorption during a catalytic process.[172] Further, CTFs prepared by following liquid-phase reaction conditions, *i.e.* either in molten $ZnCl_2$ or TFMS solution, are not proficient in constructing highly ordered hollow nanostructures. Therefore, the development of milder reaction conditions along with morphology tailoring is of immense interest.

A facile solid phase synthetic method was developed to fabricate nanoporous CTFs by trimerization of nitrile functionalized aromatic precursors catalyzed by TFMS vapor at an elevated temperature (100 °C) in the solid phase. Uniformly packed silica nanoparticles (SiO_2 NPs) were used as sacrificial templates to obtain ordered and defined hollow nanostructures with defined optical and electronic properties.[34,173] The enhanced photocatalytic activity of these nanoporous CTFs in nitrophenol and aminophenol reduction reactions showed that it was not only the chemical composition but also the morphological structure that had a considerable influence on the photocatalytic performance of the CTFs. Figure 3.9 illustrates the synthesis of nanostructured CTFs with interconnected hollow spherical pores

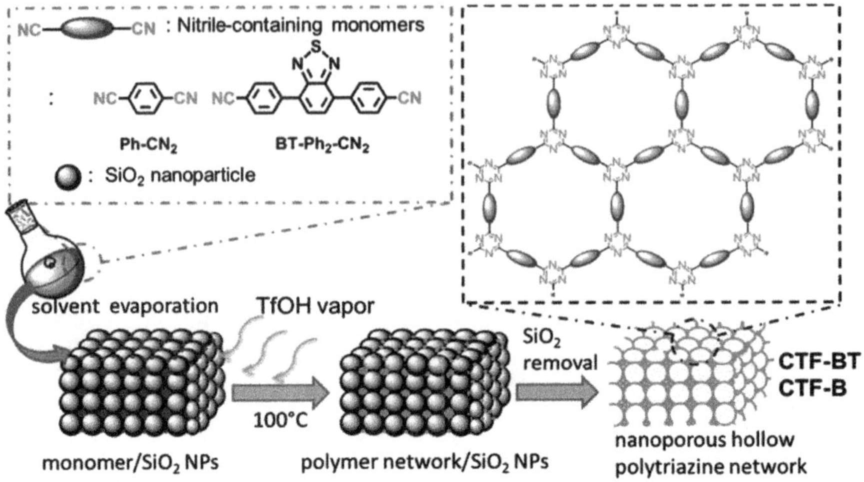

Figure 3.9 A diagram showing a systematic solid vapor approach for the synthesis of nanoporous hollow polytriazine networks. Reprinted from ref. 34 with permission from The Royal Society of Chemistry.

by a solid-state synthetic method using TFMS acid vapor. A nanoporous polymer CTF-BT was synthesized following a similar method using a photoactive and electron-donor–acceptor-type monomer, 4,4′-(benzothiadiazole-4,7-diyl) dibenzonitrile (BT-Ph_2-CN_2), where the donor–acceptor structure would improve the electron delocalization, lower the energy level and thus result in a high visible-light response.[174,175]

3.4 Miscellaneous

Apart from the above-mentioned methods, other methods have also been exploited for the synthesis of g-C_3N_4. The choice of precursors and synthesis method significantly influences the texture, morphology, optical and electronic properties.

A simple direct heating of inexpensive melamine hydrochloride could obtain porousification in the final g-C_3N_4 with a higher surface area and an increase in band gap by 0.13 eV compared to a traditionally prepared bulk C_3N_4. Their physicochemical characteristics have a direct relation with photooxidation activity in rhodamine B and an inverse relation has been found in CO_2 photoreduction activity.[176] Wang and co-workers reported a two-stage programmed heating approach, by which a rationally controlled polymerization of melamine takes place to obtain distorted g-C_3N_4 nanosheets. The 2D buckled structure and higher polycondensation promote charge separation because of a structure distortion-induced extension in the optical absorption by taking advantage of activating more $n \rightarrow \pi^*$ transitions. Experimental and theoretical studies have given evidence of the optical absorption dependency on the structural distortion and number of layers fabricated by the delamination process.[177] These properties of buckled carbon nitride samples have made them a wide visible light (>495 nm) responsive photocatalyst in water splitting.[177]

3.4.1 Heteroatom Doping

3.4.1.1 *Non-metal Doping*

A suitable redox potential as well as a strong light absorption capacity are prerequisites for a semiconductor material to be an active (photo)catalyst. Most of the time, many semiconductors are not efficient in meeting these requirements. An effective strategy to modify the photo-, electro- and chemical-properties of g-C_3N_4 could be described as a structural doping approach. In particular, intense research activity has been devoted to non-metal (nitrogen, oxygen, sulfur, phosphorus and halogen) doping to design and develop low-cost, green and effective photocatalysts that directly split water, degrade environmental pollutants and can also promote organic transformations under thermal conditions and/or sunlight irradiation.[178–180] Some notable examples from the literature are discussed below briefly.

3.4.1.1.1 Nitrogen Doping. Polycondensation of precursors such as melamine with a nitrogen-rich additive of hydrazine hydrate provides nitrogen-doped g-carbon nitride (nitrogen-doped g-CN).[181] Zhou *et al.*[182] demonstrated a similar thermal polymerization synthetic approach using citric acid and urea as precursors. An increase in nitrogen content enhances the number of lone pair electrons in the graphitic N-atom, which results in a delocalized aromatic π-conjugated system. N-doping could lead to a decrease in the band gap. The band gap is 2.72 eV for pristine g-C_3N_4, whereas N-doped g-CN has a band gap of 2.65 eV.

Li *et al.*[183] demonstrated the transformation of a g-C_3N_4 phase by a one-pot, self-sacrificing templating method to form nitrogen-doped graphene, which showed excellent electrocatalytic activity toward the oxygen reduction reaction (ORR). The annealing process is capable of tuning the relative contents of graphitic and pyridinic nitrogen well.[183]

3.4.1.1.2 Oxygen Doping. Mei *et al.* described a polycondensation reaction combined with a solution mixing pathway using melamine and cyanuric acid in deionized water to synthesize an O-doped graphitic carbon nitride (O-doped-g-C_3N_4) with a layered spongy structure. This catalyst exhibited improved ORR activity due to its high electron conductivity.[184] Furthermore, a hydrothermal route using H_2O_2 as the dopant precursor has demonstrated a precursor pre-treatment method forming hydrogen-bond-induced supramolecular aggregates for the preparation of porous O-doped-g-C_3N_4.[185,186] The combination of a porous structure and O-doping made porous O-doped-g-C_3N_4 a better catalyst for hydrogen evolution compared to bulk and porous g-C_3N_4 (non-doping) materials. Guo *et al.*[187] followed photo-Fenton reactions to dope oxygen at the edges of holey-structured g-C_3N_4 with a BET surface area of 348 m^2 g^{-1} and band gap energy of 2.43 eV. In addition, theoretical studies confirmed that oxygen doping leads to improved visible-light absorption and carrier mobility, produces more active sites and reduces the recombination of electron–hole pairs.[188]

3.4.1.1.3 Sulfur Doping. Wang *et al.* demonstrated that sulfur-doped g-C_3N_4 was fabricated using thiourea by a simple calcination at 520 °C. In such materials, a photogenerated electron could easily jump from the impure state to the conduction band (CB) or from the valence band (VB) to the impure state for sulfur-doped g-C_3N_4.[189] The existence of a sulfur impurity was responsible for a high methanol yield in CO_2 reduction compared to un-doped g-C_3N_4.[189] Ke *et al.* synthesized graphene-like sulfur-doped C_3N_4 (SC_3N_4) samples by thermal condensation using urea and benzyl disulfide at 550 to 650 °C; they showed high catalytic efficiency in the photo-reductive elimination of UO_2^{2+} under visible-light illumination.[190] A porous sulfur-doped g-C_3N_4 has been prepared by direct heating using melamine and sulfuric acid, which showed higher photocatalytic activity in carbon dioxide reduction to acetaldehyde and methane.[191] He *et al.*[192] synthesized S/g-CN by a sublimated-sulfur-bubble-mediated method and many other methods have also appeared in the literature[193–195] that can shed light on general doping strategies for designing efficient sulfur-doped g-C_3N_4 materials.

3.4.1.1.4 Phosphorus Doping. The introduction of phosphorus into the structural framework of g-C_3N_4 would lead to customized electronic properties, enhanced ionic conductivity and, hence, favor catalytic performance.[196] Zhang *et al.*[196] reported a one-pot, thermal-induced polycondensation of dicyandiamide and 1-butyl-3-methylimidazolium hexafluorophosphate (BmimPF6) precursors at 550 °C to obtain homogeneously distributed phosphorus in the carbon nitride host matrix. These low-cost, easily scalable and stable polymeric materials can fulfill the requirement to be used as industrial catalysts and/or catalyst supports. However, the high degree of polycondensation of the monomers leads to a low surface area (~10 m^2 g^{-1}). Furthermore, template-assisted methods have been followed to improve the textural properties of un-doped and doped g-C_3N_4 materials.[59,83,197,198] The peculiar properties of the doped samples contribute to the promotion of photocatalytic performances for decomposing dyes (rhodamine B and methyl orange) under visible light irradiation and for reduction of H_2O to produce H_2.[94,199] Furthermore, achieving the nanostructuring and homogeneous doping of a heteroatom in g-C_3N_4 simultaneously is a very challenging task. Ran *et al.*[200] fabricated novel porous P-doped graphitic carbon nitride nanosheets by combining P doping and thermal exfoliation strategies. Mesoporous P-doped g-C_3N_4 nanostructured flowers were also prepared by a co-condensation method in the absence of any templates.[201] Template-free methodologies have been presented by Zhu *et al.*[201] and Guo *et al.*[202] using a hexagonal rod-like supramolecular assembly of a melamine–cyanuric acid adduct and phosphorous acid as a precursor for g-C_3N_4 and a P source to prepare tubes and nanostructured flowers, respectively. These materials have a typical band gap energy, optical property and degree of crystallinity depending up on the phosphorus concentration. The down-shift VB edge due to the P in these materials exhibited a much stronger oxidation ability and showed enhanced photocatalytic NO removal.[203]

3.4.1.1.5 Halogen-modified g-C_3N_4. Iodine-modified g-C_3N_4 and iodinated g-C_3N_4 nanosheets have been synthesized by following co-condensation[204] and ball-milling[205] techniques using dicyandiamide and ammonium iodide. Bromine can also be doped with g-C_3N_4 derived from urea condensation using NH_4Br. These iodine- and bromine-doped materials are employed as efficient catalysts in H_2 production and O_2 evolution, and are superior to pristine C_3N_4.[205,206] Wang *et al.* showed a simple fluorination route using ammonium fluoride (NH_4F) or hydrogen fluoride (HF) as a cheap fluorine source by injecting directly during the thermal condensation process of g-C_3N_4 to produce F-modified g-C_3N_4, and they showed effective oxidization of benzene to phenol under visible light.[207] Li and co-workers demonstrated the synthesis of boron- and fluorine-enriched mesoporous polymeric carbon nitride using an easily available ionic liquid, 1-butyl-3-methylimidazolium tetrafluoroborate ($BmimBF_4$), as a unique soft template and dicyandiamide as a precursor for g-C_3N_4.[91] It has a "morel-like" mesopore structure with a narrow pore size distribution, which showed an improved photoactivity under visible-light illumination. In addition, a surprisingly good catalytic performance in the selective oxidation of cycloalkanes to cyclohexanone was exhibited.

3.4.1.2 Metal Doping and Heterostructure Formation

A reduction in band gap and extension in the light absorption range can effectively be achieved by introducing metal dopants on g-C_3N_4 to prepare functional metal–organic hybrid materials for improved light utilization and catalytic efficiency. For the first time, Wang *et al.* demonstrated the synthesis of metal (Fe^{3+})-containing carbon nitride compounds using dicyandiamide and metal chloride as precursors. The as-obtained materials showed high photocatalytic activities for the degradation of various organic dyes.[208] In addition, g-C_3N_4 materials doped with other transition metal elements (Co, Ni, Mn, Cu,[209] Zn[210,211]) have also been reported to have high photocatalytic activity and good stability in selective hydrocarbon oxidation and hydrogen production. Fe-g-C_3N_4 was reported as an active catalyst for the oxidation of benzene to phenol using hydrogen peroxide in the presence of visible light.[212] Zr-doped and W-doped g-C_3N_4 have also been developed by following simple pyrolysis and hydrothermal methods, for the enhanced photocatalytic degradation of rhodamine B (RhB) and methyl orange (MO) by H_2O_2 addition, respectively.[213,214] Notably, alkali-doped g-C_3N_4 has been prepared by thermal condensation of dicyandiamide and a eutectic salt melt.[215]

Furthermore, the development of heterostructures has become a simple and feasible scheme to improve the photocatalytic efficiency of g-C_3N_4 by lowering the electron–hole recombination.[216] Many categories of hetero-structured-g-C_3N_4 composite photocatalysts, namely wide band gap inorganic semiconductor/g-C_3N_4,[216] narrow band gap inorganic semiconductor/g-C_3N_4,[217–221] carbon/g-C_3N_4,[222] polymer/g-C_3N_4,[223–225] and g-C_3N_4-based ternary heterostructures[226] are well explored. Moreover, modification of g-C_3N_4 with enzymes, bio-inspired species, cobalt co-catalysts,[227–229] and noble-metal-nanoparticle-supported g-C_3N_4[230,231] offer new opportunities in heterostructure- and heterojunction-based materials for heterogeneous catalysts, as well as for environmental and photocatalytic applications. However, limited literature reports are covered here as this section falls out of the scope of this chapter.

3.5 Remarks and Future Prospects

In recent years, architecture/morphological tuning and nano-structure engineering has occupied its own place in material science in fabricating a tailor-made morphology, texture, electronic structure, and optical and surface properties, particularly for g-C_3N_4 and CTF materials. Manipulation of their properties has revealed a new level of multifunctional heterogeneous organocatalyst and these materials show promising activity towards myriad transformations such as alkylation, oxidation of alcohols and amines, dye degradation, the oxygen reduction reaction (ORR), hydrogen evolution through water splitting, and CO_2 activation/reduction, *etc.*

Various synthetic techniques and physicochemical strategies, such as the hard/soft-templating strategy, exfoliation, ionothermal and microwave-assisted approaches, supramolecular approach, solvothermal technology, top-down strategy, and combined methods, have been developed to obtain desired/optimized electronic structures, nanostructures, crystal structures, and heterostructures to improve the catalytic performances of g-C_3N_4 and CTFs. To enhance the photocatalytic performance of g-C_3N_4, there has been a focus on modification pathways to reduce its band gap and introducing nanostructures that reduce exciton recombination with an increase in visible light absorption. In addition, the electron-rich properties, basic surface functionalities and H-bonding motifs due to the presence of N- and H-atoms make them effective base catalysts and catalyst supports for various organic transformations (mainly in oxidation and hydrogenation reactions). Furthermore, modifications of structural and textural features and doping by introducing foreign impurities are necessary to construct efficient (photo/electro)catalysts. However, before thinking about doping foreign elements into g-C_3N_4, one has to look into the co-polymerization process, which is a primary requirement to extend the delocalization of the π-electrons and change the intrinsic acid–base and semiconductor properties by grafting various types of aromatic groups during the formation of the $g\text{-}C_3N_4$. This will aid in far visible absorption, which can enhance their catalytic activity in a dramatic manner. In such cases, low temperature synthesis is the key, but a tremendous effort is needed to develop new synthetic strategies. However, the development of soluble heptazine monomers (basic unit of $g\text{-}C_3N_4$ materials) holds the key for tunable chemical structures and porous catalysts.

Kailasam and Thomas *et al.* synthesized heptazine-based microporous polymers (HMPs) at room temperature. This approach allowed the use of various organic linkers to tune the structural and physicochemical properties and, therefore, it expands the scope of carbon nitride polymers in the advancement of activity and stability in catalysis and other applications.[139] In extension to that, donor–acceptor (D–A)-type HMPs have been reported and these porous materials have shown exceptionally high photocatalytic hydrogen evolution activity compared to bulk polymeric carbon nitrides prepared at high temperatures.[232] Furthermore, nano-architectured $g\text{-}C_3N_4$ materials, namely nano/meso-porous materials with nanoscale geometrical shapes (*e.g.*, rods, sheets, and spheres), tend to show outstanding photocatalytic performances compared to their classical counterparts such as bulk $g\text{-}C_3N_4$.

Even though extensive efforts have been made to construct state-of-the-art synthesis approaches to obtain various functional nanostructured $g\text{-}C_3N_4$ materials and CTFs, there is scope for further improvements by exploring new pathways and through modification of existing approaches. Further, the design of the surface junction and texture engineering is required to promote charge generation and separation on the catalyst ($g\text{-}C_3N_4$ or CTF). In addition, theoretical studies for understanding the structure formation and evolution can aid efforts to direct the experimental synthesis.

Thus, the promising examples of the novel synthesis approaches of $g\text{-}C_3N_4$ and CTFs discussed above show that there will be extensive opportunities related to the fabrication of the nano-architectures and utilization of $g\text{-}C_3N_4$ and CTFs in sustainable catalysis, solar energy conversion, and related devices.

References

1. A. Thomas, A. Fischer, F. Goettmann, M. Antonietti, J.-O. Muller, R. Schlogl and J. M. Carlsson, *J. Mater. Chem.*, 2008, **18**, 4893.
2. J. Liu, H. Wang and M. Antonietti, *Chem. Soc. Rev.*, 2016, **45**, 2308.
3. W.-J. Ong, L.-L. Tan, Y. H. Ng, S.-T. Yong and S.-P. Chai, *Chem. Rev.*, 2016, **116**(12), 7159.
4. X. Jiang, P. Wang and J. Zhao, *J. Mater. Chem. A*, 2015, **3**, 7750.

5. G. Dong, Y. Zhang, Q. Pan and J. Qiu, *J. Photochem. Photobiol., C*, 2014, **20**, 33.
6. J. V. Liebig, *Ann. Pharm.*, 1834, **10**, 10.
7. L. Gmelin, *Ann. Pharm.*, 1835, **15**, 252.
8. D. M. Teter and R. J. Hemley, *Science*, 1996, **271**, 53.
9. Y. Zheng, L. Lin, B. Wang and X. Wang, *Angew. Chem., Int. Ed.*, 2015, **54**, 12868.
10. E. Kroke, M. Schwarz, E. H. Bordon, P. Kroll, B. Noll and A. D. Norman, *New J. Chem.*, 2002, **26**, 508.
11. A. Y. Liu and M. L. Cohen, *Science*, 1989, **245**, 841.
12. F. Goettmann, A. Fischer, M. Antonietti and A. Thomas, *Chem. Commun.*, 2006, 4530.
13. F. Su, S. C. Mathew, G. Lipner, X. Fu, M. Antonietti, S. Blechert and X. Wang, *J. Am. Chem. Soc.*, 2010, **132**, 16299.
14. F. Su, S. C. Mathew, L. Mohlmann, M. Antonietti, X. Wang and S. Blechert, *Angew. Chem.*, 2011, **123**, 683.
15. F. Goettmann, A. Thomas and M. Antonietti, *Angew. Chem., Int. Ed.*, 2007, **46**, 2717.
16. M. B. Ansari, B.-H. Min, Y.-H. Mo and S.-E. Park, *Green Chem.*, 2011, **13**, 1416.
17. J. Xu, K.-Z. Long, Y. Wang, B. Xue and Y.-X. Li, *Appl. Catal., A*, 2014, **496**, 1.
18. G. Zhang, Z.-A. Lan and X. C. Wang, *Angew. Chem., Int. Ed.*, 2016, **55**, 15712.
19. V. S. Vyas, V. W. Lau and B. V. Lotsch, *Chem. Mater.*, 2016, **28**(15), 5191.
20. A. P. Cote, A. I. Benin, N. W. Ockwig, M. O'Keeffe, A. J. Matzger and O. M. Yaghi, *Science*, 2005, **310**, 1166.
21. H. M. El-Kaderi, J. R. Hunt, J. L. Mendoza-Cortes, A. P. Cote, R. E. Taylor, M. O'Keeffe and O. M. Yaghi, *Science*, 2007, **316**, 268.
22. G. Ferey, *Chem. Soc. Rev.*, 2008, **37**, 191.
23. O. M. Yaghi, M. O'Keeffe, N. W. Ockwig, H. K. Chae, M. Eddaoudi and J. Kim, *Nature*, 2003, **423**, 705.
24. P. Kuhn, M. Antonietti and A. Thomas, *Angew. Chem., Int. Ed.*, 2008, **47**, 3450.
25. P. Kuhn, A. Forget, J. Hartmann, A. Thomas and M. Antonietti, *Adv. Mater.*, 2009, **21**, 897.
26. G. Decher, *Science*, 1997, **277**, 1232.
27. B. Voit, *J. Polym. Sci., Part A: Polym. Chem.*, 2000, **38**, 2505.
28. J. Sakamoto, J. van Heijst, O. Lukin and A. D. Schluter, *Angew. Chem., Int. Ed.*, 2009, **48**, 1030.
29. O. M. Yaghi, H. L. Li, C. Davis, D. Richardson and T. L. Groy, *Acc. Chem. Res.*, 1998, **31**, 474.
30. F. Goettmann, A. Fischer, M. Antonietti and A. Thomas, *Angew. Chem., Int. Ed.*, 2006, **45**, 4467.
31. X. Wang, K. Maeda, A. Thomas, K. Takanabe, G. Xin, J. M. Carlsson, K. Domen and M. Antonietti, *Nat. Mater.*, 2009, **8**, 76.

32. S. M. J. Rogge, A. Bavykina, J. Hajek, H. Garcia, A. I. O. Suarez, A. S. Escribano, A. Vimont, G. Clet, P. Bazin, F. Kapteijn, M. Daturi, E. V. R. Fernandez, F. X. L. Xamena, V. V. Speybroeck and J. Gascon, *Chem. Soc. Rev.*, 2017, **46**, 3134.
33. C. R. DeBlase, K. E. Silberstein, T. T. Truong, H. D. Abruna and W. R. Dichtel, *J. Am. Chem. Soc.*, 2013, **135**, 16821.
34. W. Huang, Z. J. Wang, B. C. Ma, S. Ghasimi, D. Gehrig, F. Laquai, K. Landfester and K. A. I. Zhang, *J. Mater. Chem. A*, 2016, **4**, 7555.
35. X. Wang, G. Sun, P. Routh, D.-H. Kim, W. Huang and P. Chen, *Chem. Soc. Rev.*, 2014, **43**, 7067.
36. X. Ma, Y. Lv, J. Xu, Y. Liu, R. Zhang and Y. Zhu, *J. Phys. Chem. C*, 2012, **116**, 23485.
37. X.-H. Li and M. Antonietti, *Chem. Soc. Rev.*, 2013, **42**, 6593.
38. C. E. Chan-Thaw, A. Villa, P. Katekomol, D. S. Su, A. Thomas and L. Prati, *Nano Lett.*, 2010, **10**, 537.
39. Y. S. Jun, E. Z. Lee, X. Wang, W. H. Hong, G. D. Stucky and A. Thomas, *Adv. Funct. Mater.*, 2013, **23**, 3661.
40. M. Shalom, S. Inal, C. Fettkenhauer, D. Neher and M. Antonietti, *J. Am. Chem. Soc.*, 2013, **135**, 7118.
41. S. C. Yan, Z. S. Li and Z. G. Zou, *Langmuir*, 2009, **25**, 10397.
42. J. Liu, T. Zhang, Z. Wang, G. Dawson and W. Chen, *J. Mater. Chem.*, 2011, **21**, 14398.
43. F. Dong, L. Wu, Y. Sun, M. Fu, Z. Wu and S. C. Lee, *J. Mater. Chem.*, 2011, **21**, 15171.
44. Y. Sakata, K. Yoshimoto, K. Kawaguchi, H. Imamura and S. Higashimoto, *Catal. Today*, 2011, **161**, 41.
45. Y. Zhang, J. Liu, G. Wu and W. Chen, *Nanoscale*, 2012, **4**, 5300.
46. Y. P. Yuan, W. T. Xu, L. S. Yin, S. W. Cao, Y. S. Liao, Y. Q. Tng and C. Xue, *Int. J. Hydrogen Energy*, 2013, **38**, 13159.
47. D. J. Martin, K. Qiu, S. A. Shevlin, A. D. Handoko, X. Chen, Z. Guo and J. Tang, *Angew. Chem., Int. Ed.*, 2014, **53**, 9240.
48. J. Xu, L. Zhang, R. Shi and Y. Zhu, *J. Mater. Chem. A*, 2013, **1**, 14766.
49. R. C. Pawar, S. Kang, J. H. Park, J. Kim, S. Ahn and C. S. Lee, *Sci. Rep.*, 2016, **6**, 31147.
50. F. Zhao, H. Cheng, Y. Hu, L. Song, Z. Zhang, L. Jiang and L. Qu, *Sci. Rep.*, 2014, **4**, 5882.
51. H. Xu, J. Yan, X. She, L. Xu, J. Xia, Y. Xu, Y. Song, L. Huang and H. Li, *Nanoscale*, 2014, **6**, 1406.
52. Y. Lu, D. Chu, M. Zhu, Y. Du and P. Yang, *Phys. Chem. Chem. Phys.*, 2015, **17**, 17355.
53. X. Du, G. Zou, Z. Wang and X. Wang, *Nanoscale*, 2015, **7**, 8701.
54. J. H. Lee, M. J. Park, S. J. Yoo, J. H. Jang, H. J. Kim, S. W. Nam, C. W. Yoon and J. Y. Kim, *Nanoscale*, 2015, **7**, 10334.
55. Z. Wang, L. Chen, X. Du, G. Zou and X. Wang, *RSC Adv.*, 2016, **6**, 15605.
56. S. Han, D. Q. Wu, S. Li, F. Zhang and X. L. Feng, *Adv. Mater.*, 2014, **26**, 849.

57. A. Thomas, F. Goettmann and M. Antonietti, *Chem. Mater.*, 2008, **20**, 738.
58. M. Groenewolt and M. Antonietti, *Adv. Mater.*, 2005, **17**, 1789.
59. X. C. Wang, K. Maeda, X. F. Chen, K. Takanabe, K. Domen, Y. D. Hou, X. Z. Fu and M. Antonietti, *J. Am. Chem. Soc.*, 2009, **131**, 1680.
60. S. S. Park, S.-W. Chu, C. Xue, D. Zhao and C.-S. Ha, *J. Mater. Chem.*, 2011, **21**, 10801.
61. J. Zhang, F. Guo and X. Wang, *Adv. Funct. Mater.*, 2013, **23**, 3008.
62. S. N. Talapaneni, S. Anandan, G. P. Mane, C. Anand, D. S. Dhawale, S. Varghese, A. Mano, T. Mori and A. Vinu, *J. Mater. Chem.*, 2012, **22**, 9831.
63. S. N. Talapaneni, G. P. Mane, A. Mano, C. Anand, D. S. Dhawale, T. Mori and A. Vinu, *ChemSusChem*, 2012, **5**(4), 700.
64. G. P. Mane, D. S. Dhawale, C. Anand, K. Ariga, Q. Ji, M. A. Waha, T. Mori and A. Vinu, *J. Mater. Chem. A*, 2013, **1**, 2913.
65. C. Anand, S. V. Priya, G. Lawrence, G. P. Mane, D. S. Dhawale, K. S. Prasad, V. V. Balasubramanian, M. A. Wahab and A. Vinu, *Catal. Today*, 2013, **204**, 164.
66. G. P. Mane, S. N. Talapaneni, K. S. Lakhi, H. Ilbeygi, U. Ravon, K. Al-Bahily, T. Mori, D.-H. Park and A. Vinu, *Angew. Chem., Int. Ed.*, 2017, **56**, 1.
67. K. S. Lakhi, D.-H. Park, G. Singh, S. N. Talapaneni, U. Ravon, K. Al-Bahily and A. Vinu, *J. Mater. Chem. A*, 2017, **5**, 16220.
68. K. Kailasam, J. D. Epping, A. Thomas, S. Losse and H. Junge, *Energy Environ. Sci.*, 2011, **4**, 4668.
69. J. H. Sun, J. S. Zhang, M. W. Zhang, M. Antonietti, X. Z. Fu and X. C. Wang, *Nat. Commun.*, 2012, **3**, 1139.
70. J. Huang, M. Antonietti and J. Liu, *J. Mater. Chem. A*, 2014, **2**, 7686.
71. J. Zhang, M. Zhang, C. Yang and X. Wang, *Adv. Mater.*, 2014, **26**, 4121.
72. Q. Han, B. Wang, Y. Zhao, C. Hu and L. Qu, *Angew. Chem., Int. Ed.*, 2015, **54**, 11433.
73. D. D. Zheng, C. J. Huang and X. C. Wang, *Nanoscale*, 2015, **7**, 465.
74. G. Jiang, C.-H. Zhou, X. Xia, F. Yang, D. Tong, W. Yu and S. Liu, *Mater. Lett.*, 2010, **64**, 2718.
75. J. T. Kloprogge, J. Breukelaar, J. B. H. Jansen and J. W. Geus, *Clays Clay Miner.*, 1993, **41**, 103.
76. J. Zhuang, W. Lai, M. Xu, Q. Zhou and D. Tang, *ACS Appl. Mater. Interfaces*, 2015, **7**(15), 8330.
77. K. Homan, J. Chen, A. Schiano, M. Mohamed, K. Willets, S. Murugesan, K. Stevenson and S. Emelianov, *Adv. Funct. Mater.*, 2011, **21**, 1673.
78. A. Ermushev, B. Mchedlishvili, V. Oleinikov and A. Petukhov, *Quantum Electron.*, 1993, **23**, 435.
79. Q. Xiang, J. Yu and M. Jaroniec, *Chem. Soc. Rev.*, 2012, **41**, 782.
80. J. Liu, H. Wang, Z. P. Chen, H. Moehwald, S. Fiechter, R. v. d. Krol, L. Wen, L. Jiang and M. Antonietti, *Adv. Mater.*, 2015, **27**, 712.
81. J. Bian, Q. Li, C. Huang, J. Li, Y. Guo, M. Zaw and R.-Q. Zhang, *Nano Energy*, 2015, **15**, 353.

82. L. Jia, H. Wang, D. Dhawale, C. Anand, M. A. Waha, Q. Ji, K. Ariga and A. Vinu, *Chem. Commun.*, 2014, **50**, 5976.
83. X.-H. Li, J. Zhang, X. Chen, A. Fischer, A. Thomas, M. Antonietti and X. Wang, *Chem. Mater.*, 2011, **23**, 4344.
84. J. Liu, R. Cazelles, H. Zhou, A. Galarneau and M. Antonietti, *Phys. Chem. Chem. Phys.*, 2014, **16**, 14699.
85. J. Liu, J. Huang, D. Dontsova and M. Antonietti, *RSC Adv.*, 2013, **3**, 22988.
86. Y. Zhao, Z. Liu, W. Chu, L. Song, Z. Zhang, D. Yu, Y. Tian, S. Xie and L. Sun, *Adv. Mater.*, 2008, **20**, 1777.
87. S.-W. Bian, Z. Ma and W.-G. Song, *J. Phys. Chem. C*, 2009, **113**, 8668.
88. Y. Zheng, L. H. Lin, X. J. Ye, F. S. Guo and X. C. Wang, *Angew. Chem., Int. Ed.*, 2014, **53**, 11926.
89. J. Liu, J. Huang, H. Zhou and M. Antonietti, *ACS Appl. Mater. Interfaces*, 2014, **6**, 8434.
90. Y. Wang, J. S. Zhang, X. C. Wang, M. Antonietti and H. R. Li, *Angew. Chem., Int. Ed.*, 2010, **49**, 3356.
91. Y. Wang, J. S. Zhang, X. C. Wang, M. Antonietti and H. R. Li, *Angew. Chem., Int. Ed.*, 2010, **122**, 3428.
92. H. J. Yan, *Chem. Commun.*, 2012, **48**, 3430.
93. Z. Z. Lin and X. C. Wang, *ChemSusChem*, 2014, **7**, 1547.
94. L. G. Zhang, X. F. Chen, J. Guan, Y. J. Jiang, T. G. Hou and X. D. Mu, *Mater. Res. Bull.*, 2013, **48**, 3485.
95. J. S. Xu, T. J. K. Brenner, Z. P. Chen, D. Neher, M. Antonietti and M. Shalom, *ACS Appl. Mater. Interfaces*, 2014, **6**, 16481.
96. Y. Ishida, L. Chabanne, M. Antonietti and M. Shalom, *Langmuir*, 2014, **30**(2), 447.
97. Y. S. Jun, J. Park, S. U. Lee, A. Thomas, W. H. Hong and G. D. Stucky, *Angew. Chem., Int. Ed.*, 2013, **52**, 11083.
98. Y. S. Jun, J. Park, S. U. Lee, A. Thomas, W. H. Hong and G. D. Stucky, *Angew. Chem., Int. Ed.*, 2013, **125**, 11289.
99. Z. Zhao, Y. Sun and F. Dong, *Nanoscale*, 2015, **7**, 15.
100. Y. Ham, K. Maeda, D. Cha, K. Takanabe and K. Domen, *Chem.–Asian J.*, 2013, **8**, 218.
101. H. L. Gao, S. C. Yan, J. J. Wang, Y. A. Huang, P. Wang, Z. S. Li and Z. G. Zou, *Phys. Chem. Chem. Phys.*, 2013, **15**, 18077.
102. G. Algara-Siller, N. Severin, S. Y. Chong, T. Bjçrkman, R. G. Palgrave, A. Laybourn, M. Antonietti, Y. Z. Khimyak, A. V. Krasheninnikov, J. P. Rabe, U. Kaiser, A. I. Cooper, A. Thomas and M. J. Bojdys, *Angew. Chem., Int. Ed.*, 2014, **53**, 7450.
103. G. A. Siller, N. Severin, S. Y. Chong, T. Bjçrkman, R. G. Palgrave, A. Laybourn, M. Antonietti, Y. Z. Khimyak, A. V. Krasheninnikov, J. P. Rabe, U. Kaiser, A. I. Cooper, A. Thomas and M. J. Bojdys, *Angew. Chem., Int. Ed.*, 2014, **126**, 7580.
104. E. Wirnhier, M. Dçblinger, D. Gunzelmann, J. Senker, B. V. Lotsch and W. Schnick, *Chem.–Eur. J.*, 2011, **17**, 3213.

105. S. Y. Chong, J. T. A. Jones, Y. Z. Khimyak, A. I. Cooper, A. Thomas, M. Antonietti and M. J. Bojdys, *J. Mater. Chem. A*, 2013, **1**, 1102.
106. M. J. Bojdys, J. O. Mîller, M. Antonietti and A. Thomas, *Chem.–Eur. J.*, 2008, **14**, 8177.
107. K. Schwinghammer, M. B. Mesch, V. Duppel, C. Ziegler, J. Senker and B. V. Lotsch, *J. Am. Chem. Soc.*, 2014, **136**, 1730.
108. M. K. Bhunia, K. Yamauchi and K. Takanabe, *Angew. Chem., Int. Ed.*, 2014, **53**, 11001.
109. M. K. Bhunia, K. Yamauchi and K. Takanabe, *Angew. Chem., Int. Ed.*, 2014, **126**, 11181.
110. K. Schwinghammer, B. Tuffy, M. B. Mesch, E. Wirnhier, C. Martineau, F. Taulelle, W. Schnick, J. Senker and B. V. Lotsch, *Angew. Chem., Int. Ed.*, 2013, **52**, 2435.
111. K. Schwinghammer, B. Tuffy, M. B. Mesch, E. Wirnhier, C. Martineau, F. Taulelle, W. Schnick, J. Senker and B. V. Lotsch, *Angew. Chem., Int. Ed.*, 2013, **125**, 2495.
112. Y. J. Cui, Z. X. Ding, X. Z. Fu and X. C. Wang, *Angew. Chem., Int. Ed.*, 2012, **51**, 11814.
113. Y. J. Cui, Z. X. Ding, X. Z. Fu and X. C. Wang, *Angew. Chem., Int. Ed.*, 2012, **124**, 11984.
114. H. Montigaud, B. Tanguy, G. Demazeau, S. Courjault, M. Birot and J. Dunogues, *C. R. Acad. Sci., Ser. IIb*, 1997, **325**, 229.
115. Y. Zheng, L. Lin and X. Wang, in *Nanocarbons for Advanced Energy Conversion*, ed. X. Feng, Wiley-VCH Verlag GmbH & Co. KGaA, 2015, Chapter 11: Nanostructured Carbon Nitrides for Photocatalytic Water Splitting, pp. 281–300.
116. Y. Yang, J. Chen, Z. Mao, N. An, D. Wang and B. D. Fahlman, *RSC Adv.*, 2017, **7**, 2333.
117. X. Zhang, X. Xie, H. Wang, J. Zhang, B. Pan and Y. Xie, *J. Am. Chem. Soc.*, 2012, **135**, 18.
118. S. Yang, Y. Gong, J. Zhang, L. Zhan, L. Ma, Z. Fang, R. Vajtai, X. Wang and P. M. Ajayan, *Adv. Mater.*, 2013, **25**, 2452.
119. Y. Zhang, A. Thomas, M. Antonietti and X. Wang, *J. Am. Chem. Soc.*, 2008, **131**, 50.
120. P. Niu, L. Zhang, G. Liu and H. M. Cheng, *Adv. Funct. Mater.*, 2012, **22**, 4763.
121. Z. Zhou, Y. Shen, Y. Li, A. Liu, S. Liu and Y. Zhang, *ACS Nano*, 2015, **9**, 12480.
122. X. Wang, G. Sun, N. Li and P. Chen, *Chem. Soc. Rev.*, 2016, **45**, 2239.
123. W. Wang, J. C. Yu, Z. Shen, D. K. L. Chan and T. Gu, *Chem. Commun.*, 2014, **50**, 10148.
124. S. S. Yamijala, M. Mukhopadhyay and S. K. Pati, *J. Phys. Chem. C*, 2015, **119**, 12079.
125. S. S. Yamijala, A. Bandyopadhyay and S. K. Pati, *J. Phys. Chem. C*, 2013, **117**, 23295.

126. A. Bandyopadhyay, D. Ghosh, N. M. Kaley and S. K. Pati, *J. Phys. Chem. C*, 2017, **121**(3), 1982.
127. Z. Li, B. Li, S. Peng, D. Li, S. Yang and Y. Fang, *RSC Adv.*, 2014, **4**, 35144.
128. X. Chen, Q. Liu, Q. Wu, P. Du, J. Zhu, S. Dai and S. Yang, *Adv. Funct. Mater.*, 2016, **26**, 1719.
129. X. Wang, L. Wang, F. Zhao, C. Hu, Y. Zhao, Z. Zhang, S. Chen, G. Shi and L. Qu, *Nanoscale*, 2015, **7**, 3035.
130. Y. H. Xu, S. B. Jin, H. Xu, A. Nagai and D. L. Jiang, *Chem. Soc. Rev.*, 2013, **42**, 8012.
131. S. Wan, J. Guo, J. Kim, H. Ihee and D. L. Jiang, *Angew. Chem., Int. Ed.*, 2008, **47**, 8826.
132. J. X. Jiang, F. B. Su, A. Trewin, C. D. Wood, N. L. Campbell, H. J. Niu, C. Dickinson, A. Y. Ganin, M. J. Rosseinsky, Y. Z. Khimyak and A. I. Cooper, *Angew. Chem., Int. Ed.*, 2007, **46**, 8574.
133. S. Y. Ding and W. Wang, *Chem. Soc. Rev.*, 2013, **42**, 548.
134. K. Schwinghammer, S. Hug, M. B. Mesch, J. Senkerd and B. V. Lotsch, *Energy Environ. Sci.*, 2015, **8**, 3345.
135. R. Dawson, A. I. Cooper and D. J. Adams, *Prog. Polym. Sci.*, 2012, **37**, 530.
136. M. G. Rabbani, T. E. Reich, R. M. Kassab, K. T. Jackson and H. M. El-Kaderi, *Chem. Commun.*, 2012, **48**, 1141.
137. P. Kaur, J. T. Hupp and S. T. Nguyen, *ACS Catal.*, 2011, **1**, 819.
138. S. Wan, J. Guo, J. Kim, H. Ihee and D. L. Jiang, *Angew. Chem., Int. Ed.*, 2009, **48**, 5439.
139. K. Kailasam, J. Schmidt, H. Bildirir, G. G. Zhang, S. Blechert, X. C. Wang and A. Thomas, *Macromol. Rapid Commun.*, 2013, **34**, 1008.
140. K. Y. Wu, J. Guo and C. C. Wang, *Chem. Commun.*, 2014, **50**, 695.
141. X. Zhu, C. C. Tian, S. M. Mahurin, S. H. Chai, C. M. Wang, S. Brown, G. M. Veith, H. M. Luo, H. L. Liu and S. Dai, *J. Am. Chem. Soc.*, 2012, **134**, 10478.
142. A. Bhunia, I. Boldog, A. Moller and C. Janiak, *J. Mater. Chem. A*, 2013, **1**, 14990.
143. W. Zhang, C. Li, Y. P. Yuan, L. G. Qiu, A. J. Xie, Y. H. Shen and J. F. Zhu, *J. Mater. Chem.*, 2010, **20**, 6413.
144. X. Zhu, S. H. Chai, C. C. Tian, P. F. Fulvio, K. S. Han, E. W. Hagaman, G. M. Veith, S. M. Mahurin, S. Brown, H. L. Liu and S. Dai, *Macromol. Rapid Commun.*, 2013, **34**, 452.
145. L.-M. Tao, F. Niu, D. Zhang, T.-M. Wang and Q.-H. Wang, *New J. Chem.*, 2014, **38**, 2774.
146. A. Bhunia, D. Esquivel, S. Dey, R. Fernández-Terán, Y. Goto, S. Inagaki, P. V. Der Voort and C. Janiak, *J. Mater. Chem. A*, 2016, **4**, 13450.
147. T. Wang, K. Kailasam, P. Xiao, G. S. Chen, L. Q. Chen, L. H. Wang, J. L. Li and J. J. Zhu, *Microporous Mesoporous Mater.*, 2014, **187**, 63.
148. R. Palkovits, M. Antonietti, P. Kuhn, A. Thomas and F. Schuth, *Angew. Chem., Int. Ed.*, 2009, **48**, 6909.

149. P. Kuhn, A. Thomas and M. Antonietti, *Macromolecules*, 2009, **42**, 319.
150. S. Kuecken, J. Schmidt, L. Zhi and A. Thomas, *J. Mater. Chem. A*, 2015, **3**, 24422.
151. L. Hao, B. Luo, X. L. Li, M. H. Jin, Y. Fang, Z. H. Tang, Y. Y. Jia, M. H. Liang, A. Thomas, J. H. Yang and L. J. Zhi, *Energy Environ. Sci.*, 2012, **5**, 9747.
152. L. Hao, X. L. Li and L. J. Zhi, *Adv. Mater.*, 2013, **25**, 3899.
153. K. Sakaushi, G. Nickerl, F. M. Wisser, D. Nishio-Hamane, E. Hosono, H. S. Zhou, S. Kaskel and J. Eckert, *Angew. Chem., Int. Ed.*, 2012, **51**, 7850.
154. K. Sakaushi, E. Hosono, G. Nickerl, T. Gemming, H. Zhou, S. Kaskel and J. Eckert, *Nat. Commun.*, 2013, **4**, 1485.
155. K. Sakaushi, E. Hosono, G. Nickerl, H. S. Zhou, S. Kaskel and J. Eckert, *J. Power Sources*, 2014, **245**, 553.
156. C. E. Chan-Thaw, A. Villa, L. Prati and A. Thomas, *Chem.–Eur. J.*, 2011, **17**, 1052.
157. F. Niu, L. Tao, Y. Deng, H. Gao, J. Liu and W. Song, *New J. Chem.*, 2014, **38**, 5695.
158. J. Roeser, K. Kailasam and A. Thomas, *ChemSusChem*, 2012, **5**, 1793.
159. J. Zhu, K. Kailasam, A. Fischer and A. Thomas, *ACS Catal.*, 2011, **1**, 342.
160. S. Kuecken, A. Acharjya, L. Zhi, M. Schwarze, R. Schomäcker and A. Thomas, *Chem. Commun.*, 2017, **53**, 5854.
161. K. Sakaushi, G. Nickerl, H. C. Kandpal, L. C. Cortés, T. Gemming, J. Eckert, S. Kaskel and J. V. den Brink, *J. Phys. Chem. Lett.*, 2013, **4**, 2977.
162. S. Hug, M. E. Tauchert, S. Li, U. E. Pachmayr and B. V. Lotsch, *J. Mater. Chem.*, 2012, **22**, 13956.
163. A. Bhunia, V. Vasylyeva and C. Janiak, *Chem. Commun.*, 2013, **49**, 3961.
164. P. Kuhn, A. l. Forget, D. Su, A. Thomas and M. Antonietti, *J. Am. Chem. Soc.*, 2008, **130**, 13333.
165. M. J. Bojdys, J. Jeromenok, A. Thomas and M. Antonietti, *Adv. Mater.*, 2010, **22**, 2202.
166. S. Ren, M. J. Bojdys, R. Dawson, A. Laybourn, Y. Z. Khimyak, D. J. Adams and A. I. Cooper, *Adv. Mater.*, 2012, **24**, 2357.
167. B. A. Roberts and C. R. Strauss, *Acc. Chem. Res.*, 2005, **38**, 653.
168. N. B. Shustova, B. D. McCarthy and M. Dincă, *J. Am. Chem. Soc.*, 2011, **133**, 20126.
169. S. Xiong, X. Fu, L. Xiang, G. Yu, J. Guan, Z. Wang, Y. Du, X. Xiong and C. Pan, *Polym. Chem.*, 2014, **5**, 3424.
170. F. Jeremias, A. Khutia, S. K. Henninger and C. Janiak, *J. Mater. Chem.*, 2012, **22**, 10148.
171. F. Jeremias, V. Lozan, S. Henninger and C. Janiak, *Dalton Trans.*, 2013, **42**, 15967.
172. S. Keskin, T. M. van Heest and D. S. Sholl, *ChemSusChem*, 2010, **3**, 879.
173. A. Ajayaghosh, *Chem. Soc. Rev.*, 2003, **32**, 181.
174. S. Himeno, T. Komatsu and S. Fujita, *J. Chem. Eng. Data*, 2005, **50**, 369.

175. K. B. Lee, M. G. Beaver, H. S. Caram and S. Sircar, *Ind. Eng. Chem. Res.*, 2008, **47**, 8048.
176. G. Dong and L. Zhang, *J. Mater. Chem.*, 2012, **22**, 1160.
177. Y. Chen, B. Wang, S. Lin, Y. Zhang and X. Wang, *J. Phys. Chem. C*, 2014, **118**, 29981.
178. X. B. Chen, S. H. Shen, L. J. Guo and S. S. Mao, *Chem. Rev.*, 2010, **110**, 6503.
179. A. Kudo and Y. Miseki, *Chem. Soc. Rev.*, 2009, **38**, 253.
180. Y. Wang, H. Li, J. Yao, X. Wang and M. Antonietti, *Chem. Sci.*, 2011, **2**, 446.
181. J. W. Fang, H. Q. Fan, M. M. Li and C. B. Long, *J. Mater. Chem. A*, 2015, **3**, 13819.
182. Y. J. Zhou, L. X. Zhang, W. M. Huang, Q. L. Kong, X. Q. Fan, M. Wang and J. L. Shi, *Carbon*, 2016, **99**, 111.
183. J. Li, Y. Zhang, X. Zhang, J. Han, Y. Wang, L. Gu, Z. Zhang, X. Wang, J. Jian, P. Xu and B. Song, *ACS Appl. Mater. Interfaces*, 2015, **7**(35), 19626.
184. R. Mei, L. Ma, L. An, F. Wang, J. Xi, H. Sun, Z. Luo and Q. Wu, *J. Electrochem. Soc.*, 2017, **164**(4), F354.
185. J. H. Li, B. A. Shen, Z. H. Hong, B. Z. Lin, B. F. Gao and Y. L. Chen, *Chem. Commun.*, 2012, **48**, 12017.
186. Z. F. Huang, J. J. Song, L. Pan, Z. M. Wang, X. Q. Zhang, J. J. Zou, W. B. Mi, X. W. Zhang and L. Wang, *Nano Energy*, 2015, **12**, 646.
187. S. N. Guo, Y. Zhu, Y. Y. Yan, Y. L. Min, J. C. Fan and Q. J. Xu, *Appl. Catal., B*, 2016, **185**, 315.
188. J. Cui, S. H. Liang, X. S. Wang and J. M. Zhang, *Mater. Chem. Phys.*, 2016, **161**, 194.
189. K. Wang, Q. Li, B. Liu, B. Cheng, W. Ho and J. Yu, *Appl. Catal., B*, 2015, **176–177**, 44.
190. L. Ke, P. Li, X. Wu, S. Jiang, M. Luo, Y. Liu, Z. Le, C. Sun and S. Song, *Appl. Catal., B*, 2017, **205**, 319.
191. N. D. Shcherban, S. M. Filonenko, M. L. Ovcharov, A. M. Mishura, M. A. Skoryk, A. Aho and D. Y. Murzin, *ChemistrySelect*, 2016, **1**, 4987.
192. F. He, G. Chen, Y. G. Yu, Y. S. Zhou, Y. Zheng and S. Hao, *Chem. Commun.*, 2015, **51**, 425.
193. J. S. Zhang, J. H. Sun, K. Maeda, K. Domen, P. Liu, M. Antonietti, X. Z. Fu and X. C. Wang, *Energy Environ. Sci.*, 2011, **4**, 675.
194. G. Liu, P. Niu, C. H. Sun, S. C. Smith, Z. G. Chen, G. Q. Lu and H. M. Cheng, *J. Am. Chem. Soc.*, 2010, **132**, 11642.
195. S. Stolbov and S. Zuluaga, *J. Phys.: Condens. Matter*, 2013, **25**, 085507.
196. Y. J. Zhang, T. Mori, J. H. Ye and M. Antonietti, *J. Am. Chem. Soc.*, 2010, **132**, 6294.
197. X. F. Chen, Y. S. Jun, K. Takanabe, K. Maeda, K. Domen, X. Z. Fu, M. Antonietti and X. C. Wang, *Chem. Mater.*, 2009, **21**, 4093.
198. X. X. Zou, G. D. Li, Y. N. Wang, J. Zhao, C. Yan, M. Y. Guo, L. Li and J. S. Chen, *Chem. Commun.*, 2011, **47**, 1066.

199. Y. Zhou, L. Zhang, J. Liu, X. Fan, B. Wang, M. Wang, W. Ren, J. Wang, M. Li and J. Shi, *J. Mater. Chem. A*, 2015, **3**, 3862.
200. J. R. Ran, T. Y. Ma, G. P. Gao, X. W. Du and S. Z. Qiao, *Energy Environ. Sci.*, 2015, **8**, 3708.
201. Y.-P. Zhu, T.-Z. Ren and Z.-Y. Yuan, *ACS Appl. Mater. Interfaces*, 2015, **7**, 16850.
202. S. E. Guo, Z. P. Deng, M. X. Li, B. J. Jiang, C. G. Tian, Q. J. Pan and H. G. Fu, *Angew. Chem., Int. Ed.*, 2016, **55**, 1830.
203. W. Zhang, J. Zhang, F. Dong and Y. Zhang, *RSC Adv.*, 2016, **6**, 88085.
204. G. G. Zhang, M. W. Zhang, X. X. Ye, X. Q. Qiu, S. Lin and X. C. Wang, *Adv. Mater.*, 2014, **26**, 805.
205. Q. Han, C. G. Hu, F. Zhao, Z. P. Zhang, N. Chen and L. T. Qu, *J. Mater. Chem. A*, 2015, **3**, 4612.
206. Z. A. Lan, G. G. Zhang and X. C. Wang, *Appl. Catal., B*, 2016, **192**, 116.
207. Y. Wang, Y. Di, M. Antonietti, H. Li, X. Chen and X. Wang, *Chem. Mater.*, 2010, **22**, 5119.
208. X. Wang, X. Chen, A. Thomas, X. Fu and M. Antonietti, *Adv. Mater.*, 2009, **21**, 1609.
209. Z. Ding, X. Chen, M. Antonietti and X. Wang, *ChemSusChem*, 2011, **4**(2), 274.
210. B. Yue, Q. Li, H. Iwai, T. Kako and J. Ye, *Sci. Technol. Adv. Mater.*, 2011, **12**, 034401.
211. Y. Gong, M. Li, H. Li and Y. Wang, *Green Chem.*, 2015, **17**, 715.
212. X. Chen, J. Zhang, X. Fu, M. Antonietti and X. Wang, *J. Am. Chem. Soc.*, 2009, **131**(33), 11658.
213. Y. Wang, Y. Wang, Y. Li, H. Shi, Y. Xu, H. Qin, X. Li, Y. Zuo, S. Kang and L. Cui, *Catal. Commun.*, 2015, **72**, 24.
214. X. Rong, F. Qiu, J. Rong, X. Zhu, J. Yan and D. Yang, *Mater. Lett.*, 2016, **164**, 127.
215. M. Zhang, X. J. Bai, D. Liu, J. Wang and Y. F. Zhu, *Appl. Catal., B*, 2015, **164**, 77.
216. C. S. Pan, J. Xu, Y. J. Wang, D. Li and Y. F. Zhu, *Adv. Funct. Mater.*, 2012, **22**, 1518.
217. S. C. Yan, S. B. Lv, Z. S. Li and Z. G. Zou, *Dalton Trans.*, 2010, **39**, 1488.
218. Y. D. Hou, A. B. Laursen, J. S. Zhang, G. G. Zhang, Y. S. Zhu, X. C. Wang, S. Dahl and I. Chorkendorff, *Angew. Chem., Int. Ed.*, 2013, **52**, 3621.
219. Y. D. Hou, A. B. Laursen, J. S. Zhang, G. G. Zhang, Y. S. Zhu, X. C. Wang, S. Dahl and I. Chorkendorff, *Angew. Chem., Int. Ed.*, 2013, **125**, 3709.
220. J. D. Hong, Y. S. Wang, Y. B. Wang, W. Zhang and R. Xu, *ChemSusChem*, 2013, **6**, 2263.
221. S. W. Cao, Y. P. Yuan, J. Barber, S. C. J. Loo and C. Xue, *Appl. Surf. Sci.*, 2014, **319**, 344.
222. Q. J. Xiang, J. G. Yu and M. Jaroniec, *J. Phys. Chem. C*, 2011, **115**, 7355.
223. J. S. Zhang, M. W. Zhang, R. Q. Sun and X. C. Wang, *Angew. Chem., Int. Ed.*, 2012, **51**, 10145.

224. J. S. Zhang, M. W. Zhang, R. Q. Sun and X. C. Wang, *Angew. Chem., Int. Ed.*, 2012, **124**, 10292.
225. S. W. Zhang, L. P. Zhao, M. Y. Zeng, J. X. Li, J. Z. Xu and X. K. Wang, *Catal. Today*, 2014, **224**, 114.
226. S. Gawande and S. R. Thakare, *ChemCatChem*, 2012, **4**, 1759.
227. C. A. Caputo, M. A. Gross, V. W. Lau, C. Cavazza, B. V. Lotsch and E. Reisner, *Angew. Chem., Int. Ed.*, 2014, **126**, 11722.
228. G. G. Zhang, S. Zang and X. C. Wang, *ACS Catal.*, 2015, **5**, 941.
229. G. G. Zhang, C. J. Huang and X. C. Wang, *Small*, 2015, **11**, 1215.
230. Y. Wang, J. Yao, H. Li, D. Su and M. Antonietti, *J. Am. Chem. Soc.*, 2011, **133**, 2362.
231. L. Ge, C. Han, J. Liu and Y. Li, *Appl. Catal., A*, 2011, **409–410**, 215.
232. K. Kailasam, M. B. Mesch, L. Möhlmann, M. Baar, S. Blechert, M. Schwarze, M. Schröder, R. Schomäcker, J. Senker and A. Thomas, *Energy Technol.*, 2016, **4**, 744.

Section II

Characterization

CHAPTER 4

Raman, IR and INS Characterization of Functionalized Carbon Materials

E. GROPPO[a], F. BONINO[a], F. CESANO[a], A. DAMIN[a] AND M. MANZOLI*[b]

[a]Università degli Studi di Torino, Department of Chemistry, NIS and INSTM Reference Centre, Via G. Quarello 15, I-10135 and Via P. Giuria 7, Torino, I-10125, Italy; [b]Università degli Studi di Torino, Department of Drug Science and Technology, NIS and INSTM Reference Centre, Via P. Giuria 9, 10125 Torino, Italy
*E-mail: maela.manzoli@unito.it

4.1 Raman Spectroscopy Applied to Carbons

Raman spectroscopy is based on the inelastic light scattering of a laser source in the near IR–UV range. Backscattered photons come out with a lower ($\nu_0 - \nu_{vibr}$) or higher ($\nu_0 + \nu_{vibr}$) energy with respect to the incoming laser photons (ν_0). The energy difference (ν_{vibr}) corresponds to the vibrational mode. A Raman active vibrational mode should involve a change in the electric polarizability. This powerful technique is widely adopted in the characterization

Catalysis Series No. 31
Metal-free Functionalized Carbons in Catalysis: Synthesis, Characterization and Applications
Edited by Alberto Villa and Nikolaos Dimitratos

Published by the Royal Society of Chemistry, www.rsc.org

of carbon-derived materials, both crystalline [*e.g.*, HOPG (highly oriented pyrolytic graphite), nano-graphite] and amorphous (*e.g.*, turbostratic carbon, carbon fibers *etc.*).

4.1.1 Raman Spectroscopy of Perfect and Defective Graphite

The Raman spectrum of perfect graphite (*e.g.*, HOPG) with large crystalline domains is dominated by a sharp peak centred at 1580 cm^{-1} (historically indicated as the G peak, first order, see Figure 4.1(a), top), which is attributed to a E_{2g} mode (Figure 4.1(c), part A). In addition, a complex envelope of bands appears in the 2800–2600 cm^{-1} region (second order), mostly due to combination modes. In particular, two main peaks are observed in the spectrum of perfect graphite, originally indicated as G'_2 (high frequency side, sharp) and G'_1 (low frequency side, broad).[1–3]

Moving to less perfect systems (Figure 4.1(a), from top to bottom), a new peak (historically indicated as the D peak, A_{1g} symmetry) located around 1350 cm^{-1} starts to appear. This band is usually assigned to a lattice breathing mode that is forbidden in graphite, but becomes active in the presence of structural disorder (Figure 4.1(c), part B). For a long time, the intensity ratio between the D and G bands (I_D/I_G) has been inversely correlated with the carbon cluster dimension according to the Tuinstra–Koenig correlation (TK) and considered as an indicator of the size of the graphitic domains in sp^2 carbonaceous materials (see Figure 4.2(b), right part). Therefore, the I_D/I_G ratio can be usefully employed to classify the carbon materials in terms

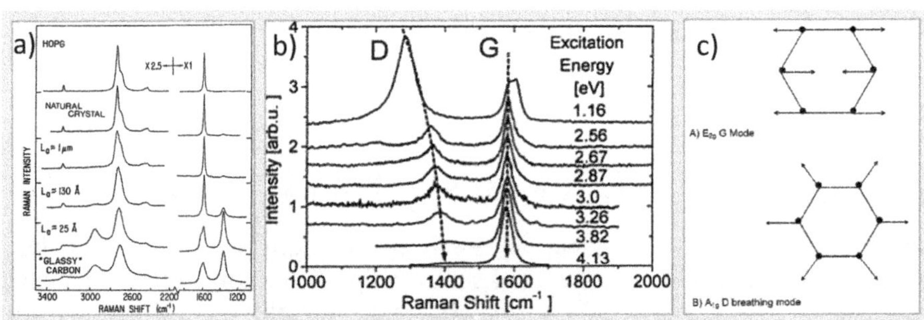

Figure 4.1 (a) First- (right side) and second-order (left side) Raman spectra of different forms of graphite, from perfect (top) to very defective up to glassy carbon (bottom). Reprinted (Figure 8) with permission from R. J. Nemanich, S. A. Solin, *Phys. Rev. B*, **20**, 392, 1979. Copyright (1979) by the American Physical Society. (b) Raman spectra of graphite excited by different laser lines. Reprinted from *J. Non. Cryst. Solids*, **227–230**, I. Pócsik, M. Hundhausen, M. Koós, L. Ley, Origin of the D peak in the Raman spectrum of microcrystalline graphite, 1083–1086, copyright (1998) with permission from Elsevier. (c) Pictorial representation of the G and D vibrational modes. Reprinted (Figure 4) with permission from A. C. Ferrari, J. Robertson, *Phys. Rev. B*, **61**, 14095, 2000. Copyright (2000) by the American Physical Society.

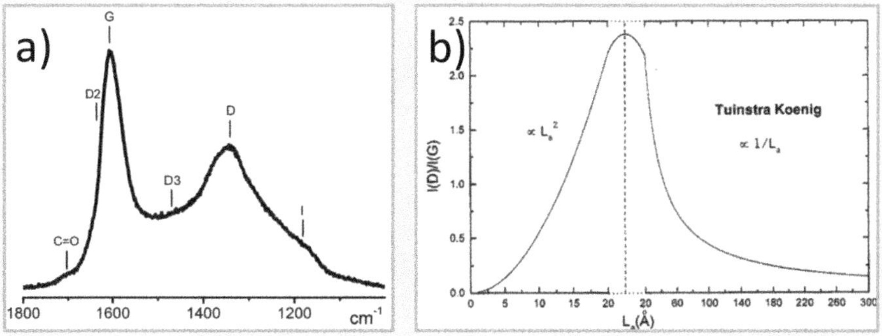

Figure 4.2 (a) Typical Raman spectrum (collected with an excitation of λ = 514 nm) of a chemically activated carbon and assignment of the main absorption bands. *Rend. Lincei*, Activated carbons for applications in catalysis: the point of view of a physical-chemist, 28, 2017, 29, A. Lazzarini (copyright Accademia Nazionale dei Linceioriginal 2017). With permission of Springer. (b) Dependence of the I_D/I_G ratio as a function of the lateral size (L_a) of the sp^2 domains. Reprinted (Figure 5) with permission from A. C. Ferrari, J. Robertson, *Phys. Rev. B*, **61**, 14095, 2000. Copyright (2000) by the American Physical Society.

of crystallinity and/or domain dimensions. Due to the fact that G'_2 and G'_1 fall quite near to the overtone region of the D peak, they have now been reclassified as second order D peak and are typically indicated as 2D, instead of the original nomenclature. Another interesting aspect characterizing the D peak is the strong dependence of its position (and of its intensity too) with respect to the exciting laser line adopted to collect the Raman spectrum.[4] Such an effect is clearly shown in Figure 4.1(b), that reports the Raman spectra of a defective graphite collected by adopting different exciting laser lines falling in the 1.16–4.13 eV (1069–300 nm) range. This behaviour suggests a strong coupling between the phonon dispersion and the electronic structure of carbon-derived materials.

4.1.2 Raman Spectroscopy of Activated Carbons

Raman spectroscopy is among the most powerful techniques to study the structure of activated carbons. These materials are entirely composed of aromatic C–C bonds without any dipole moments, but with a good polarizability. Typically, the laser lines at 514 nm and 244 nm are adopted as exciting sources. The former gives information mainly on the sp^2 domains, while the latter conveys information on both sp^2 and sp^3 carbon species. However, the λ = 514 nm laser line is the most adopted in this type of study as it is the most available in Raman laboratories.

Figure 4.2(a) displays a typical Raman spectrum of an activated carbon. It is dominated by two intense bands, attributed to vibrational modes involving sp^2 carbon species belonging to disordered microcrystalline domains.[5]

The signal around 1600 cm^{-1} is the G band: the +20 cm^{-1} shift with respect to the G band of graphite is typical of disordered graphitic systems and due to the overlap of a second band (D2, around 1620 cm^{-1}), which involves surface "graphene-like" layers or sp^2 nanostructures with a strained geometry. Furthermore, the position of the G band indicates that the graphene stacking order is very low in activated carbons and that graphite-like platelets are also partially defective. The second dominant feature around 1350 cm^{-1} is the D band. Recently, it has been proposed that the TK correlation is applicable only to graphitic domains with a lateral dimension larger than about 2 nm (Figure 4.2(b), right side).[5] In contrast, for sp^2 platelets with dimensions below 2 nm, the development of a D band in the Raman spectrum indicates ordering (Figure 4.2(b), left side).

Two additional bands are observed in the Raman spectrum shown in Figure 4.2(a): (1) the D3 band, observed between the G and the D bands around 1450 cm^{-1} and ascribed to randomly distributed amorphous carbon on the interstitial position of the graphitic islands; and (2) the I band around 1150 cm^{-1}, originated from the coexistence of the sp^3 phase and conjugated non-aromatic polyenes, present as pending groups of graphitic platelets. Besides, the very weak feature present around 1700 cm^{-1} is due to the ν(C=O) vibrations of oxygenated groups at the activated carbon surface. As expected, its intensity is very weak as it is due to a slightly active Raman vibrational mode and to surface species, hardly detectable by means of this spectroscopy, which is essentially devoted to bulk investigations.

4.1.3 Raman Spectroscopy of Graphene, Graphene Oxide and Reduced Graphene Oxide

Raman spectroscopy is largely employed to characterize graphene in terms of number of layers, defect density, edge chirality, strain, thermal properties, and amount of doping.[6–8] Indeed, the G band at around 1580 cm^{-1} is doubly-degenerate: the degeneration is lost when graphene is rolled to give a carbon nanotube. The second order 2D or G′ band at around 2700 cm^{-1} (assigned to the in-plane breathing-like mode of the carbon rings) is double resonant, and splits into four peaks in bilayer graphene. If defects are present in graphene, several additional symmetry-breaking features are found. In these cases, the band with the highest intensity is usually the D one. Also, the D′ band is influenced by symmetry breaking. Symmetry breaking is commonly due to the presence of vacancies, interstitial and substitutional atoms or to the introduction of interfaces at the borders of crystalline areas.

Concerning graphene oxide (GO, Figure 4.3(a)),[9] a typical Raman spectrum is shown in Figure 4.3(b), compared to that of reduced GO (RGO) and graphene. With respect to graphene, the spectrum of GO shows both the D and G peaks (much broader) and a very broad 2D peak. It is well known in the literature[10,11] that RGO has a Raman spectrum very similar to that of GO. Only a small increase of the I_D/I_G ratio is observed.[12,13] Finally, the Raman spectra of G1 (one-layer graphene) with a degree of functionalization of

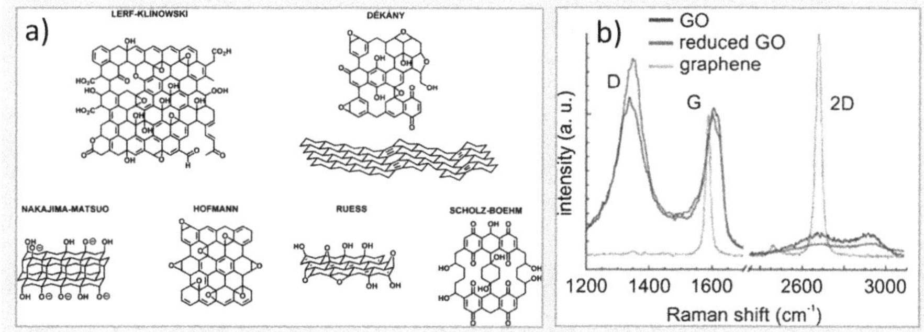

Figure 4.3 (a) Structural models of GO. Reproduced from ref. 9 with permission from the Royal Society of Chemistry. (b) Raman spectra collected from: a single sheet of graphene oxide deposited from water solution (GO), a thermally reduced single sheet of graphene oxide (R-GO), and a single sheet of pristine graphene. Reprinted (adapted) with permission from I. Jung, D. A. Dikin, R. D. Piner and R. S. Ruoff, *Nano Lett.*, 2008, **8**, 4283–4287. Copyright (2008) American Chemical Society.

about 4 and 50%, respectively, display no significant differences (data not shown), indicating that Raman spectroscopy is scarcely sensitive to surface functionalization.[11]

Graphene oxide can also be intentionally doped by means of several methods (*e.g.*, electrostatic, contact doping, chemical, substitution—mainly nitrogen or boron—and optical doping). Nitrogen-doped graphene shows better catalytic properties than the intrinsic graphene, which can be exploited in nano-devices and fuel cells. Since also amino-functionalized carbon materials display good catalytic activity, it becomes interesting to combine the amino-modification and N-doping in graphene. In this regard, Zhang *et al.*[8] proposed a simple and effective solvo-thermal method to synthesize amino-functionalized graphene, as schematically shown in Figure 4.4(a), from graphite oxide only in the presence of ammonia solution. The resultant material is an efficient metal-free catalyst, exhibiting enhanced electro-catalytic properties for oxygen reduction reactions.

Figure 4.4(b) shows the Raman spectrum (collected with an excitation of $\lambda = 632.8$ nm) of the amino-functionalised graphene prepared by Zhang *et al.*[8] compared to that of GO. The two spectra are very similar, proving once again the insensitivity of Raman spectroscopy to surface functionalization.

4.1.4 Raman Spectroscopy of Carbon Nanotubes (CNTs)

Single-walled carbon nanotubes (SWCNTs), discovered in 1993,[14] can be considered as nanoscale graphene cylinders that are closed at each end by half a fullerene. Multi-walled carbon nanotubes (MWCNTs) contain two or more concentric graphene cylinders.[15] Depending on the arrangement along the tube of the hexagon rings, SWCNTs can be metallic or semiconducting,[16]

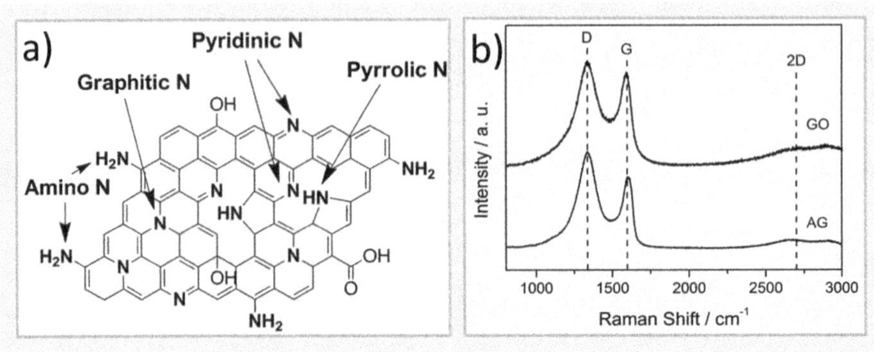

Figure 4.4 (a) Schematic representation of amino-functionalized graphene. (b) Raman spectra of amino-functionalized graphene (bottom) and graphite oxide (top). Reprinted from *Nano Energy*, **2**, C. Zhang, R. Hao, H. Liao, Y. Hou, Synthesis of amino-functionalized graphene as metal-free catalyst and exploration of the roles of various nitrogen states in oxygen reduction reaction, 88–97, Copyright (2013) with permission from Elsevier.

while MWCNTs are expected to be metallic. Raman spectroscopy is one of the most extensively employed tools to characterize CNTs. The tangential modes are the most intense high frequency modes of SWCNTs and give rise to the G band (always around 1600 cm^{-1}). When SWCNTs are functionalized, the defect-induced D band in the 1400–1300 cm^{-1} range becomes more prominent. A dependence on the diameter of the SWCNTs is frequently observed.

The most characteristic features in the Raman spectrum of SWCNTs are the low frequency 'radial breathing modes' (RBM) in the 350–150 cm^{-1} range, which are observed when all the carbon atoms of the tube vibrate radially in phase. The frequency of these vibrational modes depends inversely on the diameter *d* of the SWCNTs, and not on the way the SWCNT is rolled up to form a cylinder. In the case of functionalization, the RBMs undergo to a frequency shift. Figure 4.5 shows typical Raman spectra of SWCNTs collected with two different laser lines, in both the RBM range ((a) and (c)) and G and D range ((b) and (d)). Comparison is set to ozonized SWCNTs. Small nanotubes are preferentially ozonized. Hence, the differences observed between the two spectra can be ascribed to the functionalization of smaller nanotubes.[17]

4.1.5 Raman Spectroscopy of Other C-containing Materials

Graphitic carbon nitride (g-C_3N_4) composed of C and N elements demonstrated great potential in sustainable processes employing solar energy; furthermore, showing great stability in both very acidic and basic conditions, it could be a good candidate for a wide variety of applications in metal-free catalysis.[18] g-C_3N_4 can be obtained by a step process starting from dicyandiamide and heating: during such a process, NH_3 molecules are released and condensation occurs forming *melamine* → *melem* → *melon*- and finally porous g-C_3N_4, according to Figure 4.6(a).

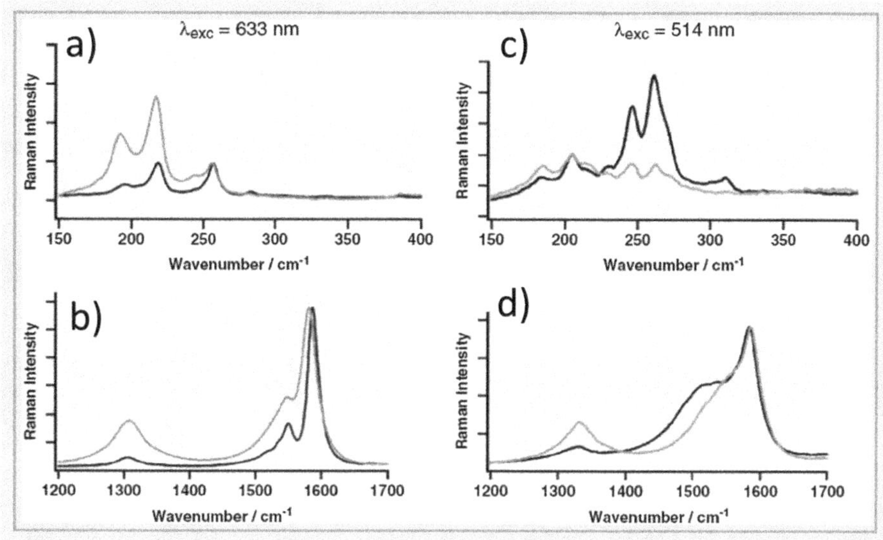

Figure 4.5 Raman spectra of as such (black) and ozonized (blue) SWCNTs collected with excitation λ = 633 nm [(a) and (b)], and λ = 514 nm [(c) and (d)] in the RBM range [(a) and (c)] and G and D bands range [(b) and (d)]. Reprinted with permission from S. Banerjee, S. S. Wong, *Nano Lett.*, 2004, **4**, 1445–1450. Copyright (2004) American Chemical Society.

Interestingly, Raman spectroscopy can be employed to follow the g-C_3N_4 formation process, as shown in Figure 4.6(b) (exciting laser line 780 nm).[19] The spectral region of 1250–400 cm^{-1} is typical for skeletal vibrations of nitrogen containing aromatic rings.[20] A triazine group[21] is present at all temperatures as demonstrated by the presence of a band falling at 980 cm^{-1}, even if starting from 550 °C, it becomes broader and finally hardly recognizable from the background at 700 °C. *Melem* bands, situated respectively at 1153 cm^{-1} and 548 cm^{-1}, are both well visible in the 400–500 °C temperature range, the second one disappearing at a higher temperature. The peak attributable to the *melon* product (located at 1233 cm^{-1}) starts to appear at 500 °C, becoming even more visible and well defined as the temperature increases up to 600 °C. The typical peak of g-C_3N_4, located at 708 cm^{-1},[22] is well visible in the 550–700 °C temperature range.

4.2 IR Spectroscopy Applied to Carbons: Principles, Problems and Solutions

IR spectroscopy relies on the use of photons in the IR region (usually in the mid-IR interval, *i.e.*, 400–4000 cm^{-1}) to excite atomic vibrations. In order for a vibrational mode in a molecule (or a molecular fragment) to be IR active, it must be associated with changes in the dipole moment. It derives that not all the vibrational modes of a molecule are IR active, but only those that satisfy the dipole selection rule. Moreover, according to the Lambert–Beer law,

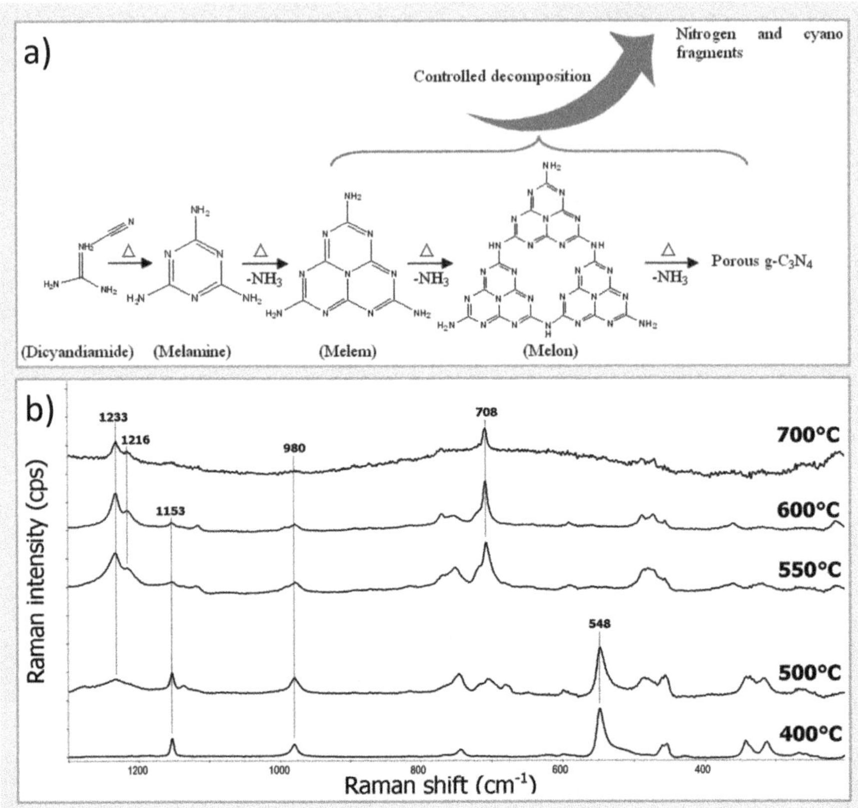

Figure 4.6 (a) Reaction processes of porous g-C_3N_4 during the thermal treatment of dicyandiamide. Reproduced from ref. 18 with permission from the Royal Society of Chemistry. (b) Raman spectra of melamine products obtained at different temperatures. Reprinted from *Mater. Chem. Phys.*, **193**, P. Praus, L. Svoboda, M. Ritz, I. Troppová, M. Šihor, K. Kočí, Graphitic carbon nitride: Synthesis, characterization and photocatalytic decomposition of nitrous oxide, 438–446, Copyright (2017) with permission from Elsevier.

the intensity of an IR absorption band generally depends on the concentration of the corresponding species, but also on the molar extinction coefficient, which is not constant along the energy interval and is a function of the type of vibration. Hence, IR spectroscopy can be used for quantitative evaluation of the absorbing species only when the extinction coefficient has been previously determined.

As far as carbonaceous materials are concerned, the ν(C–C) vibrational modes of conjugated double bonds in graphitic sp^2 domains are usually very weak in IR (opposite to what is observed in Raman), because they are not associated with a change in the dipole moment. These vibrations can be clearly observed in the IR spectrum of a carbon only in the presence of defects (surface terminations, heteroatoms, functional groups, radical

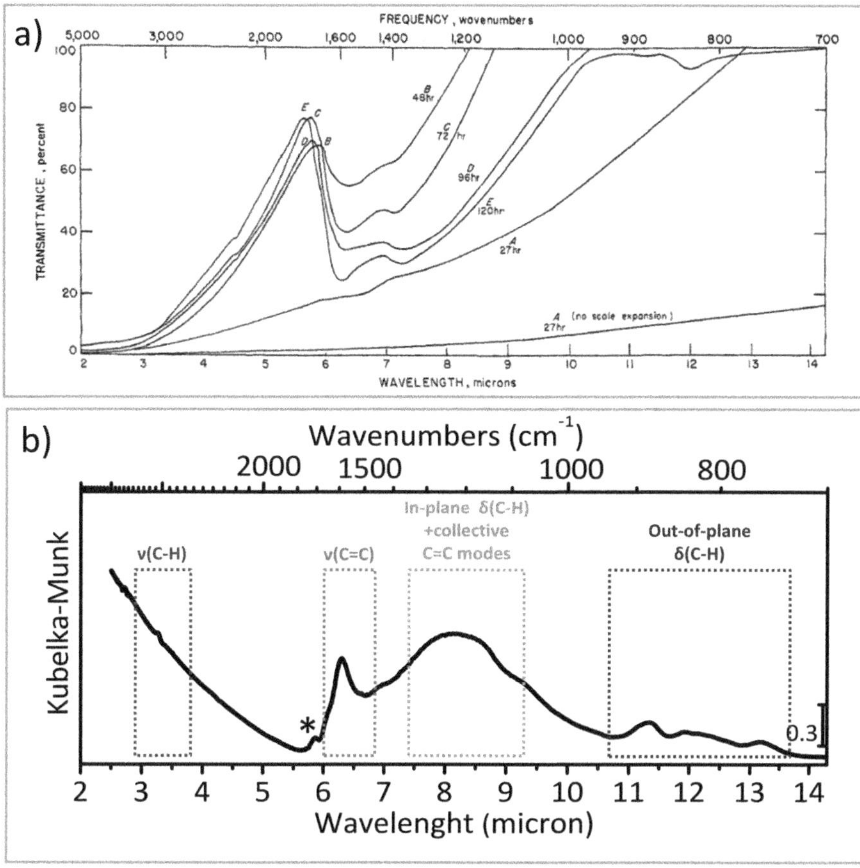

Figure 4.7 (a) IR spectrum (collected in transmission mode) of ball-milled graphite. The ball-milling time is reported for each spectrum. Reprinted with permission from R. A. Friedel, G. L. Carlson, *J. Phys. Chem.*, 1971, 75, 1149–1151. Copyright (1971) American Chemical Society. (b) IR spectrum (collected in DRIFT mode and reported in Kubelka–Munk units) of a chemically activated carbon and assignment of the main absorption bands. Unpublished spectrum.

carbon species and others), which are responsible for an increase in the dipole moment associated with the ring vibrations.[23–28] This is well exemplified by looking at the Fourier transform (FT)-IR spectrum of graphite and grounded graphite, reported in the early 1970s (Figure 4.7(a)).[29] The FT-IR spectrum of pristine graphite is completely flat, although weak bands may appear that can be assigned to adsorbed water molecules.[30] The FT-IR spectrum of graphite could be observed only by means of an exhaustive and efficient grinding, *i.e.*, when graphite loses the long-range structure. The ν(C=C) vibrations of the sp^2 domains are observed around ~1587 cm^{-1} and ~1362 cm^{-1}, very close to the Raman fingerprints for the crystalline graphite.

Two important consequences derive from the above-mentioned statements:

(1) IR spectroscopy give only indirect information on the bulk properties of carbon-based materials, while it conveys direct information on the surface species, including the C–H groups at the periphery of the sp^2 clusters (usually characterized by weak absorptions) and the functional groups containing heteroatoms (such as oxygen, nitrogen, *etc.*). The vibrational modes associated with the functional groups are usually characterized by very intense extinction coefficients. This is the reason why IR spectroscopy has been extensively used to investigate the surface species in different carbon-based materials, including activated carbons,[27,28,31,32] carbon blacks,[33] char,[34,35] coal,[36] graphite,[29] carbon fibres,[37] carbons films,[38] nano-carbons and other systems.[30] In this respect, IR spectroscopy is perfectly complementary to Raman spectroscopy.
(2) The quantitative determination of the surface species on carbons is not trivial, since the overall intensity of the IR spectrum is greatly affected by the presence of both chemical and physical defects, which enhance the dipole moment associated with the ring vibrations.

Despite the large amount of literature in this field, very often the IR spectra of carbon-based materials are characterized by undesirable spectral artefacts or distortions that complicate the data interpretation, especially when collected in the conventional transmission mode. This is mainly due to the fact that the electronic structure of foremost carbon materials results in a wide absorption through the visible and the infrared regions. As a consequence, "dark carbons" generally have a poor transmission and uneven light scattering. Dilution in KBr does not help so much, since the sample preparation is time-consuming, and the grinding conditions affect the IR spectrum. Moreover, it has been demonstrated that KBr may alter the surface properties of carbon-based materials. At least part of the problem can be avoided by performing the measurements in diffuse reflectance mode (DRIFT), whereby the scattered (rather than transmitted) light is collected. Figure 4.8 shows the two most employed methods for collecting the IR spectrum of a carbon-based material.

4.2.1 FT-IR Spectroscopy of Activated Carbons

A typical DRIFT spectrum of a carbon of wood origin chemically activated is shown in Figure 4.7(b), along with the assignment of the main absorption bands. The spectrum is dominated by a narrow band around 1600 cm^{-1}, which is due to the ν(C=C) vibrational modes of conjugated sp^2 bonds belonging to graphitic islands[26–28,39] and by a very broad absorption in the 1300–1000 cm^{-1} region due to the overlapped contribution of in-plane δ(C–H) modes and of the collective skeletal C–C vibrational modes.[26–28] These two bands are very similar to those observed in the spectrum of intensively grinded graphite (Figure 4.7(a)).[29] Different from graphite, the spectrum of

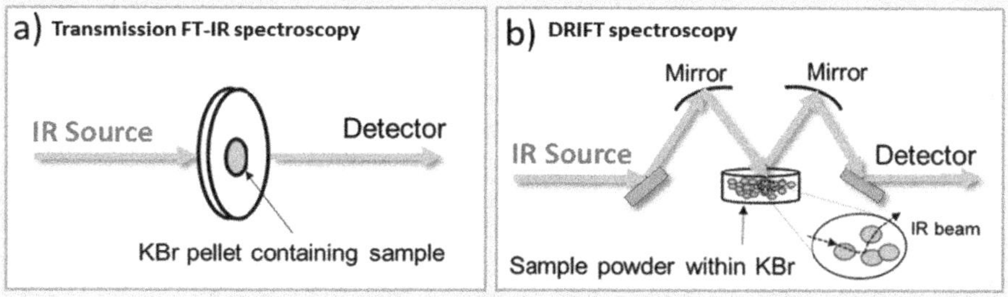

Figure 4.8 Schematic representation of the most common instrumental set-ups available for the characterization of carbon-based materials by IR spectroscopy. (a) Transmission mode; (b) DRIFT mode.

the chemically activated carbon shows also a weak band around 1710 cm^{-1} (Figure 4.7(b)), assigned to ν(C=O) vibrations of carboxyl and carbonyl groups, and a series of bands in the 900–750 cm^{-1} range, which are assigned to out-of-plane vibrations of C–H species at the borders of the sp^2 domains, differing in the number of adjacent H atoms.[27,28] A more extended discussion on this spectral region will be done in Section 4.3.1.

Not all the activated carbons display a DRIFT spectrum as rich in detail as that reported in Figure 4.7(b). The natural raw material (*e.g.*, wood, coconut shell, peat), the type of activation (physical or chemical) and the eventual post-activation treatments are all factors affecting the intensity of the DRIFT spectrum,[27,28,40] as illustrated in Figure 4.9(a). Generally speaking, physically activated (*i.e.*, by steam) carbons of peat origin display a very weak DRIFT spectrum where the absorption bands are hardly recognizable. Chemically activated carbons of wood origin usually show an intense DRIFT spectrum, indicating less regular sp^2 domains, with several bands ascribable to oxygen-containing functional groups.

The DRIFT spectrum of an activated carbon greatly changes as a consequence of post-activation treatments. Figure 4.9(b) shows the effect of an oxidative treatment (room temperature, concentrated HNO_3) on the DRIFT spectrum of a chemically activated carbon of wood origin. The whole spectrum is almost doubled in intensity, which is a direct consequence of the increased dipole moment associated with the sp^2 ring vibrations, as expected in the presence of oxygenated electron-withdrawing groups. In addition, new absorption bands are observed, denoting the formation of new functional groups:

(1) A first group of bands appears in the ν(C=O) vibrational region, with absorption maxima at 1766 and 1726 cm^{-1}, due to acidic functionalities. A precise assignment of these bands is not straightforward since the ν(C=O) vibrations of acid groups in an aromatic ring structure are affected by the presence of different peripheral groups in close proximity. Hence, the same band can be potentially ascribed to two different groups, or to the same one but in two different environments.[27,28,41,42]

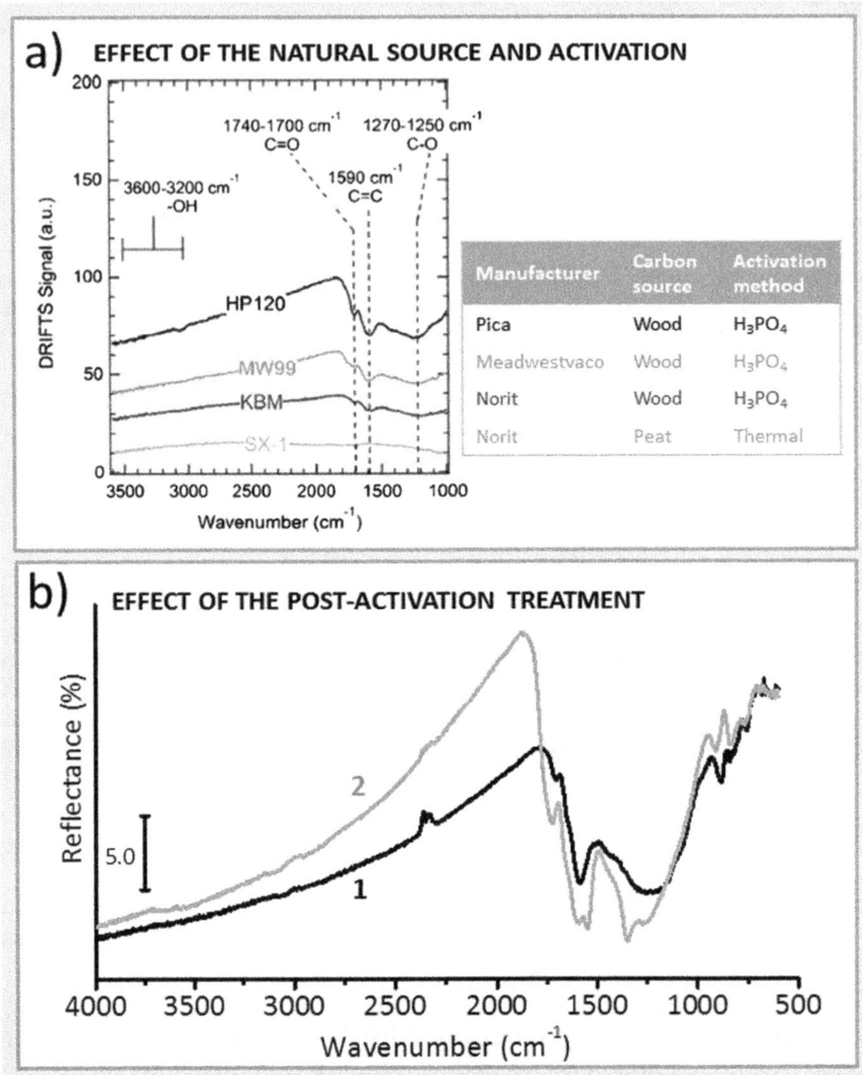

Figure 4.9 (a) DRIFT spectra of several carbons with a different origin (wood or peat) and activated either in phosphoric acid (chemical activation) or in steam (physical activation). The assignment of the main absorption bands is also reported. Reprinted from *Fuel*, **163**, M. T. Timko, J. A. Wang, J. Burgess, P. Kracke, L. Gonzalez, C. Jaye, D. A. Fischer, Roles of surface chemistry and structural defects of activated carbons in the oxidative desulfurization of benzothiophenes, 223–231, copyright (2016) with permission from Elsevier. (b) DRIFT spectrum of a chemically activated carbon (wood origin, spectrum 1) compared to that of the same sample subjected to an oxidative treatment (conc. HNO_3, room temperature, spectrum 2). Unpublished spectra.

(2) A second group of bands appears at 1550 cm^{-1} and 1350 cm^{-1}, which are assigned to the asymmetric and symmetric stretching modes of carboxylate groups.[27] The formation of carboxylates in acidic conditions is made possible because of the extended conjugation of the π electrons, which allows a redistribution of the electron density within the graphene layers. It has been demonstrated that for aromatic acids, both inductive effects (*i.e.*, the influence of neighbours on the acidic strength of carboxylic groups) and resonance effects contribute to enhancing the stability of carboxylate anions.[31] This phenomenon has been recently discussed[43,44] as responsible for the partial graphitization of activated carbons by HNO_3 treatments.

(3) Finally, the 900–750 cm^{-1} region, due to terminal C–H groups is also affected (see Section 4.3.1 for further discussion).

In this context, it is relevant to notice the role played by the activating agents ($ZnCl_2$, H_3PO_4, *etc.*),[45] which are often responsible for the formation of defective regions characterized by the presence of heteroatoms and dopants (B, N, O, P, S).[46,47] The presence of dangling groups, hydrogen saturation and polar groups has been widely demonstrated to alter the local density of the π-electrons, thus increasing the chemical reactivity. For carbons activated with H_3PO_4, absorption bands in the 1300–1000 cm^{-1} range have sometimes been attributed to phosphorus-containing groups (P=O, P–O–P, P=OOH, P–O–C and ionized $P^+–O^-$).[48] Nitrogen-doped porous carbons, which have been demonstrated to be very effective in environmental issues (fuel cells, CO_2 capture and segregation), have been reported to show absorption bands at 1263 cm^{-1} and at 3300 cm^{-1} associated with ν(C–N) and ν(NH), respectively.[49]

4.2.2 FT-IR Spectroscopy of GO, RGO, and Doped GO

The exfoliation process on graphite gives graphene, which is a flat crystalline single-atomic-layer of sp^2 carbon atoms within a two-dimensional hexagonal honeycomb lattice. As for graphite, no significant absorption bands are observed in the FT-IR spectrum of graphene.[50–52] In graphene technology, graphite is employed to synthesise GO, which is a chemically functionalized material consisting of oxygen-rich carbonaceous layers, where the graphite structure is maintained. On the contrary, the interlayer distance is increased due to the presence of oxygen-containing functional groups. GO is composed by both sp^3- sp^2-hybridized carbons, with hydroxyl and ether/epoxy groups arranged on the top and bottom of the surfaces, and with carbonyl and carboxyl groups in the sheet and on the edges, respectively. Such functional groups can be detected by FT-IR spectroscopy. Typical absorption bands are observed at 1230–1215 cm^{-1} (C–O bonds), at 1120–1110 cm^{-1} (epoxy species), at 1415 cm^{-1}, 1160 cm^{-1} (carbonyl or carboxyl groups) and at 1080–1040 cm^{-1} (alkoxy or epoxy moieties). These spectroscopic features indicate that the extended conjugated π-orbital system of the pristine graphite is destroyed and that C=O groups, with stretching modes appearing at 1740–1720 cm^{-1}, are introduced into the graphite pristine framework.

The GO oxidation degree can be evaluated by the increase in intensity of the ν(C=O) band along with the decrease in intensity of other C–O peaks due to C–OH and/or O–O–C bonds. Adsorbed water can be detected by checking for the presence of an absorption band around 3450 cm^{-1} due to the interaction with OH groups, which renders the GO hydrophilic. Generally, the band at 1640–1620 cm^{-1}, due to the presence of OH groups, overlaps with the in-plane vibrations of the skeletal C=C band related to the hexagonal aromatic rings of the graphene sheet or to the residual sp^2 character after oxidation of the graphitic network. The vibrational stretching mode of aryl C=C is observed at 1420 cm^{-1}.[13,53]

GO is used to prepare either chemically modified graphene or reduced graphene oxide (RGO) by simply heating or by chemical reduction. These processes lead to a decrease in the intensity of the absorption bands associated with the C=O, C–O and C–H groups, which are totally depleted upon deoxygenation.[12] Therefore, in contrast to what happens for Raman spectroscopy (see paragraph 2.2.1.3), the presence of a band centred in the 1585–1565 cm^{-1} range and related to the C=C aromatic group of the graphene sheet is expected when GO is completely reduced at temperatures below 1100 °C. On the contrary, an increase in intensity of the bands observed at 2950–2850 cm^{-1} and related to C–H bonds stretching is observed when chemical reduction is performed.[54]

N-doped GO can be obtained by heating GO in the presence of different amounts of N-containing molecules, such as ammonium nitrate,[55] urea,[56] thiourea, ammonium dicyanamide, cyanamide, melamine,[57] as well aminopropyltrimethoxysilane.[58] Among these, melamine gives interesting results, due to the uniform distribution of N.[59] With this in mind, N-doped graphene can be effectively characterised by FT-IR spectroscopy, which reveals the presence of bands at 1244, 1321 and 1427 cm^{-1} due to the aromatic C=N stretching mode and of one band centred at 807 cm^{-1}, due to the triazine ring mode. These absorption bands increase in intensity and blue shift with increasing the N content.[57] It is worth noting that FT-IR spectroscopy can be helpful to simultaneously investigate the effectiveness of N doping in the graphene framework and the reduction of the O-containing species in GO.[55,56] After reduction, the band at 3411 cm^{-1} attributed to the stretching vibration of hydroxyl groups decreases in intensity and a weak band at 1732 cm^{-1}, assigned to the C=O stretching vibration in conjugated graphene, indicates the presence of residual C=O groups surviving after the reduction process. Most important is the red shift of the band corresponding to the in-plane vibration of C=C (from 1624 to 1556 cm^{-1}), as a consequence of the superposition of the absorptions related to C=C and C=N, giving evidence of effective embedding of N-containing groups.[60,61]

Besides nitrogen, boron and fluorine can also be used to dope graphene. Boron-doped 2D carbons with photoluminescence properties, obtained by a hot-filament CVD technique, have been reported recently by Wang *et al.*[62] An absorption band at about 790 cm^{-1} in the IR spectrum, associated with the C–B bonds, provides evidence of boron incorporation in the material. Halogenated graphenes, such as fluorographene and chlorographene, usually exhibit much stronger IR bands owing to a higher transparency, and show prominent absorption features at ~1260 cm^{-1} or at ~1110 cm^{-1}, corresponding to *ν*(C–F) or *ν*(C–Cl), respectively. FT-IR studies conducted at increasing fluorination times have revealed an absorption band at 1112 cm^{-1}, assigned to "semi-ionic" C–F, which gradually disappears, giving rise to a strong band at 1211 cm^{-1} due to covalent C–F bonding. Features at 600 cm^{-1} or at 745 cm^{-1} are representative of the *ν*(C–Br) and *ν*(C–I) bonds.[63]

4.2.3 FT-IR Spectroscopy of CNTs

Both SWCNTs and MWCNTs have been extensively investigated by IR spectroscopy. In this regard, IR active modes of CNTs can be found in the 1580–1530 cm^{-1} range (sp^2 carbon) and 1200–1100 cm^{-1} interval (more disordered sp^3 carbon). These skeletal vibrational modes depend on the CNT symmetry and their expected wavenumbers in the IR spectrum are influenced by the nanotube diameters (~0.4–2.0 nm for SWNTs and ~5–1000 nm for MWNTs). This explains why the corresponding absorption bands are quite broad. For MWCNTs, the broadness is the result of the presence of multiple concentric tubes with different diameters, while for SWCNTs, the broadness of the *ν*(C=C) absorption band can be explained by the polydispersity of the nanotubes.[30] Additional IR bands around 2800–3000 cm^{-1} are due to the $\nu(CH_x)$ modes.[64]

The purification, functionalization and doping of CNTs require strong chemistry, involving the use of mineral acids, oxidizing agents,[65] cavitation and microwave treatments[66] or high temperature conditions. These treatments are often required for doping or co-doping CNTs with heteroatoms (*i.e.*, B, N, O, S, I). In this way, the catalytic activity of CNTs can be enhanced for a wide range of reactions including the ORR (oxygen reduction reaction), OER (oxygen evolution reaction) and HER (hydrogen evolution reaction).[47] It is believed that the doping-induced electron modulation at defective edges is responsible for the high catalytic activity of doped CNTs.[46] Generally speaking, the effects of the doping on a single nanotube are qualitatively and quantitatively very difficult to determine. These difficulties mainly arise from the intrinsic properties of nanotubes (*i.e.*, sample heterogeneity, bundling), impurities and dopant concentrations (even low). The FT-IR spectra of functionalized/doped CNTs may show new absorption bands due to the presence of oxygen-rich functional groups formed during the oxidation processes, such as C=O or –OH (absorption bands located at about 1800–1700 cm^{-1} and at 3200–3600 cm^{-1}, respectively), which are attached to the tubes' outer wall or to their open endings.[67] Furthermore, the presence of a band located at about 1560–1630 cm^{-1} may be attributed to the interaction of C=C conjugation with C=O groups at the surface. Nitrogen-doped CNTs can show bands at 1375 cm^{-1} (N–N) and 2200 cm^{-1} (N–C) bonds,[68] while for I- or Br-doped CNTs, a band around 600–650 cm^{-1} or at 580–590 cm^{-1} can be associated with C–I and C–Br bonds, respectively.[69,70] As for doping of CNTs with boron species, Preston *et al.*[71] reported the FT-IR spectra of B–(O)-doped few-walled CNTs, showing characteristic bands at 1024 cm^{-1} (B–O–C bend.), 1196 cm^{-1} (B–O bend.) and 1407 cm^{-1} (B–O stretch.).

4.2.4 FT-IR Spectroscopy of Fullerenes

Fullerenes are molecules with a closed cage structure exclusively formed by sp^2-hybridized C atoms. Each carbon is bonded to three neighbours, with 12 pentagons and a variable number of hexagons, which leads to the existence of molecules with different shapes, such as spheres, ellipsoids, tubes, *etc.*, and with different sizes, depending on the number of C atoms, going from buckyballs to giant fullerenes. The most popular and the most symmetric molecules are the spherical C60 fullerenes with the most stable molecular structure composed of 60 carbon atoms, 12 pentagons and 20 hexagons, with an icosahedral symmetry where all the rings are fused and all the double bonds are conjugated in a soccer-ball type arrangement. The C60 molecule acts as a deficient electron alkene and readily reacts with electron-rich species. Theoretically, the C60 fullerene molecule has 3N-6 = 174 vibrational degrees of freedom but, due to the high symmetry of the molecules, only four modes are IR-active.[30] Indeed, the C60 molecule shows a very simple FT-IR spectrum with only four characteristic peaks at about 1428 cm^{-1}, 1182 cm^{-1} associated with a tangential motion of the carbon atoms, and 576 cm^{-1} and 527 cm^{-1} from a primarily radial motion of the carbon atoms.[72]

For fullerene C60 nanotubes and fullerene dimers, the FT-IR spectra are the same as for pristine C60, confirming that the FNTs or dimers are composed of C60 molecules. Modification of the surface in order to add epoxide groups to the fullerene cage, for example for obtaining fullerene epoxides, leads to a shift of the main absorption bands to lower wavenumbers and to the appearance of a new band assigned to epoxy groups (1140 cm^{-1}). The C60 characteristic bands are maintained also for fullerene dimer oxides, indicating that the fullerene cages are intact in these dimers. In addition, new bands appear that may be assigned to various intra-cage vibrations due to dimerisation and epoxy groups.[73] In the presence of free chlorine and UVA irradiation, extensive oxidation of nC60 clusters occurs,[74] as indicated by the relevant increase in intensity of the bands at about 1700 cm^{-1} (due to the presence of carbonyl groups), at 1400 cm^{-1} (assigned to C–OH in-plane bending), and at 1100 cm^{-1} (C–O stretching mode).[75,76] Besides oxygen-based functional groups, additional bands are also observed in the range from 500 to 900 cm^{-1}, which are assigned to C–Cl stretching modes.[77] It is likely that reactive chlorine radicals produced *via* chlorine decomposition under photo-irradiation can bond with nucleophilic C60 *via* a free-radical addition reaction. Such a reaction has also been demonstrated to be the primary mechanism during the photochlorination of graphene.[78]

4.2.5 FT-IR Spectroscopy of Other Carbon-based Materials

As anticipated in Section 4.1.5, graphitic carbon nitride (g-C_3N_4) can be synthetized by condensation of low-cost melamine at different temperatures (from 400 °C up to 700 °C).[22,79–81] IR spectroscopy provides a fast technique to obtain information on the chemical structures obtained at different temperatures.[19] The FT-IR spectra of g-C_3N_4 display the typical C–N and C=N stretching vibrations of N-containing heterocycles in the 1640–1240 cm^{-1} range. A typical sequence of FT-IR spectra collected at increasing temperatures is shown in Figure 4.10(a). The spectrum related to the material synthesised at 400 °C shows two intense bands at 1608 and 1471 cm^{-1} and two bands at 1307 and 1100 cm^{-1} with a lower intensity due to *melem*.[22] At 500 °C, new bands appear at 1331 and 1255 cm^{-1}, assigned to *melon* and g-C_3N_4.[22,82] Melon and g-C_3N_4 are also characterized by bands at 1635, 1458, 1410, 1540, 1570, 1320 and 1242 cm^{-1}, observed in the spectra related to the samples prepared at higher temperatures. The band centred at 806 cm^{-1} is related to the out-of-plane bending of triazine rings and is observed in all the spectra without particular modification despite the increase in temperature.[83]

At higher frequencies, above 3000 cm^{-1}, the N–H stretching vibrations of primary and secondary amines are detected.[20] In particular, bands at 3487 cm^{-1}, 3427 and 3315 cm^{-1}, due to asymmetric and symmetric –NH_2 stretching, and 3150 cm^{-1}, attributed to the –NH stretching mode, are observed for the materials obtained at 400 and 500 °C. Secondary amines are still present at 700 °C, as revealed by the bands at 3252, 3160 and 3091 cm^{-1}. These features indicate that the *melem* present at 400 °C condenses, forming *melon* at 500 °C, which further produces g-C_3N_4 rigid planes. In the rigid planes, the dipole moments decrease and, therefore, the intensity of the absorption bands diminishes. Moreover, the bands observed in the 1640–1240 cm^{-1} interval increase in intensity up to 700 °C, indicating the fragmentation of the g-C_3N_4 planes into nanosheets.[19]

FT-IR spectroscopy can be employed to investigate the effect of the addition of different atoms or pre-treatments on the final structure of g-C_3N_4. Indeed, the photocatalytic activity of g-C_3N_4 can be effectively enhanced, for example, by adding chlorine,[84–86] phosphorus,[87] sulfur,[88] alkali metal halides,[89–93] or by hydrochloric acid treatment.[83] In this regard, due to its high sensitivity to the local structure, FT-IR spectroscopy can be employed to study the framework of *melon* strands of g-C_3N_4 by monitoring in-plane long-range periodic ordering.[94] Figure 4.10(b) shows the FT-IR spectra of several graphitic carbon nitride (GCN) samples doped with different alkali halides. All the spectra display the typical bands at 808 cm^{-1} and in the 1200–1900 cm^{-1} spectroscopic range, which are ascribed to the breathing mode of triazine units and to the stretching vibration modes of aromatic CN heterocycles,[21,95] indicating that the effect of alkali halides is to interrupt the intra-layer long-range periodic arrangement without modifying the basic atomic building blocks. Moreover, new components at 1000, 1155, 2147 and 2180 cm^{-1} are observed for GCN-Na, GCN-K, GCN-Rb and GCN-Cs, which can be assigned to –OH groups located at the surface of g-C_3N_4.[92,96,97] These features are due to the presence of alkali

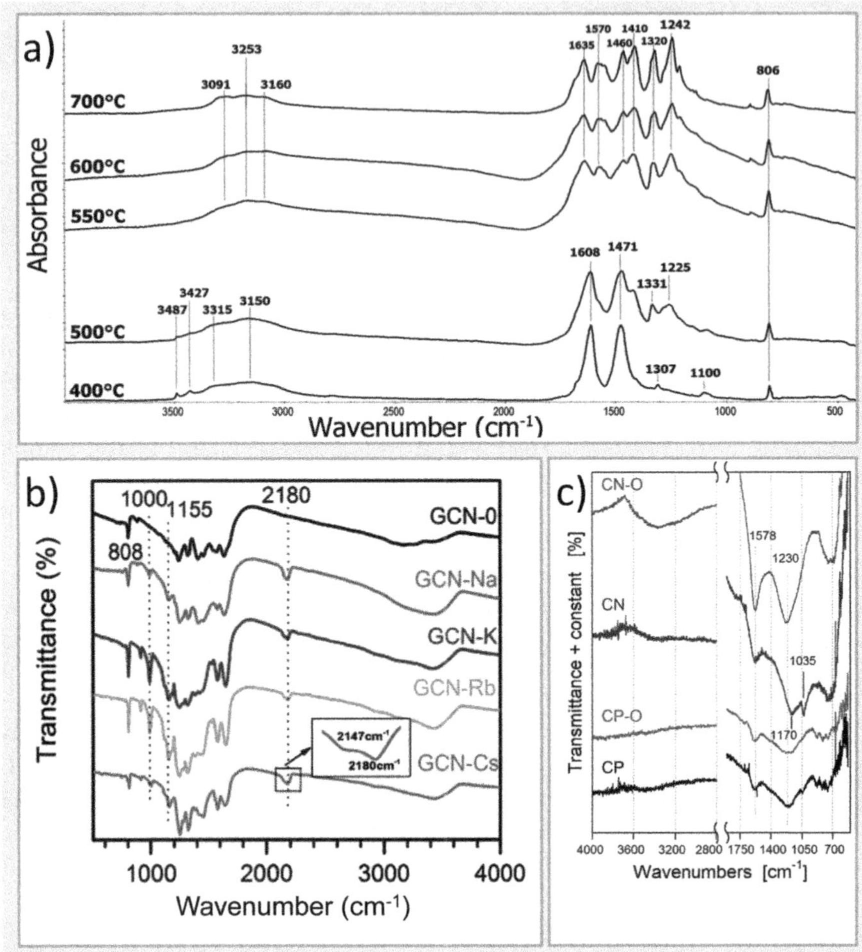

Figure 4.10 (a) FT-IR spectra of melamine g-C_3N_4 obtained at different temperatures. Reprinted from *Mater. Chem. Phys.*, **193**, P. Praus, L. Svoboda, M. Ritz, I. Troppová, M. Šihor, K. Kočí, Graphitic carbon nitride: Synthesis, characterization and photocatalytic decomposition of nitrous oxide, 438–446, copyright (2017) with permission from Elsevier. (b) g-C_3N_4 *via* MCl (M = Na, K, Rb, Cs) modification. Results of pristine g-C_3N_4 are plotted together for comparison. Reproduced from ref. 94 with permission from the Royal Society of Chemistry. (c) FT-IR spectra of the studied materials. Reprinted from *Carbon*, **107**, M. Seredych, T. J. Bandosz, Nitrogen enrichment of S-doped nanoporous carbon by g-C_3N_4: Insight into photosensitivity enhancement, 895–906, Copyright (2016) with permission from Elsevier.

ions preventing the formation of hydrogen bonds. The hydroxyls could be bonded to the amino groups of g-C_3N_4.[98] The broad absorption bands in the 3000–3600 cm^{-1} region are due to the terminal amino groups.[99] No vibration related to alkali-related groups is detected, indicating that the *melon* strand structure of carbon nitride is not modified upon the salt treatment.

To further distinguish changes in the surface chemistry of the pristine mixture of g-C_3N_4 and a sulfur-containing polymer and the final S-doped nanoporous carbon/g-C_3N_4 composite, FT-IR spectra can be collected, and the results are shown in Figure 4.11(c). The bands at 1578 and 1230 cm^{-1} assigned to the aromatic ring stretching mode vibrations of C=C, related to the carbons, and of C=N (for the CN and CN–O) conjugated with carbonyl/carboxylate groups and nitroso (N–O) configurations[100] indicate the occurrence of CN oxidation. In addition, the band at 1230 cm^{-1} can also be attributed to aromatic C–N and C–O and might be related to S–O stretching vibrations.[100,101] The two bands are increased in intensity after air oxidation, due to the formation of carbonyl/carboxylate groups or N–O configurations.[100] The weak peak at 1035 cm^{-1}, is assigned to the C–O stretching mode and can be related to the presence of phenol, ether and ester groups in the carbon structure. The band at 1170 cm^{-1} is related to the stretching vibration of the C–O bond. However, the former band can also be due to N–O stretching vibrations[100] and both absorption bands overlap with those due to nitrogen groups originating from the graphitic carbon nitride used for the synthesis of the sample,[101] showing evidence of the complexity of the spectra for a correct analysis.

FT-IR spectroscopy can be employed to define the chemical structure of complex systems, such as g-C_3N_4/acidified g-C_3N_4 composites used in photocatalysis.[102] In this case, the intensity of the bands at 1318 and 1240 cm^{-1}, due to the stretching modes of C–NH–C connected units,[103–105] gradually decreases with the increase of acidified g-C_3N_4 content. Moreover, with increasing amounts of acidified g-C_3N_4 in the composite, a red shift of the bands at 1640 cm^{-1}, related to C–N stretching, and at 1405 cm^{-1}, due to C–N stretching vibrations of heterocycles, is observed. The breathing mode of tri-*s*-triazine rings at 795 cm^{-1} follows the same trend, suggesting an electrostatic interaction among tri-*s*-triazine units of acidified g-C_3N_4 and those of g-C_3N_4.

The FT-IR spectra of a ternary metal-free hetero-structure made by sheet-on-sheet graphene/C_3N_4/carbon dots for pollutant abatement under visible-light photo-catalysis maintain bands at 1241, 1323, 1411, 1573, and 1644 cm^{-1} ascribed to the C–N stretching of heterocycles[106] and at 806 cm^{-1} assigned to the breathing mode of triazine units.[83] This result demonstrates that solvent-assisted exfoliation and the subsequent loading of carbon dots do not destroy the lattice structure of C_3N_4.

g-C_3N_4 can be modified with graphene oxide and reduced graphene oxide nanosheets with different lateral sizes to obtain a hybrid composed of RGO nanoparticles with sizes in the 4–135 nm range with enhanced photoactivity with respect to similar size RGO- and GO-containing systems.[107] The influence of GO and RGO lateral sizes can be investigated by FT-IR spectroscopy, which reveals the depletion of the bands of the oxygen-functional groups for the nanocomposites, indicating that GO undergoes reduction. At the same time, it is possible to observe the appearance of new peaks corresponding to chemical bonding between the C and N of g-C_3N_4 carbon nitride: a band at 807 cm^{-1} due to the breathing mode of triazine units as well as bands at 1245,

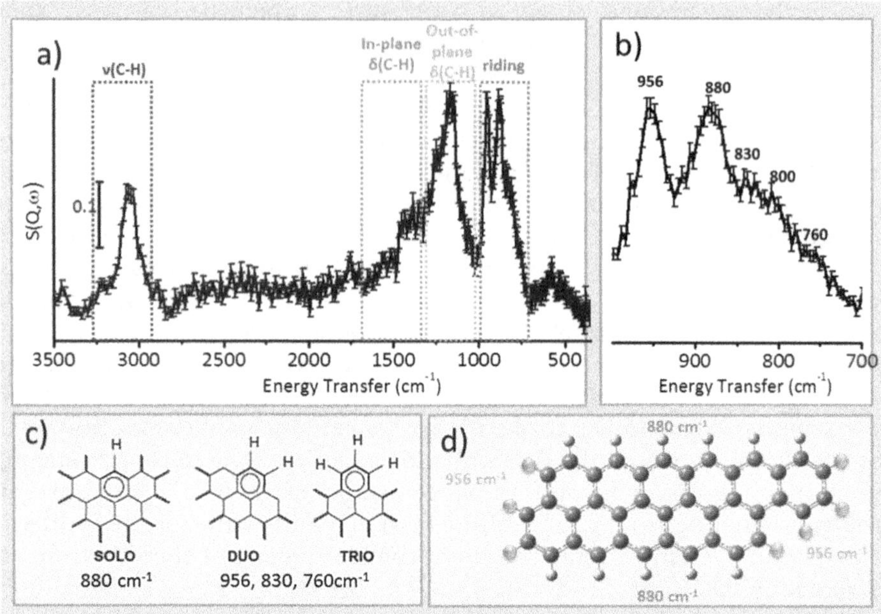

Figure 4.11 (a) INS spectrum of an activated carbon of wood origin and assignment of the main absorption bands. (b) Magnification of the 1000–700 cm^{-1} region, containing the characteristic out-of-plane δ(C–H) vibrational modes. Reproduced from ref. 28 with permission from the Royal Society of Chemistry. (c) Description of *solo*, *duo*, and *trio* structures and corresponding vibrational modes. Reprinted from *Carbon*, **43**, A. Centrone, L. Brambilla, T. Renouard, L. Gherghel, C. Mathis, K. Mullen, G. Zerbi, Structure of new carbonaceous materials: The role of vibrational spectroscopy, 1593–1609, copyright (2005) with permission from Elsevier. (d) Schematic representation of the out-of-plane δ(C–H) vibrational modes for regular and irregular borders. Carbon atoms are shown in grey, hydrogen belonging to regular borders in white, and hydrogen belonging to irregular borders in yellow. The contribution at 880 cm^{-1} is connected to an in-phase oscillation of *solo* species on well-extended borders, while the intense mode at 958 cm^{-1} is connected to out-of-phase modes at *duo* and *trio* sites that characterize a defective and not extended border. Reprinted from A. Piovano, A. Lazzarini, R. Pellegrini, G. Leofanti, G. Agostini, S. Rudić, A. L. Bugaev, C. Lamberti and E. Groppo, *Adv. Condens. Matter Phys.*, 2015, **2015**, 803267. Published under the terms of the CC BY 3.0 licence, https://creativecommons.org/licenses/by/3.0/.

1320, 1420 and 1560 cm^{-1} related to C–N stretching, a peak at 1640 cm^{-1} of the C=N stretching mode and finally a broad band at 3160 cm^{-1} assigned to the stretching modes of secondary and primary amines.[108,109]

Finally, B_4C merits a mention. Despite the fact that it does not find application in catalysis, it is an advanced material with excellent properties, such as low density, high hardness and high corrosion resistance. Moreover, due to the huge thermal neutron-absorbing cross section, B_4C is capable of capturing thermal neutrons and it finds application in the nuclear shielding field.[110] The FT-IR spectrum of B_4C shows bands related to C–C stretching at 1150–1450 cm^{-1}, B–C stretching at about 1200 cm^{-1} and B–O at 1450 cm^{-1}, respectively.[110]

4.3 Neutron Properties, and Inelastic Neutron Scattering Applied to Carbons

The inelastic neutron scattering (INS) technique probes the motion of atoms and molecules within a material. The technique relies on the use of neutrons, which are uncharged particles with a mass very similar to that of a hydrogen atom and a negligible electric dipole moment. For this reason, neutrons interact only with the nucleus of atoms—which is typically only ~0.1% of the diameter of the atom and 10^6 times smaller than the distance between two atoms—and can penetrate the matter for large distances before being scattered or absorbed.[111–113] This means that INS spectra are representative of an entire sample, although large amounts of samples (grams) are necessary for the measurements.

Opposite to Raman and FT-IR spectroscopy, INS has no symmetry selection rules, which means that all the vibrational modes are allowed.[111–114] The intensity of a band at the characteristic vibrational frequency of a certain chemical species is proportional to: (1) the amount of the species under the neutron beam; (2) the incoherent cross section of the species involved (which is element- and isotope-dependent); and (3) the amplitude of motion of the atoms in the vibrational mode (*i.e.*, how much they displace with respect to the equilibrium position), which is larger for light atoms. For most of the elements, both the incoherent cross section and the amplitude of motion are very small, except for hydrogen ^{1}H, which has an incoherent scattering cross section more than one order of magnitude larger than that of all the other elements, and the smallest mass of any element. This is the reason why INS is extremely powerful in detecting vibrations involving hydrogenous species, and explains also why an INS spectrum is usually accessible in the whole 0–4000 cm^{-1} range without exhibiting a cut-off due to lattice absorption, which is typically observed in the FT-IR spectra of most of the materials.[111–113]

In carbon-based materials, hydrogen atoms are present mainly at the edges of the sp^2 graphitic domains constituting the basic structural units, and to a lesser extent in functional groups containing other heteroatoms (such as carboxylic groups, –COOH, or hydroxyl groups, –OH). The type and relative amount of these species influence the surface properties of all the carbon-based materials, are responsible for the catalytic activity of metal-free carbons as well as for the adsorption of organic reactants during the catalysis, and also influence the deposition of metal nanoparticles during the preparation of carbon-supported catalysts.[115–117] Albeit not as diffuse as IR and Raman spectroscopies, INS has been widely applied to the investigation of carbon-based materials[118–127] and carbon-supported catalysts,[128–135] offering unprecedented details on surface properties at the atomic level. Most of the work is related to activated carbons or coal,[118–122,135–137] while there are very few reports on INS applied to other carbonaceous materials such as graphite, graphene and carbon nanotubes, the latter motivated mainly by their potential in H_2 absorption.[123–125,127] The most prolific research groups in this field are those of Albers and Parker,[111,113,118,122,129–131,133,136] who published the first INS spectra of activated carbons in the late 1990s. Although the performances of the neutron facilities have been largely improved in the last decade with a consequent progress in the overall quality of INS spectra, these are still the reference papers in the field of INS spectroscopy applied to carbonaceous materials.

4.3.1 INS Spectra of Activated Carbons

A typical INS spectrum of an activated carbon is shown in Figure 4.11(a). The absorption bands centred around 3060 cm^{-1}, 1200 cm^{-1} and in the 800–1000 cm^{-1} region are assigned, respectively, to the stretching, in-plane and out-of-plane bending modes of the hydrogen species belonging to the condensed rings' edges.[26,28] The weak bands in the 700–400 cm^{-1} region are assigned to so-called riding vibrations, which are substantial movements of the hydrogen atoms indirectly induced by C–C torsion modes of the carbon atoms at the edges of the graphitic fragments. In other words, the H atoms move in phase with the vibrations of the surface C atoms, as a rider moves with their horse.[28,138]

It has been demonstrated that the relative intensity of the above-mentioned bands depends on the origin of the activated carbon (*e.g.*, wood, peat or coconut),[118] on the activation procedure,[28] and on the eventual post-treatment. The most characteristic region is that of the out-of-plane C–H bending modes (Figure 4.11(b)), where well-defined bands are observed at 956, 880, 830, 800 and 760 cm^{-1}. Most of these bands are also observed in the DRIFT spectra of the same carbons (Figure 4.7 (b)), and can be assigned to specific structures at the "borders" of the sp^2 domains, which are defined as *solo*, *duo* and *trio* referring to the number of adjacent aromatic C–H groups that vibrate out-of-plane in the fused rings (Figure 4.11(c)). In particular, according to the literature[26,28,118] and to recent calculations,[139] the band at 880 cm^{-1} is associated with *solo* species belonging to regular borders at extended sp^2 domains, which vibrate in phase (Figure 4.11(d)). The band at 956 cm^{-1}, which is usually absent in the DRIFT spectra, was assigned only recently[28,139] to *duo* and *trio* species belonging to irregular borders of the sp^2 domains, which vibrate not in phase (Figure 4.11(d)). Hence, the relative intensity of the two bands at 880 and 956 cm^{-1} indicates the degree of regularity of the sp^2 domains.

It is worth noticing here that a full assignment of the bands observed by INS is possible only by employing quantum chemistry methods to calculate vibrational frequencies of a suitable model. While for IR spectroscopy the quantitative evaluation of the band intensities is not an easy task (because they depend on the vibration polarization), for INS a quantitative analysis is usually easy and robust,[114] since the intensity of an INS spectrum depends on the nucleus' incoherent cross section and on the square of the normal mode displacements. The limiting step in the accuracy of calculation for carbonaceous materials is the choice of the model, which should be big enough to be representative of the system, but sufficiently small to be computable in a reasonable timescale.

The potential of INS spectroscopy in the analysis of activated carbons is illustrated in Figure 4.12, which reports a few INS spectra of activated carbons differing in the origin of the natural raw material[118] (a), in the activation procedure[28] (b), or in the post-activation treatment[27] (c). Figure 4.12(a) compares the INS spectra of three activated carbons derived from different natural sources (namely beech, coconut shell and peat).[118] The three spectra show a different absolute intensity (that scales with the amount of hydrogen) and distinct differences in the out-of-plane C–H bending region and in particular in the relative intensity of the band at 880 cm^{-1}, suggesting that the raw material affects the extension and regularity of the graphitic domains in the corresponding activated carbon. It is worth noticing that discrepancies between the intensity of the INS spectra and the hydrogen content in carbon materials have also been reported. In particular, Fillaux *et al.*[140] suggested that some hydrogen species in graphitic carbons behave like free protons and, therefore, they respond in a different way.

Figure 4.12(b) shows the INS spectra of two carbons with the same origin (wood) but a different activation treatment (in steam or in the presence of phosphoric acid). They differ again in the absolute intensity—the chemically activated carbon contains more hydrogen—and in the relative abundance of the hydrogen termination—the chemically activated carbon presents less extended and more irregular borders.[28]

Figure 4.12(c) shows the INS spectra of a chemically activated carbon of wood origin and of the same carbon subjected to a post-activation treatment in HNO_3,[27] which is known to introduce oxidized functional groups.[43,44,141–143] The latter has a lower intensity, which indicates that a fraction of the hydrogen terminations has been consumed in favour of new oxygenated functional groups. In addition, the absorption bands at 956, 830 and 760 cm^{-1} significantly decrease in intensity, while that at 880 cm^{-1} upward shifts to 912 cm^{-1}. These results indicate that the treatment in HNO_3 affects mainly the irregular borders of the graphitic layers, which are likely attacked by oxygenated functional groups. The presence of these electron-withdrawing species at the irregular borders has indirect consequences also on the regular extended edges, as reflected by the substantial shift of the band associated with the vibrations of the corresponding C–H terminations.

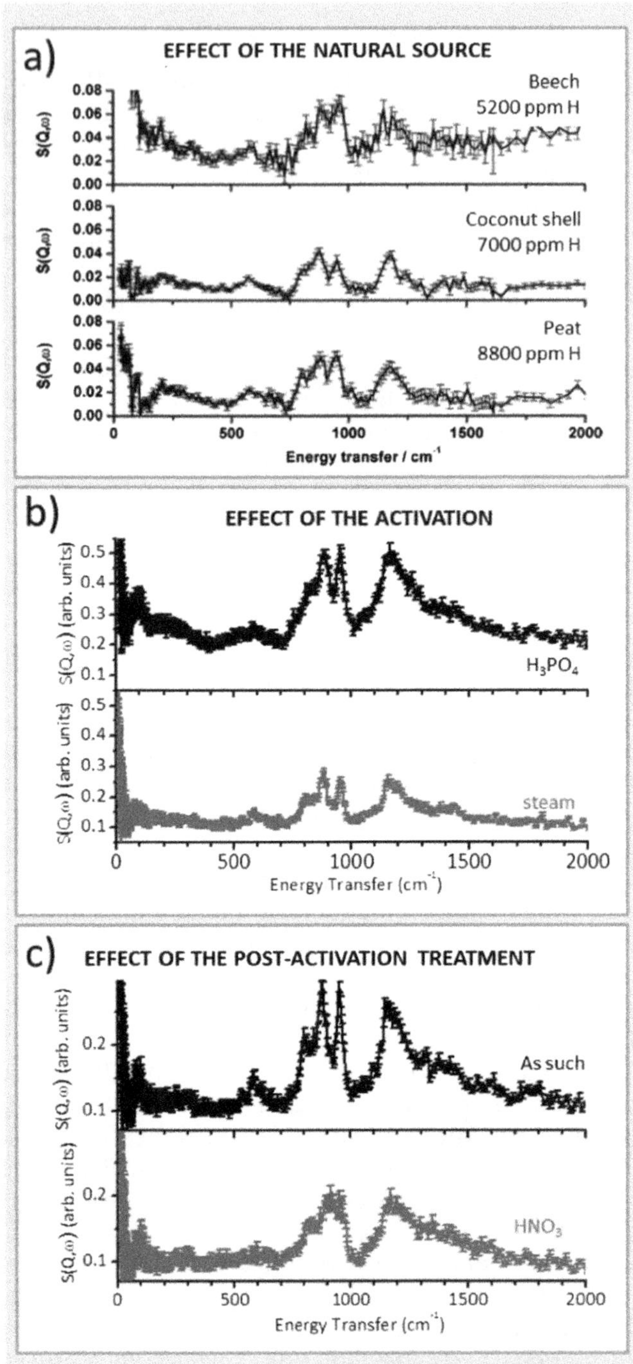

Figure 4.12 INS spectra of activated carbons differing in the origin of the natural raw material ((a) reproduced from ref. 118 with permission from The Royal Society of Chemistry), in the activation procedure ((b) reproduced from ref. 28 with permission from the PCCP Owner Societies) and in the post-activation treatment ((c) reproduced from ref. 27 with permission from The Royal Society of Chemistry).

4.3.2 INS Spectroscopy of Other Carbonaceous Materials

Also other carbon-based materials (diamond, various forms of graphite, fullerenes, nanotubes and so on) have been extensively studied by INS spectroscopy, when readily available. The INS spectra of highly oriented pyrolytic graphite and C70 fullerene[144,145] are shown in Figure 4.13(a) as examples of carbon-based materials with formally no hydrogen termination and characterized by a completely different vibrational behaviour. The vibrational bands in the spectrum of graphite (spectrum a1 in Figure 4.13(a)) are very broad and separated into four main groups, which can be approximately described as:[114,124,146,147] local in-plane stretches at 1400–1600 cm^{-1}, local in-plane bends at 600–900 cm^{-1}, sheet deformations at 400–500 cm^{-1} and rigid sheet modes below 100 cm^{-1} (the latter are not shown in Figure 4.13). Opposite to graphite, the INS spectrum of the C70 fullerene (spectrum a2 in Figure 4.13(a)) has a molecular-like character, and contains a multitude of narrow vibrational bands with similar intensities. Such a rich spectrum is in contrast to both Raman and IR spectra, which contain a limited number of vibrational bands with substantially different intensities. This is a consequence of the fact that all the modes are allowed in INS spectroscopy, and that all the atoms have the same scattering cross section and, consequently, similar amplitudes of vibration. All the bands have been recently re-assigned based on periodic density functional theory calculations.[144]

The INS spectrum of graphite progressively broadens upon reducing the size of the sp^2 domains (*e.g.*, by ball-milling). For example, the INS spectrum of graphene (obtained by thermal exfoliation of graphite oxide) does not contain any distinguishable feature, but only an increasing background in the low frequency part (not shown). Things drastically change upon hydrogenation. Figure 4.13(b) compares the experimental INS spectra of hydrogenated nano-graphite (obtained by mechanical ball-milling of graphite) and hydrogenated graphene.[124,146] The INS spectrum of H-nanographite (spectrum b1 in Figure 4.13(b)) is quite similar to those of activated carbons (Figure 4.11), especially concerning the position in frequency of the bands, which can be attributed to common dynamical features. The spectrum is satisfactorily reproduced by a large PAH molecule (inset and full line in Figure 4.13(b)), with all the edges terminated with hydrogen. This means that the ball-milling and successive hydrogenation procedures cause a disruption of the regular graphitic domains and that single hydrogen saturation per edge carbon atoms applies. This demonstrates the effectiveness of defects in dissociating the molecule of hydrogen and trapping atomic hydrogen in covalent bonds, a phenomenon of potential interest in catalysis.

By contrast, the spectrum of H-graphene (spectrum b2 in Figure 4.13(b)) is different, and dominated by a peak at 108 meV (870 cm^{-1}), with a broad tail around 118 meV (950 cm^{-1}). This band has been indicated as the spectroscopic fingerprint for H atoms chemisorbed at sites specific for H-graphene samples, as clearly indicated by subtracting the spectrum of H-nanographite from that of H-graphene (spectrum c1 in Figure 4.13(c)). This extra dynamical

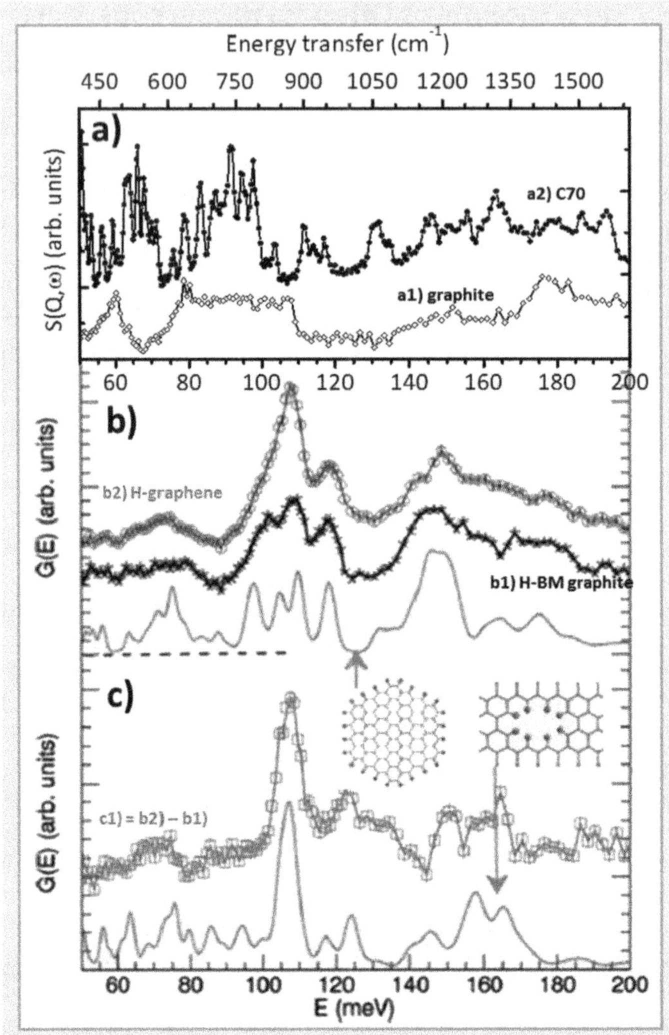

Figure 4.13 (a) INS spectra of highly oriented pyrolytic graphite (spectrum a1, reprinted from J. K. Walters, R. J. Newport, S. F. Parker and W. S. Howells, *J. Phys. Condens. Matter*, 1995, 7, 10059–10073. Acknowledgements for reproduction of IOP Publishing material in RSC publications) and C70 fullerene (spectrum a2, reprinted from K. Refson and S. F. Parker, *Chem. Open*, 2015, **4**, 620–625. Published under the terms of the CC BY 4.0 licence, https://creativecommons.org/licenses/by/4.0/) as examples of carbon-based materials formally having no hydrogen termination. (b) Experimental INS spectra of hydrogenated nano-graphite (obtained by mechanical ball-milling of graphite, spectrum b1) and hydrogenated graphene (spectrum b2), the latter satisfactorily reproduced by a large PAH molecule (inset and red line in (b)), with all the edges terminated with hydrogen. (c) Difference between spectra b2 and b1 (spectrum c1) and simulated spectrum for H chemisorbed inside small voids (inset and full line in (c)). (b) and (c) have been reproduced from ref. 146 with permission from the PCCP Owner Societies.

feature in H-graphene has been attributed to H covalently bonded to in-plane defects (*e.g.*, small voids), which are produced during the synthesis process. Indeed, the experimental INS spectrum of the H-graphene has been reproduced satisfactorily by accounting for a proportion of H chemisorbed in equal proportions, at the edges and inside small voids (inset and full line in Figure 4.13(c)).

4.4 Conclusions and Future Perspectives

Vibrational spectroscopies are among the most powerful techniques applied to the characterisation of metal-free carbon materials. However, carbon-based materials usually show very complex spectra that can generate a lot of problems for the interpretation, especially when considering new composite/hybrid C-containing systems.

In this chapter, different types of materials have been examined in order to correlate the spectroscopic features with the modification produced by treating the materials with opportune molecules or upon addition of dopants. The goal was to unravel the changes occurring on the materials during the synthesis and to establish structure–activity relationships. Efforts have been made to provide different case studies by which it is possible to have a panorama of the kind of information provided by Raman, IR and INS spectroscopies. It is shown that these techniques give complementary insights at the atomic scale on the samples examined.

In the future, to improve atom and energy efficiency for industrial applications, catalyst design will further gain strategic importance. In this regard, intense efforts are being dedicated to the possibility of obtaining green metal-free materials required in the area of scaling up carbonaceous catalysts. Therefore, the results presented here highlight the pivotal significance of the spectroscopic characterisation of metal-free carbonaceous nanomaterials. Indeed, a deep understanding of the structure at a nanometric level, which can be gained by performing an extended characterisation, will help in unravelling the nature of the active sites leading to the design and synthesis of more performant and stable catalysts.

References

1. A. C. Ferrari and J. Robertson, *Phys. Rev. B: Condens. Matter Mater. Phys.*, 2000, **61**, 14095–14107.
2. R. P. Vidano, D. B. Fischbach, L. J. Willis and T. M. Loehr, *Solid State Commun.*, 1981, **39**, 341–344.
3. R. J. Nemanich and S. A. Solin, *Phys. Rev. B: Condens. Matter Mater. Phys.*, 1979, **20**, 392–401.
4. I. Pócsik, M. Hundhausen, M. Koós and L. Ley, *J. Non-Cryst. Solids*, 1998, **227–230**, 1083–1086.
5. A. Lazzarini, *Rend. Lincei*, 2017, **28**, 29–42.

6. R. Beams, L. Gustavo Cançado and L. Novotny, *J. Phys.: Condens. Matter*, 2015, **27**, 83002.
7. A. C. Ferrari and D. M. Basko, *Nat. Nanotechnol.*, 2013, **8**, 235–246.
8. C. Zhang, R. Hao, H. Liao and Y. Hou, *Nano Energy*, 2013, **2**, 88–97.
9. D. R. Dreyer, A. D. Todd and C. W. Bielawski, *Chem. Soc. Rev.*, 2014, **43**, 5288.
10. S. Eigler, *Chem.-Eur. J.*, 2016, **22**, 7012–7027.
11. I. Jung, D. A. Dikin, R. D. Piner and R. S. Ruoff, *Nano Lett.*, 2008, **8**, 4283–4287.
12. D. Mandal, P. Routh and A. K. Nandi, *ChemistrySelect*, 2017, **2**, 3163–3171.
13. L. F. Dumée, C. Feng, L. He, F.-M. Allioux, Z. Yi, W. Gao, C. Banos, J. B. Davies and L. Kong, *Appl. Surf. Sci.*, 2014, **322**, 126–135.
14. S. Iijima and T. Ichihashi, *Nature*, 1993, **363**, 603–605.
15. S. Iijima, *Nature*, 1991, **354**, 56–58.
16. R. Graupner, *J. Raman Spectrosc.*, 2007, **38**, 673–683.
17. S. Banerjee and S. S. Wong, *Nano Lett.*, 2004, **4**, 1445–1450.
18. K. K. Han, C. C. Wang, Y. Y. Li, M. M. Wan, Y. Wang, J. H. Zhu, H. M. Cheng and M. Antonietti, *RSC Adv.*, 2013, **3**, 9465.
19. P. Praus, L. Svoboda, M. Ritz, I. Troppová, M. Šihor and K. Kočí, *Mater. Chem. Phys.*, 2017, **193**, 438–446.
20. M. K. Marchewka, *Mater. Sci. Eng., B*, 2002, **95**, 214–221.
21. Q. Xiang, J. Yu and M. Jaroniec, *J. Phys. Chem. C*, 2011, **115**, 7355–7363.
22. I. Papailias, T. Giannakopoulou, N. Todorova, D. Demotikali, T. Vaimakis and C. Trapalis, *Appl. Surf. Sci.*, 2015, **358**, 278–286.
23. K.-D. Henning and H. von Kienle, *Ullmann's Encyclopedia of Industrial Chemistry*, Wiley-VCH, 2010.
24. R. Schlogl, in *Handbook of Heterogeneous Catalysis*, ed. G. Ert, H. Knozinger, F. Schuth and J. Weitkamp, Wiley-VCH, 2008, p. 357.
25. H. Marsh and F. Rodriguez-Reinoso, *Activated Carbon*, Elsevier B.V, 2006, p. 13.
26. A. Centrone, L. Brambilla, T. Renouard, L. Gherghel, C. Mathis, K. Mullen and G. Zerbi, *Carbon*, 2005, **43**, 1593–1609.
27. A. Lazzarini, R. Pellegrini, A. Piovano, S. Rudić, C. Castan-Guerrero, P. Torelli, M. R. Chierotti, R. Gobetto, C. Lamberti and E. Groppo, *Catal. Sci. Technol.*, 2017, **7**, 4162–4172.
28. A. Lazzarini, A. Piovano, R. Pellegrini, G. Leofanti, G. Agostini, S. Rudić, M. R. Chierotti, R. Gobetto, A. Battiato, G. Spoto, A. Zecchina, C. Lamberti and E. Groppo, *Catal. Sci. Technol.*, 2016, **6**, 4910–4922.
29. R. A. Friedel and G. L. Carlson, *J. Phys. Chem.*, 1971, **75**, 1149–1151.
30. V. Ţucureanu, A. Matei and A. M. Avram, *Crit. Rev. Anal. Chem.*, 2016, **46**, 502–520.
31. C. Moreno-Castilla, M. López-Ramón, F. Carrasco-Marín, M. V. Lopez-Ramon and F. Carrasco-Marin, *Carbon*, 2000, **38**, 1995–2001.
32. E. Fuente, J. A. Menéndez, M. A. Díez, D. Suárez and M. A. Montes-Morán, *J. Phys. Chem. B*, 2003, **107**, 6350–6359.
33. F. Rositani, P. L. Antonucci, M. Minutoli, N. Giordano and A. Villari, *Carbon*, 1987, **25**, 325–332.

34. A. O. Odeh, *J. Fuel Chem. Technol.*, 2015, **43**, 129–137.
35. J. Lee, K. H. Kim and E. E. Kwon, *Renewable Sustainable Energy Rev.*, 2017, **77**, 70–79.
36. T. Zerlia, *Fuel*, 1985, **64**, 1310–1312.
37. C. Yang, *Fuel*, 1995, **74**, 543–548.
38. P. K. Chu and L. Li, *Mater. Chem. Phys.*, 2006, **96**, 253–277.
39. A. C. Ferrari, S. E. Rodil and J. Robertson, *Phys. Rev. B: Condens. Matter Mater. Phys.*, 2003, **67**, 155306.
40. M. T. Timko, J. A. Wang, J. Burgess, P. Kracke, L. Gonzalez, C. Jaye and D. A. Fischer, *Fuel*, 2016, **163**, 223–231.
41. T. Kar, S. Scheiner, U. Adhikari and A. K. Roy, *J. Phys. Chem. C*, 2013, **117**, 18206–18215.
42. U. Adhikari, S. Scheiner, A. K. Roy and T. Kar, *Carbon*, 2014, **73**, 194–205.
43. J. Collins, T. Ngo, D. Qu and M. Foster, *Carbon*, 2013, **57**, 174–183.
44. J. Collins, D. Zheng, T. Ngo, D. Qu and M. Foster, *Carbon*, 2014, **79**, 500–517.
45. F. Cesano, M. M. Rahman, S. Bertarione, J. G. Vitillo, D. Scarano and A. Zecchina, *Carbon*, 2012, **50**, 2047–2051.
46. C. Tang and Q. Zhang, *Adv. Mater.*, 2017, **29**, 1604103.
47. X. Liu and L. Dai, *Nat. Rev. Mater.*, 2016, **1**, 16064.
48. N. A. Fathy, M. A. Shouman and R. M. M. Aboelenin, *Asia-Pac. J. Chem. Eng.*, 2016, **11**, 836–845.
49. D. Li, Y. Chen, M. Zheng, H. Zhao, Y. Zhao and Z. Sun, *ACS Sustainable Chem. Eng.*, 2016, **4**, 298–304.
50. M. Sangermano, S. Marchi, L. Valentini, S. B. Bon and P. Fabbri, *Macromol. Mater. Eng.*, 2011, **296**, 401–407.
51. A. C. Obreja, D. Cristea, R. Gavrila, V. Schiopu, A. Dinescu, M. Danila and F. Comanescu, *Appl. Surf. Sci.*, 2013, **276**, 458–467.
52. D. W. Lee, L. V. De Los Santos, J. W. Seo, L. L. Felix, A. D. Bustamante, J. M. Cole and C. H. W. Barnes, *J. Phys. Chem. B*, 2010, **114**, 5723–5728.
53. D. W. Chang, E. K. Lee, E. Y. Park, H. Yu, H.-J. Choi, I.-Y. Jeon, G.-J. Sohn, D. Shin, N. Park, J. H. Oh, L. Dai and J.-B. Baek, *J. Am. Chem. Soc.*, 2013, **135**, 8981–8988.
54. S. Cravanzola, F. Cesano, G. Magnacca, A. Zecchina and D. Scarano, *RSC Adv.*, 2016, **6**, 59001–59008.
55. S. Indrawirawan, H. Sun, X. Duan and S. Wang, *J. Mater. Chem. A*, 2015, **3**, 3432–3440.
56. Y. Zhang, Z. Sun, H. Wang, Y. Wang, M. Liang and S. Xue, *RSC Adv.*, 2015, **5**, 10430–10439.
57. B. Shateesh, G. B. Markad and S. K. Haram, *Electrochim. Acta*, 2016, **200**, 53–58.
58. S. Rana and S. B. Jonnalagadda, *Catal. Commun.*, 2017, **92**, 31–34.
59. S. Yang, X. Feng, X. Wang and K. Müllen, *Angew. Chem., Int. Ed.*, 2011, **50**, 5339–5343.
60. M. J. Ju, J. C. Kim, H.-J. Choi, I. T. Choi, S. G. Kim, K. Lim, J. Ko, J.-J. Lee, I.-Y. Jeon, J.-B. Baek and H. K. Kim, *ACS Nano*, 2013, **7**, 5243–5250.

61. Y. Xue, J. Liu, H. Chen, R. Wang, D. Li, J. Qu and L. Dai, *Angew. Chem., Int. Ed.*, 2012, **51**, 12124–12127.
62. B. B. Wang, M. K. Zhu, K. Ostrikov, R. W. Shao and K. Zheng, *J. Mater. Chem. C*, 2015, **3**, 1106–1112.
63. F. Karlický, K. Kumara Ramanatha Datta, M. Otyepka and R. Zbořil, *ACS Nano*, 2013, **7**, 6434–6464.
64. J. H. Lehman, M. Terrones, E. Mansfield, K. E. Hurst and V. Meunier, *Carbon*, 2011, **49**, 2581–2602.
65. D. Tasis, N. Tagmatarchis, A. Bianco, M. Prato, D. Tasis, N. Tagmatarchis, A. Bianco and M. Prato, *Chem. Rev.*, 2006, **106**, 1105–1136.
66. G. Cravotto, D. Garella, E. C. Gaudino, F. Turci, S. Bertarione, G. Agostini, F. Cesano and D. Scarano, *New J. Chem.*, 2011, **35**, 915–919.
67. J. Zhang, H. Zou, Q. Qing, Y. Yang, Q. Li, Z. Liu, X. Guo and Z. Du, *J. Phys. Chem. B*, 2003, **107**, 3712–3718.
68. A. Misra, P. K. Tyagi, M. K. Singh and D. S. Misra, *Diamond Relat. Mater.*, 2006, **15**, 385–388.
69. N. F. A. Zainal, T. I. Tunku Kudin, A. Azira, A. Z. Ahmed, S. Abdullah and M. Rusop, *AIP Conf. Proc.*, 2008, **1017**, 159–163.
70. L. Moradi and I. Etesami, *Fullerenes, Nanotubes, Carbon Nanostruct.*, 2016, **24**, 213–218.
71. C. Preston, D. Song, J. Taillon, J. Cumings and L. Hu, *Nanotechnology*, 2016, **27**, 445601.
72. M. Prato, *J. Mater. Chem.*, 1997, **7**, 1097–1109.
73. M. R. Resmi, S. Ma, R. Caprioli and T. Pradeep, *Chem. Phys. Lett.*, 2001, **333**, 515–521.
74. J. Wu, W. Li, J. D. Fortner, M. Wiesner, P. Fioux, M. A. Gomez-Fatou, A. M. Benito, W. K. Maser, M. T. Martinez, F. Zerbetto, A. Koch and J. B. Hughes, *Environ. Sci.: Nano*, 2017, **4**, 117–126.
75. J. Lee, M. Cho, J. D. Fortner, J. B. Hughes and J.-H. Kim, *Environ. Sci. Technol.*, 2009, **43**, 4878–4883.
76. Y. S. Hwang and Q. Li, *Environ. Sci. Technol.*, 2010, **44**, 3008–3013.
77. F. Cataldo, *Carbon*, 1994, **32**, 437–443.
78. Y. Gao, X. Wu, X. C. Zeng, E. Wang, D. Geohagan, Z. Y. Zhang, C. Xiao, K. Deng, Z. Liu, S. Asahina, H. Kazumori, M. O'Keeffe, O. Terasaki, J. F. Stoddart and O. M. Yaghi, *J. Mater. Chem. A*, 2014, **2**, 5910–5914.
79. S. Yang, Y. Gong, J. Zhang, L. Zhan, L. Ma, Z. Fang, R. Vajtai, X. Wang and P. M. Ajayan, *Adv. Mater.*, 2013, **25**, 2452–2456.
80. Y. Zhang, Q. Pan, G. Chai, M. Liang, G. Dong, Q. Zhang and J. Qiu, *Sci. Rep.*, 2013, **3**, 1943.
81. B. Jürgens, E. Irran, J. Senker, P. Kroll, H. Müller and W. Schnick, *J. Am. Chem. Soc.*, 2003, **125**, 10288–10300.
82. P. Wu, J. Wang, J. Zhao, L. Guo and F. E. Osterloh, *J. Mater. Chem. A*, 2014, **2**, 20338–20344.
83. H. Yin, Q. Guo, D. He, J. Li and S. Sun, *Chem. Phys. Lett.*, 2017, **669**, 22–28.
84. Z. Yao, H. Nie, Z. Yang, X. Zhou, Z. Liu and S. Huang, *Chem. Commun.*, 2012, **48**, 1027–1029.

85. C. Liu, Y. Zhang, F. Dong, A. H. Reshak, L. Ye, N. Pinna, C. Zeng, T. Zhang and H. Huang, *Appl. Catal., B*, 2017, **203**, 465–474.
86. W.-J. Ong, L.-L. Tan, S.-P. Chai and S.-T. Yong, *Dalton Trans.*, 2015, **44**, 1249–1257.
87. B. Chai, J. Yan, C. Wang, Z. Ren and Y. Zhu, *Appl. Surf. Sci.*, 2017, **391**, 376–383.
88. M. Seredych and T. J. Bandosz, *Carbon*, 2016, **107**, 895–906.
89. W. Fang, J. Liu, L. Yu, Z. Jiang and W. Shangguan, *Appl. Catal., B*, 2017, **209**, 631–636.
90. S. Hu, F. Li, Z. Fan, F. Wang, Y. Zhao, Z. Lv, M. Antonietti, J. Z. Gui and X. C. Wang, *Dalton Trans.*, 2015, **44**, 1084–1092.
91. M. Zhang, X. Bai, D. Liu, J. Wang and Y. Zhu, *Appl. Catal., B*, 2015, **164**, 77–81.
92. Y. Li, H. Xu, S. Ouyang, D. Lu, X. Wang, D. Wang and J. Ye, *J. Mater. Chem. A*, 2016, **4**, 2943–2950.
93. H. Liu, D. Chen, Z. Wang, H. Jing and R. Zhang, *Appl. Catal., B*, 2017, **203**, 300–313.
94. Y. Xu, Y. Gong, H. Ren, W. Liu, L. Niu, C. Li, X. Liu, H. Xu, J. Martin-Calleja, T. Zhang and J. Yao, *RSC Adv.*, 2017, **7**, 32592–32600.
95. Y. Shang, X. Chen, W. Liu, P. Tan, H. Chen, L. Wu, C. Ma, X. Xiong and J. Pan, *Appl. Catal., B*, 2017, **204**, 78–88.
96. X. L. Wang, W. Q. Fang, H. F. Wang, H. Zhang, H. Zhao, Y. Yao, H. G. Yang and W. Schnick, *J. Mater. Chem. A*, 2013, **1**, 14089.
97. H. Yu, R. Shi, Y. Zhao, T. Bian, Y. Zhao, C. Zhou, G. I. N. Waterhouse, L.-Z. Wu, C.-H. Tung and T. Zhang, *Adv. Mater.*, 2017, **29**, 1605148.
98. A. Thomas, A. Fischer, F. Goettmann, M. Antonietti, J.-O. Müller, R. Schlögl and J. M. Carlsson, *J. Mater. Chem.*, 2008, **18**, 4893.
99. L. Lin, H. Ou, Y. Zhang and X. Wang, *ACS Catal.*, 2016, **6**, 3921–3931.
100. J. Zawadzki and M. Wisniewskí, *Carbon*, 2003, **41**, 2257–2267.
101. Y. Li, H. Zhang, P. Liu, D. Wang, Y. Li and H. Zhao, *Small*, 2013, **9**, 3336–3344.
102. X. Yang, F. Qian, G. Zou, M. Li, J. Lu, Y. Li and M. Bao, *Appl. Catal., B*, 2016, **193**, 22–35.
103. Q. Liang, Z. Li, X. Yu, Z.-H. Huang, F. Kang and Q.-H. Yang, *Adv. Mater.*, 2015, **27**, 4634–4639.
104. X. Du, G. Zou, Z. Wang and X. Wang, *Nanoscale*, 2015, **7**, 8701–8706.
105. Y.-P. Zhu, T.-Z. Ren and Z.-Y. Yuan, *ACS Appl. Mater. Interfaces*, 2015, 7, 16850–16856.
106. F. Zhang, Q. Wen, M. Hong, Z. Zhuang and Y. Yu, *Chem. Eng. J.*, 2017, **307**, 593–603.
107. M. Aleksandrzak, W. Kukulka and E. Mijowska, *Appl. Surf. Sci.*, 2017, **398**, 56–62.
108. X. Li, J. Zhang, L. Shen, Y. Ma, W. Lei, Q. Cui and G. Zou, *Appl. Phys. A*, 2009, **94**, 387–392.
109. M. Groenewolt and M. Antonietti, *Adv. Mater.*, 2005, **17**, 1789–1792.
110. Y. Zhang, Y. Zhang, H. Gong, J. Yu, J. Zhao, Z. Zhang and Y. Zhang, *J. Sol-Gel Sci. Technol.*, 2016, **80**, 683–689.
111. S. F. Parker, *Johnson Matthey Technol. Rev.*, 2016, **60**, 132–144.

112. R. Pynn, in *Neutron Scattering Applications and Techniques*, ed. L. Liang, Springer Science+Business Media, LLC, 2009, pp. 1–29.
113. S. F. Parker, D. Lennon and P. W. Albers, *Appl. Spectrosc.*, 2011, **65**, 1325–1341.
114. P. C. H. Mitchell, F. S. Parker, A. J. Ramirez-Cuesta and J. Tomkinson, *Vibrational Spectroscopy with Neutrons*, World Scientific, London, UK, 2005.
115. J. L. Figueiredo and M. F. R. Pereira, *Catal. Today*, 2010, **150**, 2–7.
116. L. R. Radovic, I. F. Silva, J. I. Ume and J. A. Menbndez, *Carbon*, 1997, **35**, 1339–1348.
117. H.-P. H. Boehm, *Carbon*, 2002, **40**, 145–149.
118. P. W. Albers, J. Pietsch, J. Krauter and S. F. Parker, *Phys. Chem. Chem. Phys.*, 2003, **5**, 1941–1949.
119. D. Lennon, D. T. Lundie, S. D. Jackson, G. J. Kelly and S. F. Parker, *Langmuir*, 2002, **18**, 4667–4673.
120. P. Papanek, W. A. Kamitakahara, P. Zhou and J. E. Fischer, *J. Phys.: Condens. Matter*, 2001, 8287.
121. R. Papoular, *Fuel*, 1995, **74**, 865–873.
122. S. F. Parker, S. Imberti, S. K. Callear and P. W. Albers, *Chem. Phys.*, 2013, **427**, 44–48.
123. M. Casartelli, S. Casolo, G. F. Tantardini and R. Martinazzo, *Carbon*, 2014, **77**, 165–174.
124. C. Cavallari, D. Pontiroli, M. Jiménez-Ruiz, A. Ivanov, M. Mazzani, M. Gaboardi, M. Aramini, M. Brunelli, M. Riccò and S. Rols, *J. Phys.: Conf. Ser.*, 2014, **554**, 12009.
125. A. I. Kolesnikov, I. O. Bashkin, V. E. Antonov, D. Colognesi, J. Mayers and A. P. Moravsky, *J. Alloys Compd.*, 2007, **446–447**, 389–392.
126. Y. Liu, C. M. Brown, D. A. Neumann, D. B. Geohegan, A. A. Puretzky, C. M. Rouleau, H. Hu, D. Styers-Barnett, P. O. Krasnov and B. I. Yakobson, *Carbon*, 2012, **50**, 4953–4964.
127. D. Pontiroli, M. Aramini, M. Gaboardi, M. Mazzani, S. Sanna, F. Caracciolo, P. Carretta, C. Cavallari, S. Rols, R. Tatti, L. Aversa, R. Verucchi and M. Riccò, *J. Phys. Chem. C*, 2014, **118**, 7110–7116.
128. P. W. Albers, M. Lopez, G. Sextl, G. Jeske and S. E. Parker, *J. Catal.*, 2004, **223**, 44–53.
129. P. Albers, R. Burmeister, K. Seibold, G. Prescher, S. F. Parker and D. K. Ross, *J. Catal.*, 1999, **181**, 145–154.
130. P. Albers, E. Auer, K. Ruth and S. F. Parker, *J. Catal.*, 2000, **196**, 174–179.
131. P. W. Albers, J. G. E. Krauter, D. K. Ross, R. G. Heidenreich, K. Köhler and S. F. Parker, *Langmuir*, 2004, **20**, 8254–8260.
132. D. R. Jones, S. Iqbal, S. A. Kondrat, G. M. Lari, P. J. Miedziak, D. J. Morgan, S. F. Parker and G. J. Hutchings, *Phys. Chem. Chem. Phys.*, 2016, **18**, 17259–17264.
133. S. F. Parker, D. Adroja, M. Jiménez-Ruiz, M. Tischer, K. Möbus, S. D. Wieland and P. Albers, *Phys. Chem. Chem. Phys.*, 2016, **18**, 17196–17201.

134. M. Suleiman, J. Faupel, C. Borchers, H. U. Krebs, R. Kirchheim and A. Pundt, *J. Alloys Compd.*, 2005, **404–406**, 523–528.
135. C. S. Tsao, Y. Liu, H. Y. Chuang, H. H. Tseng, T. Y. Chen, C. H. Chen, M. S. Yu, Q. Li, A. Lueking and S. H. Chen, *J. Phys. Chem. Lett.*, 2011, **2**, 2322–2325.
136. P. W. Albers, W. Weber, K. Möbus, S. D. Wieland and S. F. Parker, *Carbon*, 2016, **109**, 239–245.
137. D. Lennon, J. McNamara, J. R. Phillips, R. M. Ibberson and S. F. Parker, *Phys. Chem. Chem. Phys.*, 2000, **2**, 4447–4451.
138. P. C. H. Mitchell, A. J. Ramirez-Cuesta, S. F. Parker, J. Tomkinson and D. Thompsett, *J. Phys. Chem. B*, 2003, **107**, 6838–6845.
139. A. Piovano, A. Lazzarini, R. Pellegrini, G. Leofanti, G. Agostini, S. Rudić, A. L. Bugaev, C. Lamberti and E. Groppo, *Adv. Condens. Matter Phys.*, 2015, **2015**, 803267, 8 pages.
140. F. Fillaux, R. Papoular, A. Lautiè and J. Tomkinson, *Carbon*, 1994, **32**, 1325–1331.
141. J. P. Chen and S. Wu, 2004, 2233–2242.
142. J. Jaramillo, P. Modesto and V. Gómez-serrano, *Fuel Process. Technol.*, 2010, **91**, 1768–1775.
143. J. L. Figueiredo, M. F. R. Pereira, M. M. A. Freitas and J. J. M. Orfao, *Carbon*, 1999, **37**, 1379–1389.
144. K. Refson and S. F. Parker, *ChemistryOpen*, 2015, **4**, 620–625.
145. S. F. Parker, S. M. Bennington, J. W. Taylor, H. Herman, I. Silverwood, P. Albers and K. Refson, *Phys. Chem. Chem. Phys.*, 2011, **13**, 7789.
146. C. Cavallari, D. Pontiroli, M. Jimenez-Ruiz, M. Johnson, M. Aramini, M. Gaboardi, S. F. Parker, M. Riccó and S. Rols, *Phys. Chem. Chem. Phys.*, 2016, **18**, 24820–24824.
147. J. K. Walters, R. J. Newport, S. F. Parker and W. S. Howells, *J. Phys.: Condens. Matter*, 1995, 7, 10059–10073.

CHAPTER 5

XPS Characterization of Metal-free Functionalized Carbons

W. LUO[a] AND S. ZAFEIRATOS*[b]

[a]Institute of Chemical Sciences and Engineering (ISIC), École Polytechnique Fédérale de Lausanne (EPFL) Valais/Wallis, Sion, Switzerland; [b]Institut de Chimie et Procédés pour l'Energie, l'Environnement et la Santé (ICPEES), ECPM, UMR 7515, CNRS-Université de Strasbourg, Strasbourg Cedex 02, France
*E-mail: spiros.zafeiratos@unistra.fr

5.1 Introduction

One great challenge in the development of sustainable catalysts is to substitute critical raw components, like expensive noble metals. Carbon materials in their various forms feature a variety of advanced characteristics, such as low price, high chemical stability and high specific area. Thus, there has been great interest in recent years to develop methods to employ them as metal-free catalysts in several relevant catalytic processes.[1,2] Functionalization of graphitic carbon surfaces is essential to introduce active sites for catalytic reactions.[3] Defects and functional groups, as well as heteroatoms, have been introduced onto the carbon surface, showing advanced catalytic performance for several reactions. The mechanisms underlying the catalytic performance are closely related to the catalytic active centers that are located

Catalysis Series No. 31
Metal-free Functionalized Carbons in Catalysis: Synthesis, Characterization and Applications
Edited by Alberto Villa and Nikolaos Dimitratos

Published by the Royal Society of Chemistry, www.rsc.org

on the carbon surface. In this respect, specific information about the chemical and electronic state, as well as the elemental composition of the carbon surface is essential.

X-Ray photoelectron spectroscopy (XPS), which is also known as electron spectroscopy for chemical analysis (ESCA) or photoelectron spectroscopy (PES), is a spectroscopic technique to examine, with high surface sensitivity, the chemical state and composition of solids. The method was initially developed in the late '50s by the physicist Kai Siegbahn at Uppsala University (Sweden), who later obtained the Nobel Prize for this work (1981).[4] A great advantage of the technique is the possibility to detect practically all elements of the periodic table with a detection limit down to 1% or even lower. Therefore, it is not surprising that today, XPS has been well established as a standard characterization method to analyze materials both in industry and academia, in a wide range of applications extending from biology to microelectronics.

XPS is regularly used to analyze the chemical composition, purity and the nature of chemical bonds on the surface of carbons. Under specific circumstances, analysis of XPS spectra can also provide structural information for carbon materials.[5] Apart from laboratory-based XPS instruments, the application of synchrotron radiation offers particular advantages as compared to conventional laboratory X-ray tubes and extends the capabilities of the technique. Notably, the so-called hard X-ray photoemission spectroscopy (HAXPES) uses excitation energies in the hard X-ray region (up to 10 keV) and so the signal of the interior carbon layers (up to 20 nm depth) becomes accessible by the technique.[6] In addition, nowadays, XPS studies can also be performed *in situ* in a reactive gas atmosphere under several mbar pressure.[7]

5.1.1 Physical Principles and Practical Aspects of XPS

The basic principle of XPS spectroscopy is the photoelectric effect (Figure 5.1(a)), which was observed by H. Hertz in 1887 and theoretically explained by A. Einstein in 1905. The photoelectric effect describes the emission of electrons (usually referred to as photoelectrons) from the surface due to the incidence of electromagnetic radiation (*e.g.* X-rays). In a typical laboratory-based XPS spectrometer, the X-rays are produced by an X-ray source (monochromatized or achromatized) while the photoelectrons' kinetic energy is measured by a hemispherical electron analyzer combined with a detector (usually multi-channeltron). In contrast to laboratory X-ray sources, X-rays produced in synchrotron facilities can be used as the excitation source.

A wide energy scan XPS spectrum of carbon materials looks typically like the one shown in Figure 5.1(b). The spectrum consists of several narrow distinct peaks on a continuous background, which increase with the decrease of electron kinetic energy. The narrow peaks can be grouped into three main types: (i) primarily, photoelectrons that have escaped from the sample and reached the analyzer without suffering collisions, (ii) satellite peaks, which contain peaks associated with phenomena induced by the X-rays or related to

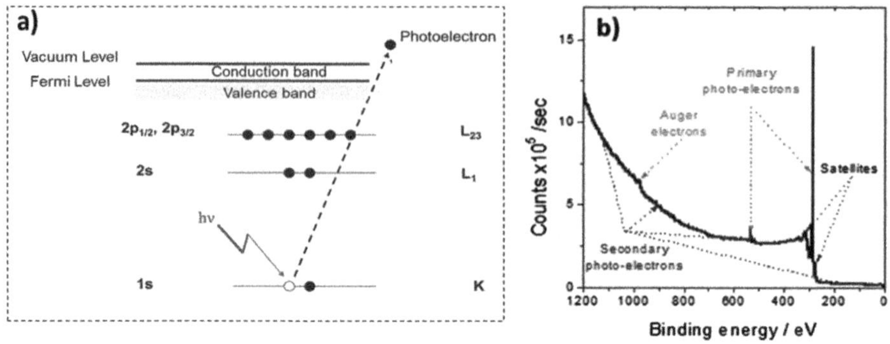

Figure 5.1 (a) Schematic representation of the photoelectric effect, (b) a typical wide energy, low-resolution XPS spectrum of a carbon material.

the electronic structure of the sample and (iii) X-ray excited Auger electrons. The background consists of secondary, scattered electrons that experience random energy losses after photoemission (Figure 5.1(b)). In general, a photoelectron spectrum contains both core-level and valence band electrons. However, the valence band is quite complex since several features are superimposed in a narrow region and is, therefore, difficult to analyze. On the other hand, in the case of the core-level spectra, the position of atoms from different chemical elements is easier to distinguish. In the case of carbon materials, the most important information is derived by the binding energy and the line shape of the C 1s core level peak, as well as the satellite peaks related to it.

The quantity measured by an electron spectrometer is the kinetic energy of the ejected photoelectrons. The binding energy of these electrons inside the atom is correlated with the measured kinetic energy of the detected photoelectrons according to eqn (5.1a):

$$E_K = h\nu - E_B - \varphi_a \tag{5.1a}$$

$$E_B = h\nu - E_K \tag{5.1b}$$

where E_K is the kinetic energy of the photoelectron as measured by the analyzer, $h\nu$ is the energy of the incident photons, E_B is the binding energy of the ejected photoelectron in the atom (which is characteristic for each and every element) and φ_a is the work function of the analyzer, which is considered as constant. Therefore, for all XPS spectrometers, the φ_a is compensated artificially and the calculation of E_B is simplified to eqn (5.1b).

A very important feature of XPS spectroscopy is that the recorded E_B of a specific element reflects its local chemical environment. Therefore, in most cases, core-level photoelectrons are used for elemental analysis, while small shifts in their E_B (known as chemical shifts) can be used to determine the oxidation state. In order to correlate the E_B with the electronic charge of the atom, a simple, but sometimes misleading, electrostatic model is often used. According to this, if some negative charge (electrons) is removed from the atom due to participation to a chemical bond, the atom is positively charged, the kinetic energy of the photoelectrons decreases (due to attraction forces) and the XPS peak shifts to a higher E_B. However, this oversimplified model does not account for other factors that can drastically affect the measured chemical shifts, such as final-state effects. Therefore, for more reliable results, it is essential to compare the chemical shifts with reference compounds or *ab initio* calculations.[8]

The reference level (zero of the E_B scale) is the Fermi level of the analyzer, which is aligned with that of the sample at least for electron conductive samples. For wide band-gap or insulating samples, a positive charge is established on the sample surface due to photoelectron emissions. As a result, the measured E_K is reduced, which results to a shift towards a higher E_B usually called "electrostatic charging". One should be very cautious, since sometimes electrostatic charging can be confused with the chemical shift. Therefore, in the case of non-conduction samples, the binding energy scale has to be calibrated based on an "internal reference" peak. The most commonly used internal reference is the C 1s peak of adventitious carbon,[9] which typically appears at a binding energy of 284.5 eV, although in various papers, this BE may vary from 284.4 to 285.0 eV.

Most carbon allotropes are electrically conductive; therefore, eqn (5.1b) can be directly applied to calculate the binding energies, without energy scale corrections. However, in the case of insulating or wide band-gap carbon allotropes (for example, diamond-like carbons), it is quite challenging to distinguish adventitious carbon peaks among the other carbon species. In that case, the use of an internal reference peak is complicated. For that reason, other, more elaborate energy scale calibration methods have been proposed.[9]

The photoelectron peak intensity is proportional to the amount of the existing elements and, therefore, core-level spectra can be used for quantitative analysis. In order to do so, the possibility for each element to produce photoelectrons should be taken into account, by using measured, empirical or theoretically calculated atomic sensitivity factors (ASF), which are sometimes called photoelectron cross sections. The ASF strongly depend on the element, the specific core-level shell and the excitation photon energy. For laboratory X-ray source experiments, empirically derived ASF are usually used,[10] while for synchrotron-based experiments, calculated values are typically employed since they take into account the photon energy dependence of the ASF.[11]

XPS is a surface-sensitive method, which probes only a very thin layer of the outermost surface in the range between 1 and 10 nm (this value strongly depends on the photoelectron kinetic energy as will be discussed later in the text). The physical origin behind the surface sensitivity of XPS is the high probability that low kinetic energy electrons will undergo inelastic scattering and lose energy when they traverse into matter. This means that photoelectrons emitted from atoms deep in the solid will lose energy due to scattering and they will probably never escape from the sample, or, at best, they will just contribute to the spectral background. The average distance that a photoelectron can travel within a solid is called the inelastic mean free path (IMFP or λ) and can be calculated by the Tanuma, Powell and Penn algorithm, designated as TPP-2M, using S. Tougaard software.[12] Usually, when referring to the sampling or analysis depth of the XPS technique, a good indicative value is to use three times the IMFP. The IMFP strongly depends on the electron kinetic energy and the material being analyzed. Therefore, when a variable photon energy source is available (for example, synchrotron radiation), it is possible to get some information about the depth distribution of the elements or the chemical species that exist over the analyzed sample.[13,14] These types of measurements are called energy-resolved depth profile (ERDP) measurements, or simply depth-profile measurements.

5.1.2 XPS Instrumentation

The two basic parts of an X-ray photoelectron spectrometer (see Figure 5.2) are the excitation source (X-ray tube or synchrotron radiation) and the photoelectron detection/analysis system (the electron lenses and the electron energy analyzer). The most commonly used X-ray sources in XPS apparatuses are magnesium (MgK_α, $h\nu$ = 1253.6 eV) and aluminium (AlK_α, $h\nu$ = 1486.6 eV), mainly due to their strong intensity and narrow band width (0.8 to 0.9 eV,

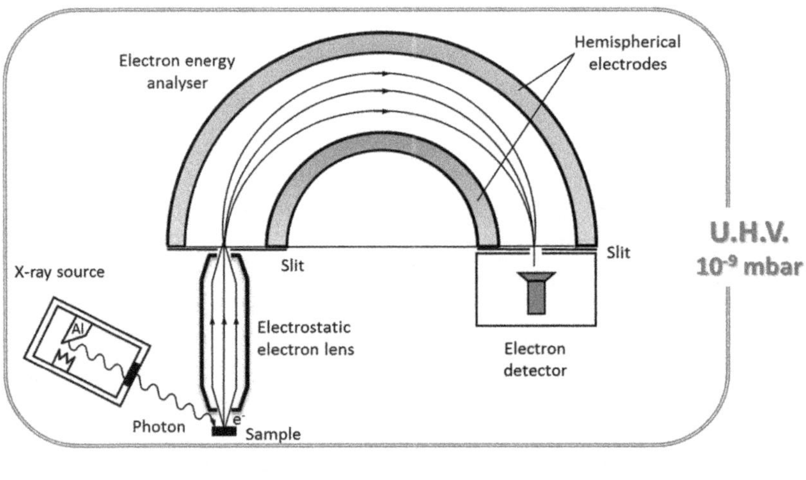

Figure 5.2 A schematic diagram of a typical laboratory-based XPS apparatus. An X-ray radiation source irradiates the sample with photons and the released photoelectrons pass through the lens system, which slows them down before they enter the hemispherical energy analyzer. Only photoelectrons with the selected kinetic energy can pass through and be detected by the electron detector.

respectively). The XPS apparatuses are maintained under ultra-high vacuum conditions (10^{-7}–10^{-10} Pa) in order to minimize gas collisions of the emitted electrons after photoionization and to maintain the stability and long lifetime requirements of the spectrometer.

During an XPS measurement, the incident photons are absorbed by the sample and the emitted photoelectrons (but also Auger and stray electrons) enter the *electron lenses*, where they are focused by means of electrostatic fields. The electron lenses physically separate the sample from the electron analyzer and their main function is to retard or accelerate the electrons to a fixed energy (*pass energy*) and focus them towards the analyzer entrance (*slit*). In laboratory experiments, the probed sample area and the electrons' acceptance angle (usually few degrees) are typically defined by the lenses' settings.

Consequently, the electrons enter the analyzer, where only those with a given passing energy can pass through the electrostatic field created between the outer and inner (hemi)spheres. Although several different types of electron energy analyzers have been developed and utilized through the years, the electrostatic hemispherical analyzer is the one most frequently used in today's commercial instruments. The analyzer defines, to a large extent, the energy resolution, which is practically the ability to resolve peaks with small energy differences. Finally, at the exit slit of the analyzer, the number of photoelectrons is multiplied and counted by the electrode detection system.

5.1.3 XPS Peak Characteristics, Chemical Shift and Quantification

An XPS spectrum (Figure 5.1(b)) is displayed as a plot of the number of electrons *versus* their kinetic energy, or, more frequently, *versus* the binding energy as this is calculated by eqn (5.1b). Typically, the binding energy is shown in electron volts (eV), while the intensity is measured in counts or counts per seconds (c/s). However, very often the intensity is shown in arbitrary units (a.u.), since the absolute spectral intensity is meaningless. Usually, the XPS analysis starts with a scan in a wide energy region (typically from 1200 to −2 eV) and with low spectral resolution (*survey spectrum*). This first spectrum is used before the detailed analysis in order to identify all elements present on the sample. Both photoelectron and Auger electron peaks are recorded in the survey scan, while other types of peaks, such as shake-up and energy loss lines, might be visible but with lower intensities. After all elements have been identified (*elemental analysis*), detailed spectra (that are sometimes called high resolution scans) of the more intense peaks of the element(s) of interest, are recorded. The detailed spectra allow the precise determination of the binding energy and the chemical shifts (*chemical state analysis*), as well as an accurate estimation of the peak areas to be used for the *quantitative analysis* of the sample. Quantification of XPS results is possible since the number of photoelectrons of an element is dependent on the atomic concentration of that element on the sample surface. The peak area after background subtraction is used to estimate the elemental concentration.

For a quantification of laboratory XPS results, the intensities of the spectra should be normalized to the atomic sensitivity factors,[10,15] which describe the probability of photoelectron emission from a particular core level.[12] There are different routes to estimate the atomic sensitivity factors for each element. This can be done by first principle calculations,[11] the use of published empirically-derived factors[10] or the use of reference samples to estimate the sensitivity factors of the particular instrument utilized. In practice, the first option is used for synchrotron-based studies, while for laboratory X-ray source instruments, published databases are employed. However, one has to be cautious that the absolute values of the sensitivity factors have no particular meaning and the quantification should always be based on the intensity ratios of the photoelectron peaks derived from different elements.

To calculate the relative concentration of the various constituents on the surface, one has to use a formula that depends on the specific morphology of the sample. This is a very challenging task, since it is very difficult to know the exact morphology of the sample; even in cases where this is feasible, the sample morphology has to be homogeneous in the analysis area.[16] Therefore, in a great number of studies, the atomic concentration (at%) is calculated based on a simplified morphology model. This model assumes that the sample is homogeneous, and all the elements are uniformly distributed within the analyzed surface region. The atomic concentration of an element x, C_x, in a sample that contains n different constituents can be expressed as:

$$C_x = \frac{I_x/\mathrm{ASF}_x}{\sum_{i=1}^{n}\left(I_i/\mathrm{ASF}_i\right)} \tag{5.2}$$

where I_x is the XPS peak area of the element x after background subtraction and ASF_x is the sensitivity factor (or photoionization cross section) for the peak x. Although this formula might not give accurate absolute values of the atomic concentration, it is safe to utilize in order to compare different samples, providing that their morphologies are not drastically different.

The quality of the XPS instrument affects the XPS spectrum very much. Two more significant factors that differentiate a "good quality" spectrum from a "bad quality" one are the signal to noise (S–N) ratio and the resolution of the spectrum. These two factors are interconnected, and a good compromise should be found between them in each experiment. As expected for the same accumulation time, the S–N ratio of a peak is directly related to the amount of the examined element and its sensitivity factor. Accordingly, it is easier to detect elements with high ASF (for example Pt), as compared to carbon, since the Pt 4f peak has a cross section almost 10 times higher than that of the C 1s. However, the sensitivity factor depends not only on the element, but also on the excitation photon energy. This might be particularly useful in

Table 5.1 Photoionization cross section or atomic sensitivity factor (ASF) and estimated IMFP of the C 1s photoelectron peak estimated for various photon energies.

Photon energy, eV	Sensitivity factor[11] (values relevant to Al Kα)	Estimated IMFP[12] (information depth), nm
1486.6 (Al Kα)	1	3.30 (9.9)
1253.6 (Mg Kα)	1.6	2.8 (8.4)
600	12	1.27 (3.8)
305	58	0.08 (0.24)

synchrotron radiation XPS experiments where the excitation photon energy can be easily varied. Of course, one has to keep in mind that when the photon energy is varied, both the ASF and the information depth are varied as well. In the particular case of C 1s, the photoionization cross section and the information depth for some characteristic excitation photon energies are included in Table 5.1. As is evident, when instead of the standard Al Kα source (1486.6 eV) one uses a synchrotron source with 300 eV photon energy, the C 1s cross section can increase by up to 60 times. This significantly improves the S–N ratio of samples with weak C 1s signals. In addition, lower excitation photon energies enhance the surface sensitivity of the XPS method and, therefore, its ability to detect surface species.

Instrumental factors can play an important role in the quality of the recorded spectra. The two primary elements of an XPS spectrometer are the analyzer/detection system and the X-ray source. Almost all modern XPS instruments use the hemispherical energy analyzer type (HEA). The quality of the analyzer is closely related to its electron transfer lenses, the diameter of the hemispheres, the detector and, of course, the electronics. The development in this field is rapid, and a detailed description is out of the scope of this chapter.

The X-ray source can also affect the quality of the X-ray spectrum and, in particular, the S–N and the resolution. In laboratory-based XPS experiments, both monochromatized and achromatized X-ray sources can be used. Although achromatized X-ray sources have higher photon fluxes, the spectra are broader than those of monochromatized X-ray sources, and also include additional peaks excited by X-ray satellites. A comparison of survey and C 1s spectra of a carbon material produced with monochromatic and non-monochromatic radiation is shown in Figure 5.3. There it is evident that apart from the presence of X-ray satellites that might complicate the analysis (for example, when the satellite of one peak overlaps with the main peak of another element), the width and shape of the C 1s peak are also influenced. It is clear that the high resolution monochromatic synchrotron beam leads to a narrower C 1s peak. This has to be considered when the C 1s peak is analyzed further, for example using a peak-fitting procedure, in order to reveal chemical information.

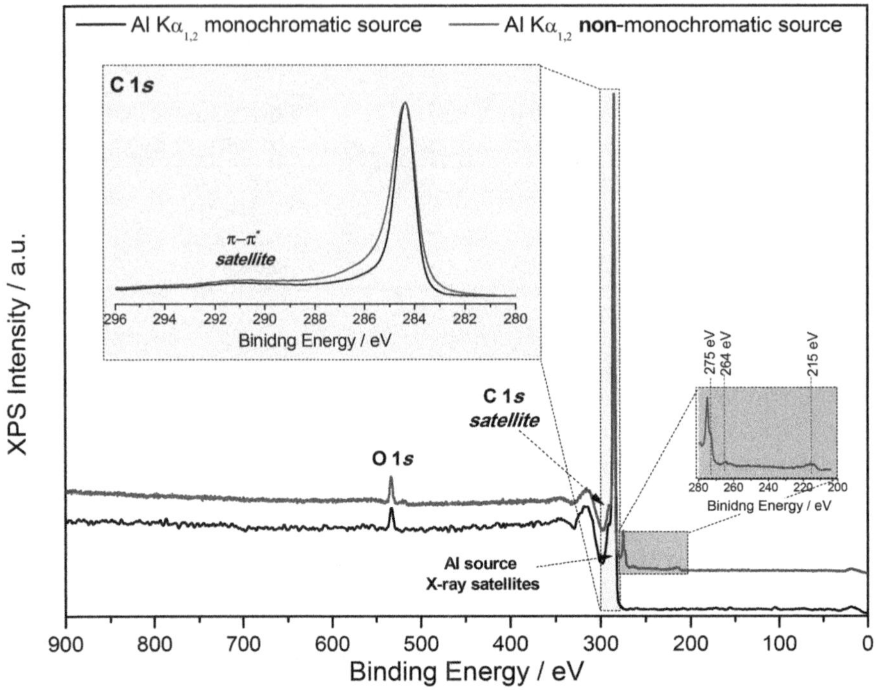

Figure 5.3 Survey and high-resolution spectra of a graphite sample recorded by a monochromatic and non-monochromatic AlK_{α} X-ray source (1486.6 eV). The spectra were measured using the same analyzer and settings, and are normalized to the maximum peak intensity.

5.2 Analysis of the C 1s Spectra of Carbonaceous Materials

The most frequently presented XPS spectrum of carbon materials is the C 1s spectrum, therefore, some emphasis is placed on the methodology for analyzing the C 1s peak. The C 1s peak shape and binding energy are determined by the nature of carbon bonds and, therefore, they may diversify in various pristine carbon materials. At a higher binding energy, the main C 1s peak is often accompanied by a π–π* inter-band transition, which is often called a plasmon satellite peak (see inset of Figure 5.3).[17] The binding energy and the line shape of the C 1s peak can be used in order to provide insights into the nature of the carbon material and reveal if variants of carbon species coexist on the sample surface. Apart from the main C 1s peak shape, the intensity of the π–π* satellite features can give additional information about the degree of graphitization of the material.[18] The identification of different chemical states of carbon only from the C 1s peak position might be possible, but usually, it should also be accompanied by careful analysis of the C 1s peak line shape. There is a plethora of reports about the E_B of the C 1s peak of various carbon materials, which do not always match one another. In Table 5.2,

Table 5.2 The binding energy (E_B) of the C 1s XPS peak of various pristine carbon materials. The table is a partial reproduction of the E_B reported in the review of T. Susi *et al.*[19] The reported energies have been rounded off to one decimal point.

Material	E_B of C 1s peak (eV)	Reference
Graphite	284.4	20 and 21
Supported single-layer graphene (SLG)	284.0–284.8[a]	21–25
Few-layer graphene (FLG)	284.5	26 and 27
Carbon nanotubes (CNTs)	284.4–284.5	28 and 29
C_{60}	285.2	30
Annealed nanodiamond (ND)	286.5–287.3[b]	31–33

[a]The reported binding energies are strongly influenced by the support material.
[b]The assignment of this peak in the literature has been reported in a wide range of energies varying from −1.5 to 3 eV as compared to graphitic carbon. This is because the diamond is an insulator and often exhibits electrostatic charging problems; for details, please refer to ref. 33.

we give a short overview of the reported E_B for the most common pristine carbon materials. For a more complete list, we propose that readers refer to the recent work of T. Susi *et al.*[19]

The chemical shifts in the C 1s spectrum occur over a relatively narrow energy range, usually from a few tenths of eV up to 3 eV. Thus, evaluation of the chemical state of elements from high-resolution XPS spectra often requires spectral analysis. Fitting the spectrum using model peaks is one common approach. The line shapes defined from the model spectra (*e.g.* acquired from pure standards or the starting material) can be used as synthetic components in a peak curve-fitting model. Another way to curve-fit complex C 1s spectra is to mathematically manipulate the spectra by performing a spectral subtraction. This method allows the enhancement of differences in the spectra and detection of small changes in the surface composition. However, before any attempt to analyze and interpret the C 1s peak using any of the aforementioned approaches, one has to take into account the following practical points:

- Samples that have been exposed to the atmosphere will have a detectable quantity of adventitious carbon contamination, which will overlap with the main carbon peak.
- The C 1s region overlaps with Ru 3d, Sr 3p and K 2p peaks from ruthenium, strontium and potassium, respectively. Therefore, the possible presence of these elements should be considered in the analysis.
- Some materials may be sensitive to prolonged vacuum and X-ray exposure (degradation). This is typically observed when polymer-type materials are analyzed.
- Some carbon materials might not have an adequate electrical conductivity; therefore, electrostatic charging problems may appear.
- The C 1s peak shape depends on the material and is affected by the amount of sp^2 and sp^3 carbons (double- and triple-bonded carbons).

5.2.1 Peak Fitting or Peak Deconvolution Procedure

For the analysis of the C 1s spectrum, the experimental peak can be deconvoluted into two or more peaks using a mathematical algorithm; this procedure is called peak fitting or simply deconvolution. The decomposition of the C 1s XP spectrum is complicated since there is a large number of parameters that are difficult to define with certainty (background, line-shape, binding energy, line-width *etc.*) and, therefore, this analysis might lead to uncertainties depending on the parameters used in the deconvolution.[34] An additional difficulty, which of course might also turn into an advantage, is that these parameters are dependent on the type of carbon bonds as mentioned above.

Curve-fit procedures are subjective and can result in non-unique curve fits. An important issue in the peak fitting process of C 1s peaks is the choice of the fitting curve (peak) shape. In some cases, the fitting components are chosen to have a symmetric line shape, which is usually a mixed Gaussian/Lorentzian peak profile, while in others, a peak asymmetry is introduced, commonly by using a Doniach–Šunjić (DS) curve shape. The correct choice of peak shape has important implications on the results of the deconvolution procedure, as can be seen in the example (Figure 5.4). A peak profile with a high asymmetric tail towards a higher binding energy, or without asymmetry at all, can lead to equally good mathematical fitting procedures. However, the amount of C 1s peaks used in the two cases, and, therefore, the different carbon species proposed by the fitting, is very different. Since many carbon compounds and species fall within a very narrow energy range, it is not possible to decide the reliability of the peak-fitting process just by the E_B of the C 1s peaks. Therefore, if the C 1s peak asymmetry is neglected, this can erroneously introduce more peaks (carbon species) in the fitting process leading to misinterpretation of the XPS results.[35]

Background subtraction is another important factor that affects the deconvolution of C 1s spectra. Both linear and Shirley backgrounds are widely used for background subtraction to get a reasonable fit quality.[17,20,36,37] The type of subtracted background can influence the peak fitting procedure as shown in the results (Figure 5.4). The background shape of the C 1s spectra depends strongly on the depth concentration profile of the carbon atoms.[38] In other words, the background will be different, in general, if a carbon material is embedded into a matrix or if it stands over a substrate. Failure to use the correct background profile will lead to errors in the fitting procedure. Another important point is the energy window of the background subtraction. For example, if the background is subtracted only under the main C 1s peak (288–281 eV in Figure 5.4), then several C 1s components are neglected or underestimated. Ideally, the background should be subtracted between 296 and 280 eV, which, of course, may vary for different carbon allotropes.

In order to be able to explain the choice of a suitable C 1s curve shape, one should recall the origin of the asymmetric tailing on the C 1s spectra. The line asymmetry is derived from the interaction of the positive core hole formed as a result of the primary photoemission process with the conduction

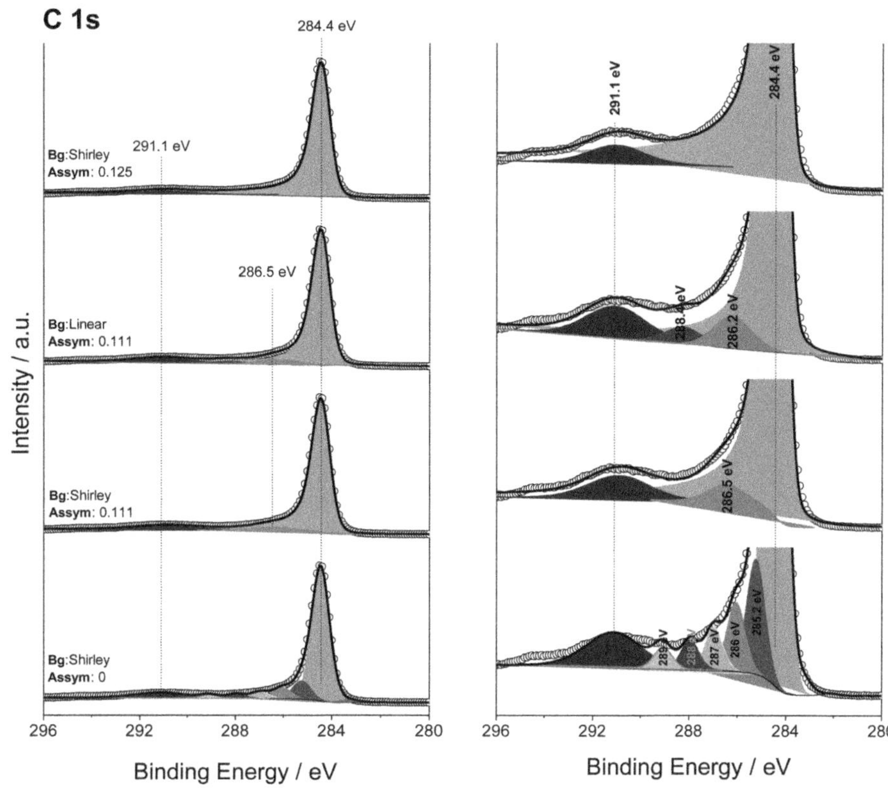

Figure 5.4 Example of a peak-fitting procedure for a graphite sample using a peak shape with different asymmetry and different background types. In the figure on the right, the *y*-axis scale has been increased for a closer examination of the energy region below 285 eV.

band states (shake-up).[39] The asymmetry depends on the band structure and is usually more pronounced when the Fermi level density of states (DOS) is high. In particular, the empty electron space (core hole) left after photoionization is screened by relaxation of the remaining electrons. A side effect of this process is the excitation of valence band electrons to unoccupied conduction band states, causing the escaping photoelectrons to lose energy, which is manifested as a high E_B peak. The resulting asymmetry is expected to be independent of the interlayer interactions, which explains why amorphous carbon materials, such as activated charcoal, can also feature asymmetric line shapes.

Apart from the shake-up-induced asymmetry, fitting components for other types of electronic losses, like the π–π* transition, should also be included in the fitting. The position and relative intensity of the π–π* transition component also depends on the nature of the carbon. For example, this satellite feature for HOPG is shifted about 6 eV compared to the main sp^2 peak. To the contrary, for other sp^2 systems, such as fullerenes, the reported π–π* satellite shift is around 1.9 or 1.4 eV, which is significantly smaller.[40] Unfortunately, the fitting process is complicated even further since one has to take into account various oxygenated carbon species on the surface, either due to exposure in the ambient atmosphere or due to the preparation process. If possible, one should also consider the O 1s spectra in order to estimate the energy and relative contribution of oxygenated carbon species in the C 1s spectrum.

As discussed above, apart from the intrinsic factors related to the chemical state of the material, the C 1s peak full width at half maximum (FWHM) is determined by the X-ray source. The FWHM of the peak directly affects the deconvolution of the C 1s peak and, consequently, the quantification of different chemical species in the carbon material. For example, if a narrow main peak is chosen to fit the broader C 1s peak generated from the lab source, a number of additional peaks at the higher binding energy side will be needed to obtain an appropriate fit, which will lead to underestimation of the graphitic carbon. To the contrary, a broad main peak will lead to overestimation of the amount of graphitic carbon.

5.2.2 Use of a Difference C 1s Spectra Peak Fitting Procedure

For studies focusing on the investigation of surface modifications, the analysis is more complicated, since new species may appear at E_B close to those of species present in the starting material. In that case, instead of peak fitting, one can use the spectra of the starting material as a reference and perform spectra subtraction.[41] The same procedure can be followed when reference C 1s spectra are available, preferentially recorded with the same XPS apparatus under identical conditions. Prior to spectra subtraction, the XPS peak intensities should be properly normalized so as to avoid artefacts in the procedure.

It is clear that using spectra subtraction, one cannot identify the carbon species of the original (reference) sample state. However, it can facilitate identification of species that appear, or even disappear, in the C 1s spectrum as compared to the reference state. This might be extremely useful when the exact C 1s curve shape is not known. One has to pay particular attention when applying the spectra subtraction analysis method that there are no peak shifts induced by charging problems, because in that case, the appearance of new spectral features will be just an artefact of the mathematical procedure. A characteristic example of spectra subtraction analysis is given in Figure 5.6.

5.3 XPS Studies of Pristine Carbon Materials

In this section, we will describe XPS studies of pristine carbon materials focusing on C 1s peak binding energies and line shapes. Carbon is known to take a great number of configurations and each of those states may have its own band structure and thus its own C 1s line shape. The XPS spectra of pristine carbons might be affected by several extrinsic and intrinsic factors

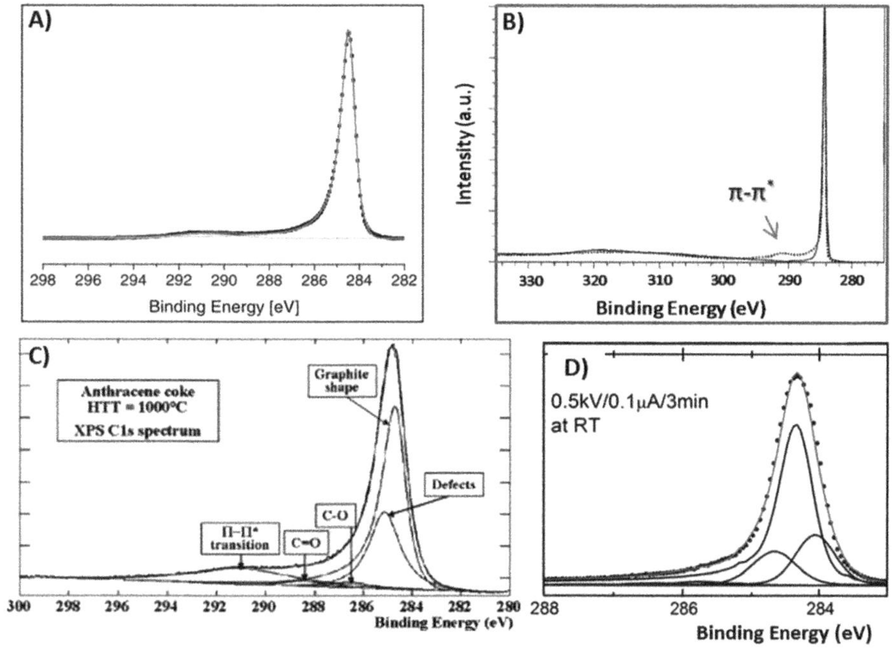

Figure 5.5 (A) XPS C 1s spectrum of HOPG. Reproduced from ref. 43, © 2003 Wiley-VCH Verlag GmnH&Co. KGaA. (B) Diamond (continuous line) and graphite (dotted line) C 1s spectra with the associated π–π* transition peak in the graphite. The two spectra have been aligned to the same EB. Reprinted from *Diamond and Related Materials*, **13**, G. Speranza and N. Laidani, Measurement of the relative abundance of sp^2 and sp^3 hybridised atoms in carbon based materials by XPS: a critical approach. Part I, 445–450, copyright (2004), with permission from Elsevier.[37] (C) C 1s spectrum of anthracene coke heat treated at a maximum temperature (HTT) of 1000 °C. Reprinted from *Carbon*, **42**, H. Estrade-Szwarckopf, XPS photoemission in carbonaceous materials: A "defect" peak beside the graphitic asymmetric peak, 1713–1721, copyright (2004), with permission from Elsevier.[42] (D) C 1s spectrum of sputtered HOPG measured using X-rays with an energy of 980 eV produced in a synchrotron facility. The sputter conditions are indicated in the figure. Reproduced from ref. 17 © 2015 Wiley-VCH Verlag GmnH&Co. KGaA.

of the material.[42] Therefore, in this part, the XPS spectra of various forms of carbon, such as highly oriented pyrolytic graphite (HOPG), carbon nanotubes (CNTs), graphene and diamond, will be presented, and the factors that influence the C 1s spectrum will be discussed.

HOPG is composed of sp^2-bonded carbon layers and serves as a model system for well-ordered graphite. It has been studied with XPS for many years and can serve as a good example to discuss the C 1s peak shape.[19] A typical HOPG C 1s spectrum is shown in Figure 5.5(A).[43] The narrow asymmetric C 1s peak typically appears at 284.4 eV E_B,[20,21,44] but in different works, may vary from 284.3 to 284.7 eV.[19,43] Apart from the main peak,

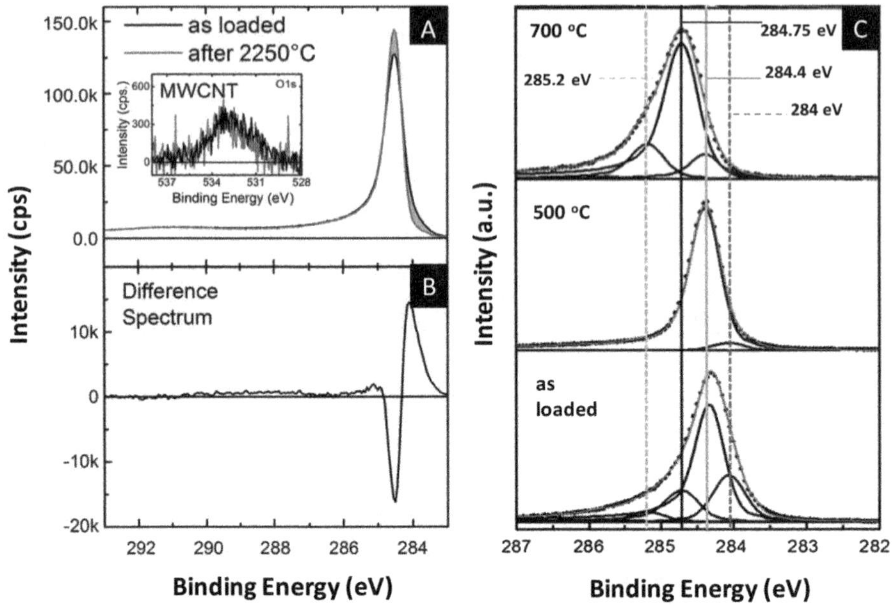

Figure 5.6 (A) C 1s spectra of as-loaded MWCNTs and after annealing at 2250 °C. Inset: corresponding O 1s spectra. (B) Difference spectrum of the two spectra in part A. Reproduced from ref. 43, © 2015 Wiley-VCH Verlag GmnH&Co. KGaA. (C) C 1s spectra of MLG on Cu after loading and subsequent annealing at 500 and 700 °C, respectively. Adapted from ref. 36 with permission from the PCCP Owner Societies.

a broad feature at about 6 eV higher in binding energy is typically present in the C 1s spectrum. This peak is characterized as the π–π^* satellite peak and it appears due to the loss of energy in the $\pi \rightarrow \pi^*$ transition for samples with high graphitic character.[37,45–48] The satellite peak grows more intense with increasing graphitization of the material and, as is evident in Figure 5.5(B), the peak is absent in the case of diamond-like carbon (mainly sp^3 type bonds).[37]

The C 1s peak has an asymmetric line shape towards higher binding energies that can be described by the Doniach–Sunjic (DS) line form.[49,50] Asymmetric XPS spectra are commonly observed for metals, and are attributed to final-state effects caused by the high DOS near the Fermi edge.[17,51] Graphite, a semi-metal with a very low Fermi level DOS, would normally induce a low asymmetry factor. However, the generation of C 1s peak asymmetry can be explained because electron–hole pairs are easily created in this specific material and induce a noticeable asymmetry of the C 1s line.[42] In the case of carbon allotropes with a metallic nature (*e.g.* graphite), an asymmetry should be introduced in the peak shape, while in the case of electrically insulating or semiconducting samples (*e.g.* diamond), the peak line shape is symmetric.[29] In addition, the C 1s peak of carbonaceous materials often contains peaks related to carbon atoms bound to oxygen. The relative intensity of the different C 1s components depends on the type of carbon material and the preparation method.

It is not only the peak shape, but also the peak width (FWHM) that should be considered in the analysis. Unreasonably broad C 1s peaks might be related with the presence of defected carbon sites, even if these sites are not clearly distinguished in the C 1s spectrum.[42] In that case, the C 1s can be fitted with a narrow peak due to graphitic sp^2 bonds, and a broader peak shifted to a higher E_B (about +1 eV) attributed to defective, sp^3 character carbons (Figure 5.5(C)).[42,52] However, the assignment of defect-like carbon to a C 1s component is controversial in the literature. Both a lower[53] and higher[42,54] E_B than the main sp^2 peak of graphite (usually at 284.4 eV) have been put forward.

This issue was recently addressed by R. Blume and co-workers[17] who attributed the two components to different types of defective carbon sites. In particular, the high E_B peak is due to disordered carbon, while the low E_B is induced by defect-like graphite. The authors realized that in order to fit the C 1s spectrum of sputtered HOPG (see Figure 5.5(D)), two broad symmetric peaks should be included in the fit, together with the asymmetric sp^2 C 1s peak measured on pristine HOPG. Using synchrotron-based energy-resolved depth profile measurements combined with *in situ* annealing, the authors assigned the two C 1s species to two different types of "less perfectly order" carbons in the well-ordered graphitic structure.

Although carbon nanotubes (CNTs) and graphene are also highly ordered graphitic carbon, their C 1s peak features cannot be simply considered the same as those of HOPG. In the case of CNTs, the differences may be caused by defects, functional groups, contamination and/or oxidation during sample preparation and exposure to air, *etc.* In the case of graphene, it is technically challenging to measure free-standing single layer graphene, therefore, most of the reported spectra are from multi-layer graphene (MLG).

Single-layer and few-layer graphene can be obtained in high qualities and in large-scale layers by chemical vapor deposition (CVD) on metal foils (such as Cu, Co *etc.*). However, exposure to air during transfer to the XPS chamber may introduce other carbon species. As shown in Figure 5.6(A), four peaks can be found on multi-layer graphene (MLG) supported on Cu after exposure to air.[36] The main asymmetric component at 284.4 eV is assigned to the C 1s peak from the ordered graphene, while the minor peak at 284 eV is due to graphene defects. The peaks at higher binding energies (BE > 285.0 eV) are assigned to C–O, C–OH bonds *etc.* due to contamination in air; all these peaks can be numerically fitted with symmetric peak components.[36] After annealing at 500 °C in a vacuum, the adsorbates are removed, and the peaks at high binding energies disappear. Since monolayer graphene is supported on copper, interactions between graphene and the substrate also play an important role in the C 1s peak E_B. As shown in Figure 5.6(A), further heating to 700 °C causes a shift of the main peak to a higher binding energy (284.75 eV), while it increases the sp^3 peak at 285.2 eV.

Figure 5.6(B) shows the C 1s spectra of commercial multi-walled carbon nanotubes (MWCNTs) before and after annealing. The mathematical difference between the two peaks is shown in Figure 5.6(C).[17] It is clear that the high binding energy side of the C 1s peaks and the O 1s peaks are barely changed after the annealing treatment. On the other hand, the feature at a low binding energy (~284 eV) disappears after annealing. This was attributed to a decrease in defective carbon during the graphitization procedure induced at high temperatures.[44,55]

Diamond is another carbon form where the carbon atoms are sp^3 hybridized. Despite many works on the fitting of the diamond C 1s peak, the absolute binding energy of this peak is still under debate.[37,56–59] While the C 1s peak of CVD-deposited diamond was observed at a lower binding energy (284.1 eV) than graphite (284.3 eV),[58] some other works report the C 1s position of diamond at a significantly higher E_B (283.7–289.2 eV).[57] This is probably related to the electronic structure of diamond, a wide band-gap semiconductor, which induces charging. Regardless of the peak position, diamond and graphite C 1s spectra are often aligned to the same binding energy (Figure 5.5) for comparison reasons.[37] The electronic structure of diamond also has an effect on the C 1s peak shape. In particular, the main peak of diamond at ~284.3 eV is symmetric and not asymmetric like that of graphite, underlying their differences in the electronic structure. Another difference in the spectra of the two carbon forms is the π–π^* satellite peak at ~291 eV, which is absent in the diamond C 1s peak.

5.4 XPS Characterization of Defects on Carbon Materials

Defects can easily form during the preparation of carbon materials; pure sp^2 carbon graphite or sp^3 carbon diamond do not exist, and even the highest quality CNTs contain various structural defects.[60] Moreover, defects will be spontaneously introduced on carbon materials when functionalization steps are carried out. In addition, defect sites might be introduced deliberately, since they can act as active centres or anchor sites for functional groups. Therefore, it is important to get detailed information of the defects on carbon materials with XPS. Unfortunately, this is not trivial, and the identification of defect carbon sites in the C 1s peak has been under debate for many years. Some works have described the defect peak at 285 eV[42] or at an even higher binding energy (285.6 eV),[54] while others have found it at a lower binding energy (~284 eV) compared to the graphite peak.[17,55,61]

In order to identify the C 1s binding energy of the defect peak, a strategy has been proposed to deliberately create defects on HOPG by Ar^+ ion sputtering.[17,55,61] With a low sputtering energy (40–80 eV), vacancy defects can be produced in a controllable manner.[62] The example in Figure 5.7(a) shows the deconvolution of the C 1s peak of HOPG after Ar^+ bombardment.[61] Density

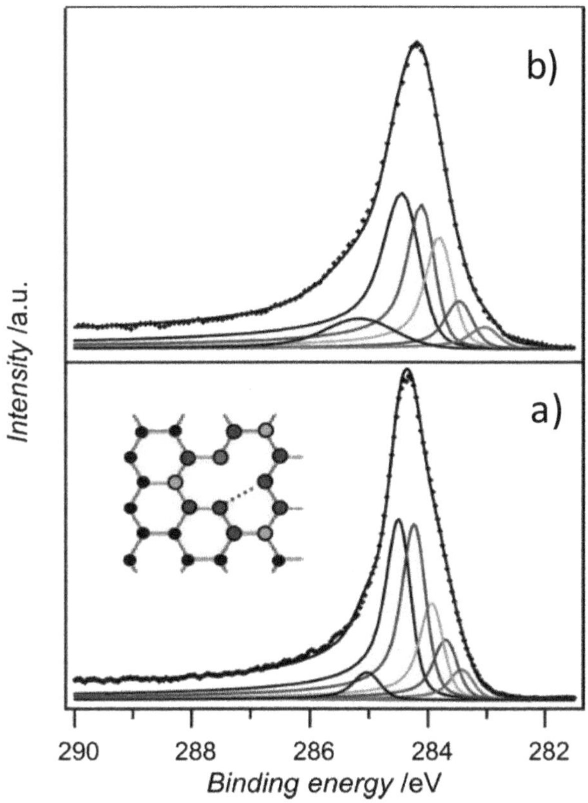

Figure 5.7 (a) C 1s spectra of Ar^+-bombarded HOPG. The inset shows the simulated defect structure of a single-atom vacancy in the graphene cage; the carbon atoms are colored as the corresponding C 1s components. (b) CNTs subjected to an oxygen dose of 2.7×10^{17} atoms cm^2 at 20 °C. Reproduced from ref. 61, © 2009 Wiley-VCH Verlag GmnH&Co. KGaA.

functional theory (DFT) simulation predicted that when a vacancy defect is created on the sp^2 carbon layer, the binding energy of all the C atoms surrounding the vacancy (inset of Figure 5.7(a)) should decrease. Based on this prediction and the carbon geometric position, four components can be deconvoluted from the C 1s peak at the lower binding energy region shown with the same colour as that of the carbon atoms. The peak at 284.5 eV is from sp^2 carbon while the peak at 285 eV can be assigned to the sp^3 carbon.[52,62] Similar fitting strategies can also be utilized for the C 1s peak of defective CNTs as shown in Figure 5.7(b).

Through these results, it can be concluded that the carbon atoms from the vacancy defects (or point-like defects[17]) result in C 1s peaks at lower binding energies. However, sp^3 carbons located at the edge of the sp^2 carbon layer are also disordered carbons and can be considered as "defects". Therefore, differences in the attribution of the term "defect" could also cause misinterpretation and debate.

5.5 XPS Characterization of Carbon Materials with Oxygen Functional Groups

In most of its usual compounds (*e.g.* with oxygen, fluorine, nitrogen), carbon appears as a donor of electrons. According to the simple electrostatic charge transfer model to explain the E_B shifts (see Section 5.1.1), this gives rise to C 1s XPS peaks shifted towards higher binding energies relatively to the graphitic C–C one. Oxidation is the most widely used method to functionalize the carbon surface. For example, chemical oxidation,[63] photo-oxidation,[64] oxygen plasma[65] or gas phase oxidation[66] have been applied to purify CNTs (remove the amorphous carbons and metallic catalyst impurities), which at the same time, decorate the graphitic surface with oxygen-containing groups (–COOH, –OH, –CO *etc.*). Most of the oxygen-containing groups can be removed by high temperature annealing under non-oxidative atmospheres,[67] but in many cases, they are deliberately left on the sample surface.

There are, in fact, a very large number of possible oxide groups on carbons and they are often mixed together on the surface. Ideally, carbon/oxygen chemical shift data should be obtained from reference samples, but, unfortunately, such standards cannot be easily prepared. Figure 5.8 shows an example of XP spectra recorded on fresh MWCNTs and the same sample after HNO_3 and ($HNO_3 + H_2SO_4$) treatments.[68] All C 1s spectra can be deconvoluted to a main sp^2 carbon peak (~284.5 eV), a sp^3 carbon peak (~285.6 eV), a π–π*

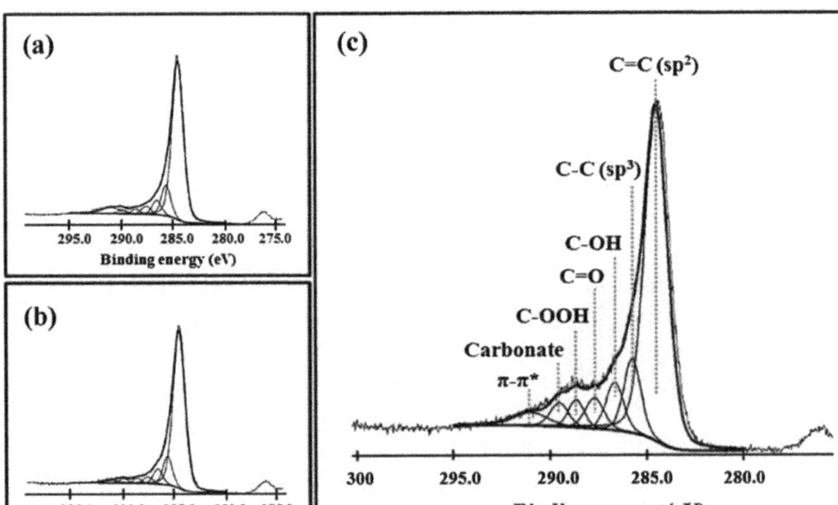

Figure 5.8 XPS C 1s spectra of MWCNTs (a) as-received, (b) HNO_3-treated, and (c) ($HNO_3 + H_2SO_4$)-treated. Reprinted from *Applied Surface Science*, **266**, H. Gong, S.-T. Kim, J. D. Lee, S. Yim, Simple quantification of surface carboxylic acids on chemically oxidized multi-walled carbon nanotubes, 219–224, Copyright (2013), with permission from Elsevier.[68]

satellite peak (~291.0 eV) and the additional oxygen functional groups.[63,68,69] Moreover, the calculated relative percentages of the oxygen-containing groups indicate that after acid treatments, the concentration of the functional groups has been increased, especially for the (HNO_3 + H_2SO_4)-treated sample.

Functionalized CNTs containing oxygen groups have shown advanced catalytic performance as metal-free catalysts in many reactions such as oxidative dehydrogenation of *n*-butane,[70] oxidative dehydrogenation of ethylbenzene,[71] and nitroarene reduction.[72] XPS was utilized to identify the functional groups and their role in the catalytic reaction. For example, Zhang *et al.*[70] found that concentrated HNO_3-treated CNTs efficiently catalysed the oxidative dehydrogenation of *n*-butane to butenes, especially butadiene. With the help of near-ambient pressure XPS, they could monitor O 1s peak evolution under the relevant temperature and reaction gas phase conditions. As shown in Figure 5.9, for both the HNO_3-oxidized CNTs sample (oCNTs) and phosphorus-passivated oCNTs sample (P-oCNTs), ketonic C=O groups (~531.2 eV)

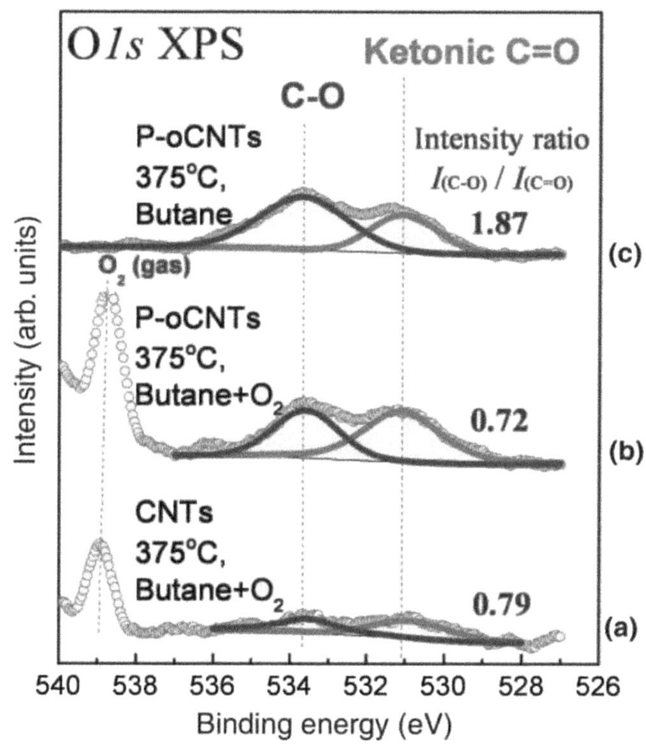

Figure 5.9 O 1s spectra taken from *in situ* XPS on working catalysts at 350 to 375 °C: (a) CNTs: butane + O_2 (1 : 1), 0.25 mbar; (b) P-oCNTs: butane + O_2 (1 : 1), 0.25 mbar; (c) P-oCNTs: butane, 0.125 mbar. From J. Zhang, X. Liu, R. Blume, A. Zhang, R. Schlögl and D. S. Su, *Science*, 2008, **322**(5898), 73–77. Reprinted with permission from AAAS.[70]

Table 5.3 C 1s and O 1s peak assignments reported in the literature.

Binding energy (eV)					
C–C (sp^2)	C–OH C–O C–O–C	C=O	C–OOH –O–C=O	CO_3	Reference
C 1s					
284.5	286.6	287.6	288.6	289.6	68
284.1	286.7		288.3	290.0	63
284.6	286.2	287.2	288.9		65
	286.1	287.6	288.4		73
284.5	286.1	287.5	288.7		74
O 1s					
	533.2	531.9		530.7	63
	533.2	531.6			74
	533.1	531.2			70
	533.5	531.2			75

and C–O groups (*i.e.* ether and hydroxyl at ~533.1 eV) are found to be stable oxygen-containing functional groups. However, under the reaction conditions (butane + O_2, 1 : 1, 0.25 atm, 350 to 375 °C), the ratio of $I_{(C-O)}/I_{(C=O)}$ values are in the range of 0.72–0.8, while after switching off O_2, this ratio drastically increases to 1.87 and is accompanied with a loss of catalytic activity. These results suggest that ketonic C=O groups are a critical ingredient of the active sites, while C–O groups may be inactive intermediates or spectators in the reaction.

The binding energies of the major carbon–oxygen components in O 1s spectra can be distinguished into four main groups (in increasing binding energy): (a) carbonate groups CO_3; around 530.5 ± 0.3 eV, (b) C=O in an acid or ester; around 531.5 ± 0.4 eV, (c) simple C–O, C–OH, C–O–C bonds; at 533.0 ± 0.5 eV, (d) C–OH in an acid or ester; at 533.4 ± 0.3 eV and (e) chemisorbed water species on carbon; at 534.0 ± 0.4 eV. Table 5.3 contains a suggested, but not exhaustive, list of the C 1s and O 1s peak binding energies of various carbon and oxygen functional groups, typically reported in the literature for oxygen species on the surface of carbons.

5.6 XPS Characterization of Carbon Materials with Heteroatom Dopants

Besides functionalization with oxygen-containing groups, replacement of lattice carbon by atoms of other elements (heteroatom doping) has attracted great interest in recent years since the heteroatoms can render desirable physical and chemical properties to the carbon materials for various catalytic processes.[76,77] The amount of heteroatoms plays an important role in modifying the properties of the carbon materials, such as the electrical conductivity, acidity, oxidation stability, as well as the catalytic activity. XPS has a great advantage compared to bulk techniques, since it can show the surface concentration and chemical nature of heteroatoms at the position where the catalytic reaction occurs. Consequently, XPS measurements are indispensable to reveal the local structure of the heteroatoms in carbon materials and correlate the catalyst chemistry with performance. Numerous atoms have been used to functionalize carbon materials, with the most prominent being nitrogen (N), boron (B), phosphorus (P) and sulfur (S). Limited studies on other heteroatoms such as Si, Al and Ni also exist, but they will not be discussed here.

5.6.1 XPS Characterization of Carbon Materials Doped with Nitrogen

Nitrogen-doped carbon materials have been intensively studied in the last few years as metal-free catalysts,[78,79] notably to replace Pt-based catalysts in the electrochemical oxygen reduction reaction (ORR).[80–84] Three types of nitrogen are usually reported in nitrogen-doped carbon materials: pyridinic (N bonded to two C atoms), pyrrolic (N in a pentagonal ring) and graphitic nitrogen (N bonded to three C atoms, also called substituted N or quaternary N). As shown in a schematic representation of the different types of nitrogen atoms on doped graphene (Figure 5.10(a)), usually all these N types coexist on N-doped carbon materials.[76,80,85] Therefore, in order to estimate the relative contribution of each N species on the surface, the N 1s spectrum should be fitted with at least three separate nitrogen peaks. The N 1s peak line shape is a convolution of a Gaussian (G) and Lorentzian (L) peak profile and is always a symmetric peak. The positions of these peaks may vary over a wide range due to the different environments of nitrogen atoms and, therefore, sometimes E_B values reported in the literature are controversial.[86]

In the example presented in Figure 5.10, C 1s spectra collected from graphene produced by reduction of graphene oxide, GO, (Figure 5.10(b)) and N-doped graphene (Figure 5.10(d)) are compared.[87] As is evident, the C 1s peak profile looks very similar in the two cases; however, according to the fitting procedure proposed by the authors, in the first case, several oxygen functional groups co-exist, while in the second, the oxygen-containing groups are absent and new nitrogen-containing groups appear. Unfortunately, the fitting procedure is far from straightforward and if it is not supported by additional experimental evidences, the assignment of the measured binding energies into specific atomic configurations might be contested.

The association of nitrogen species with reaction performance is essential for developing highly active catalysts; however, the investigation is complicated by the mixing of different types of nitrogen species in the carbon materials, which is inevitable during the nitrogen doping process.[80] Recently, Guo *et al.*[80] developed nitrogen-doped HOPG as model catalyst systems to

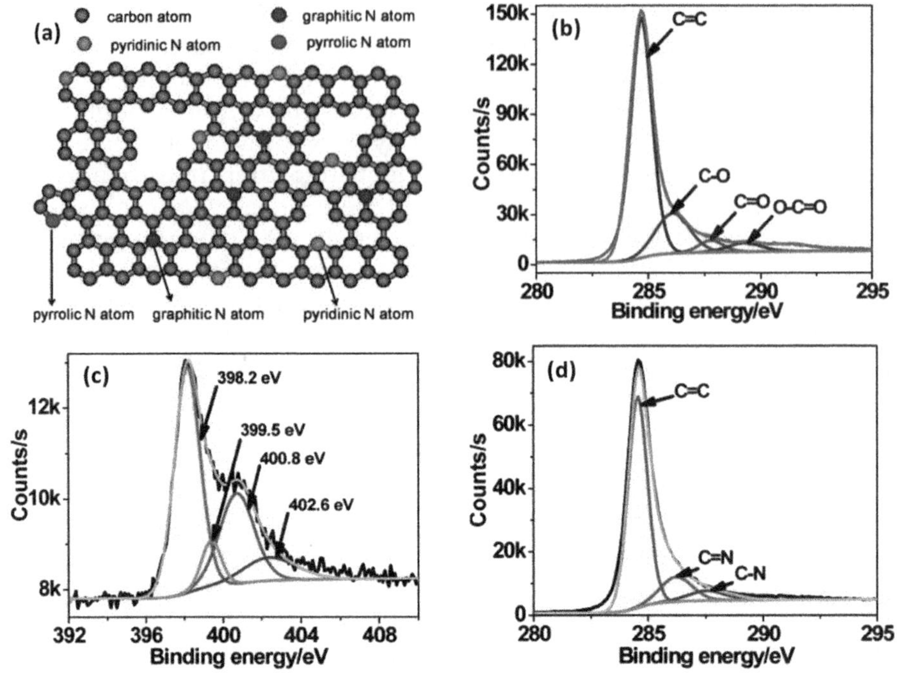

Figure 5.10 (a) Schematic structure of nitrogen-doped graphene. (b) High-resolution C 1s spectra of graphene prepared by 800 °C annealing of GO in Ar. C 1s (c) and N 1s (d) spectra of N-graphene prepared by annealing GO/melamine (1:5 mass ratio, 800 °C, 30 min). Reprinted with permission from H. Wang, T. Maiyalagan and X. Wang, *ACS Catal.*, 2012, **2**, 781–794. Copyright (2012) American Chemical Society.[82]

clarify the type of active sites in the oxygen reduction reaction (ORR). This study can serve as an example of how XPS results can be utilized in order to provide insights into the chemistry of electrocatalytic reactions involving N-doped carbon materials. In this work, the pyridinic N on HOPG (pyri-HOPG) was obtained through annealing of edge-patterned HOPG in NH_3, while the graphitic N on HOPG (grap-HOPG) was prepared by nitrogen ion beam sputtering. The highly selective methods for the preparation of specific types of nitrogen were proved by the N 1s XPS spectra as shown in Figure 5.11(A). In addition to the chemical identification, quantification of the XPS spectra was used in order to estimate the concentration of pyridinic N (95%) and graphitic N (82%) nitrogen species. In Figure 5.11(B), the electrocatalytic ORR results showed that the current density increases with the concentration of pyridinic N. A linear relationship was found at the investigated potentials (0.2 V, 0.3 V and 0.4 V),[80] indicating that pyridinic N creates the active sites.

Post-reaction analysis of the sample after ORR tests using XPS illustrated that during the reaction the N 1s component at 400.1 eV (either pyrrolic N or pyridonic N) increased from 11 to 29% (Figure 5.12(A)). This was attributed to the reaction of carbon atoms (next to pyridinic N) with OH species, which

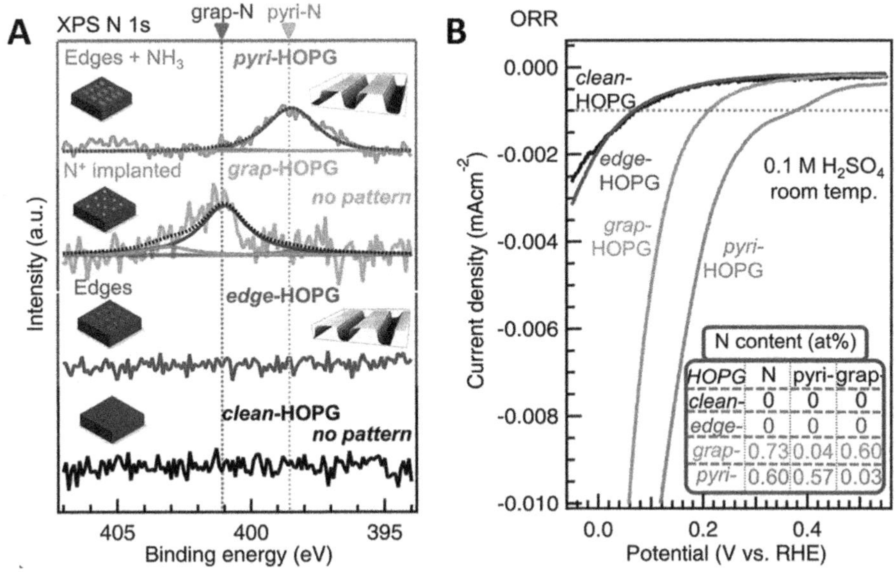

Figure 5.11 (A) N 1s XPS spectra of N-HOPG model catalysts and reference samples. (B) ORR results and N content for model catalysts corresponding to (A). From D. Guo, R. Shibuya, C. Akiba, S. Saji, T. Kondo and J. Nakamura, *Science*, 2016, **351**(6271), 361–365. Reprinted with permission from AAAS.[80]

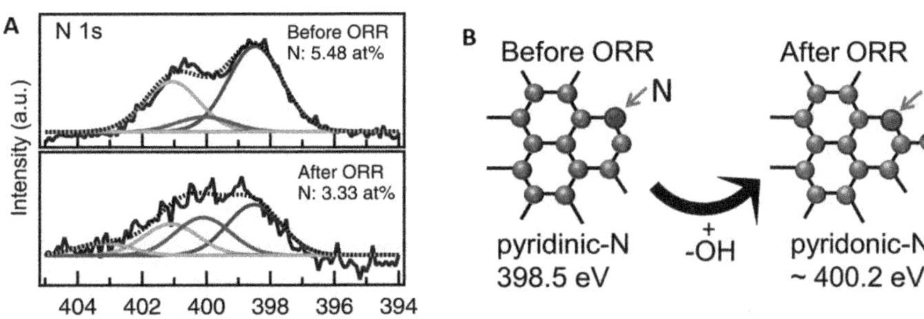

Figure 5.12 (A) N 1s XPS spectra of the N-HOPG model catalyst before and after ORR, respectively. (B) Schematic illustration of the transformation of pyridinic N to pyridonic N due to the attachment of OH to the neighboring C atom. From D. Guo, R. Shibuya, C. Akiba, S. Saji, T. Kondo and J. Nakamura, *Science*, 2016, **351**(6271), 361–365. Reprinted with permission from AAAS.[80]

converts the pyridinic N to pyridonic N as shown in Figure 5.12(B). This result suggested that the carbon atoms next to the pyridinic N are the active centres instead of the pyridinic N. With the additional support of CO_2 thermal programmed desorption (TPD) results, it was concluded that the ORR active sites are carbon atoms with Lewis basicity next to the pyridinic N. This

Table 5.4 C 1s and N 1s peak assignments according to the literature.

	Species	Binding energies (eV)	Reference
C 1s	C=C	284.6–284.8	85,87,88
	C=N	285.8–285.9	85 and 87
	C–N	285.8–287.5	85,88,89
N 1s	Pyridinic N	398.1–399.8	78–80,82,84–88
	Pyrrolic N	399.8–401.2	78–80,82,84–88
	Graphitic N	401–402.7	78–80,82,85–88
	Nitrogen oxide	402.6–405.6	78,79,82,85,88

species affects the ORR reaction by increasing the O_2 adsorption at the initial step of the reaction.[80]

Usually, for the fitting of the N 1s spectrum, the E_B windows used are: (a) pyridinic N (398.1–399.3 eV), (b) pyrrolic N (399.5–401.2 eV), and (c) graphitic (quaternary) N (401.1–402.7 eV). Apart from these three types of nitrogen, N 1s peaks corresponding to –NO and –NH_2 groups have also been observed at ~402.8 and ~399.5 eV, respectively, in several studies.[78,85,88] In addition, doping of nitrogen can also impose shifts of the C 1s spectrum. The C 1s shift is always reported at a higher E_B and its magnitude may vary from that of graphitic carbon up to 3 eV. However, since higher binding energy contributions could also be induced from other carbon bonds, *e.g.*, carbon–oxygen bonds, one should be very cautious before drawing conclusions based on literature E_B values. Table 5.4 summarizes the C 1s and N 1s binding energies of N-functionalized carbon materials typically reported in the literature.

5.6.2 XPS Characterization of Carbon Materials with Other Heteroatoms

Apart from nitrogen doping, other non-metal elements, such as boron,[90,91] sulfur,[92,93] phosphorus,[94,95] fluorine[96,97] and other halogen elements,[98] have also been doped or co-doped into carbon materials and tested as metal-free catalysts. For boron-doped carbons, several B species have been determined by XPS, which include species incorporated into the carbon structure forming boron–carbon (B–C) bonds and species originating from the surface oxidation of boron atoms.[99] Formation of B–C bonds will shift the C 1s spectra to lower binding energies and increase the FWHM of the B 1s peak.[90,99] The B 1s spectra can be separated into two regions: 186–190 eV and 190–194 eV, which mainly correspond to the characteristic peaks of B–B, B–C and O–B–N species, respectively.[91,99]

Yang *et al.*[91] developed boron-doped CNT samples (B-CNTs) through a CVD method and tested them as metal-free ORR catalysts with quite promising results. In more recent work,[100] the same group prepared co-doped B and N CNTs following two synthetic routes: (i) one-step doping of CNTs by B and

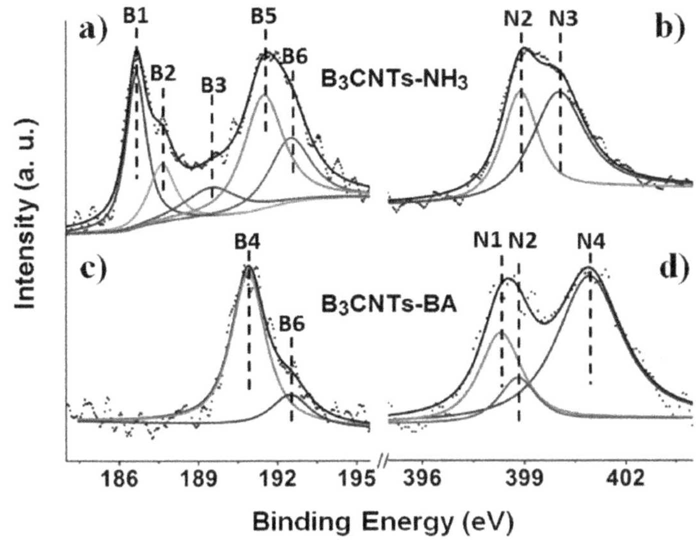

Figure 5.13 XPS spectra of (a),(c) B 1s and (b),(d) N 1s core levels recorded on CNTs co-doped with B and N using one-step (a),(b) and two-step (c),(d) preparation methods. Reprinted with permission from Y. Zhao, L. Yang, S. Chen, X. Wang, Y. Ma, Q. Wu *et al.*, *J. Am. Chem. Soc.*, 2013, **135**(4), 1201–1204. Copyright (2013) American Chemical Society.[100]

N during the CVD procedure (B_3CNTs-BA) and (ii) a two-step doping procedure by annealing B-CNTs in NH_3 at 400 °C (B_3CNTs-NH_3). Typical B 1s and N 1s XPS spectra in the two cases are shown in Figure 5.13. The assignment of the peaks in Figure 5.13 according to their E_B suggests that the B1, B2, B3, B5 and B6 species correspond to a B cluster, B_4C, BC_3, BC_2O and BCO_2 species, respectively. An extra B4 component was linked to N–B–O due to the co-doping of N with B, which can also be observed in the nitrogen spectrum noted as N1 in Figure 5.13(d). The other N components are the same as mono-doped N corresponding to pyridinic N, pyrrolic N, and graphitic N.[100] From the XPS data analysis, it became clear that the boron and nitrogen species formed on CNTs are strongly dependent on the preparation methods.

The potential application of S-doped carbon materials in the ORR has attracted much attention in recent years.[92,101–104] Sulfur's atomic radius is significantly larger than that of carbon, which is expected to cause significant perturbations in the material lattice and may hinder the doping procedure in the case of the S substitution of C atoms. For all doping elements discussed so far (C, O, N and B), the analyzed 1s photoemission peak was a single-component peak. The S 2p spectrum is an exception, since it has an intrinsic spin–orbit splitting, which produces a doublet that should be considered when the XPS spectra are analyzed. Depending on the chemical state of sulfur and/or the instrument resolution, the S 2p peak can be fitted either by two split peaks (with ~1.16 eV separation and 0.511 intensity ratio)[104] or by

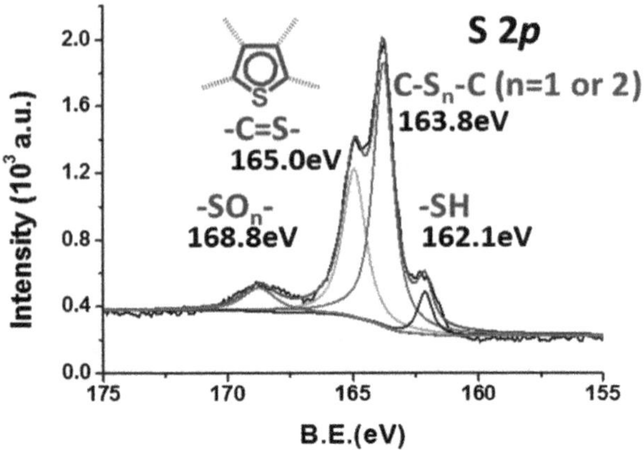

Figure 5.14 S 2p XPS spectra from nitrogen/sulfur co-doped three-dimensional graphene frameworks. Reprinted from *Carbon*, **62**, Y. Su, Y. Zhang, X. Zhuang, S. Li, D. Wu, F. Zhang, X. Feng, Low temperature synthesis of nitrogen/sulfur co-doped three-dimensional graphene frameworks as efficient metal-free electrocatalyst for oxygen reduction reaction, 296–301, Copyright (2013), with permission from Elsevier.[105]

a single asymmetric peak in the case of low-resolution instruments.[92] Moreover, the deconvolution of the sulfur peak may be complicated due to the presence of unavoidable oxygen "impurities" (CNTs, GO *etc.*) and residuals caused by the utilization of organic sulfur (such as benzyl disulfide) as the doping source.

An example of S 2p XPS spectra from nitrogen/sulfur co-doped three-dimensional graphene frameworks is shown in Figure 5.14.[105] The S 2p spectrum has been fitted with four asymmetric S 2p components. The high intensity peak at 163.8 eV is attributed to C–S$_n$–C (n = 1 or 2) bonds while the second one at 165 eV is assigned to –C=S– bonds. Additionally, oxidized (–SO$_n$–) and reduced (–SH) sulfur moieties can be found at higher (168.8 eV) and lower binding energies (162.1 eV), respectively. Some other works have also found well-separated –C–S(O)$_x$–C– (x = 2, 3 and 4) as stable sulfur species in doped carbon materials.[92,103]

Phosphorus is an element in the same group as nitrogen in the periodic table, therefore, has the same number of valence electrons,[5] which is expected to lead to similar chemical properties.[95] However, P has a larger atomic radius and lower electronegativity than C, which can efficiently modify the structure and the electronic properties of carbon materials and is expected to enhance the catalytic activity.[95,106] The P 2p peak, similar to S 2p, has closely spaced spin–orbit components (Δ = 0.87 eV), therefore, it is often presented as a single asymmetric peak. Especially, in the case of P-doped carbon materials where the P amount is very low (0–2 at%), the P 2p signal to noise ratio is also low and, therefore, it is almost impossible to resolve the spin–orbit splitting of the P 2p peak.

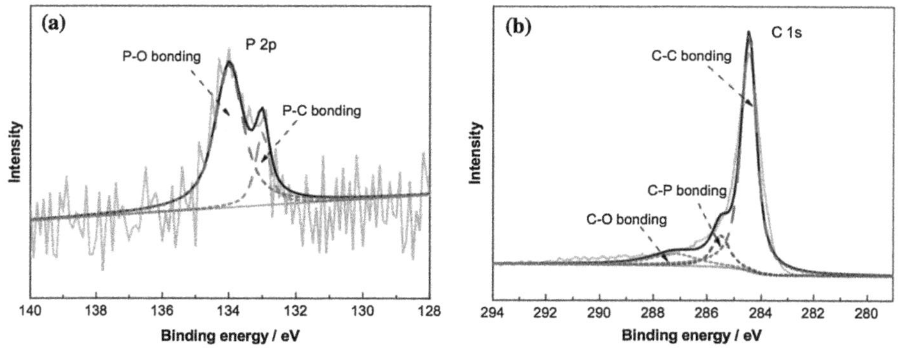

Figure 5.15 XPS spectra of (a) P 2p and (b) C 1s of phosphorus-doped carbon hollow spheres. Reprinted from *Carbon*, **82**, J. Wu, C. Jin, Z. Yang, J. Tian and R. Yang, Synthesis of phosphorus-doped carbon hollow spheres as efficient metal-free electrocatalysts for oxygen reduction, 562–571, Copyright (2015), with permission from Elsevier.[107]

Two P 2p peaks are usually observed in the XPS spectra of P-doped carbon materials, with the one at a higher binding energy assigned as P–O bonds and the other at a lower binding energy as P–C bonds, as shown in Figure 5.15(a).[95,107,108] Another feature of P doping on sp^2 carbon materials is the broadening of the C 1s peak due to the introduction of defects and sp^3 carbons.[109] Despite the lower electronegativity of phosphorus, a lower positive charge on carbon is barely observed. This is due to the fact that the main configuration of phosphorus in the carbon network is P–O–C and, thus, a more positively charged carbon is expected, as shown in Figure 5.15(b).[110]

Halogen elements, including fluorine,[96,97] chlorine,[98,111] bromine[98,111] and iodine,[98,112] have been doped onto carbon materials and tested as metal-free catalysts. Pure halogen-treated carbon black was studied in detail by XPS in the late 90s by Papirer *et al.*[113–115] Due to the complexity of the reactions between carbon black and halogen elements, the resulting materials were expected to have a polyphasic nature. In the photoelectron spectra of the carbon and halogen elements, both a semi-ionic bond and a covalent bond were present, while the additional oxygen-containing groups increased the number of XPS peaks. However, when halogen-doped carbons are used as catalysts, only a small amount of halogen elements is introduced to activate the carbon materials, to avoid severe carbon damage. Thus, identification of those elements by XPS becomes easier, and usually only one C–X (X = F, Cl, Br) bond is considered as proof of the doping of those elements. The same peak is also used to calculate the content of the element on the carbon material.[96,97,116] In some cases, additional peaks, such as partially ionized bonds ($-C-Br^+-C$ and $-C-I^+-C$) or I with different valence states, have been reported.[98,112] As a summary, Table 5.5 shows the binding energies of the above-mentioned heteroatoms in carbon materials typically reported in the literature.

Table 5.5 XPS peaks of heteroatom (B, P, S, F, Cl, Br and I) assignment according to the literature.

	Species	Binding energies (eV)	Reference
B 1s	B–B	~186.5	91,100,117
	B–C	187–190	91,99,100,117
	O–B–C	190–193	91,100,117
	B–O	~193.0	91,100,117
P 2p	P–C	130.6–132.7	95,107,118,119
	P–O	132.3–134.2	95,107,118,119
S 2p	S–H	~162.1	105
	C–S_n–C	~163.8	92,101,103,105,120
	–C=S–	~165.0	105 and 120
	S–O, C–SO_n–C	166–172	92,103,105,120
F 1s	Ionic C–F	~685.9	121
	Semi-ionic C–F	685–687.5	97 and 121
	Covalent C–F	686.1–690.0	121 and 122
Cl 2p	C–Cl	199.8–202.1	98,111,122
Br 3d	C–Br ($3d_{5/2}$)	~69.8	98 and 111
	C–Br^+–C ($3d_{5/2}$)	~69.3	98 and 111
I 3d	C–I ($3d_{5/2}$)	~619.3	98 and 123
	C–I^+–C ($3d_{5/2}$)	~619.9	98 and 123

5.7 Conclusions

Carbon-based materials have attracted great interest in recent years as metal-free catalysts due to their multiple advantages, such as low price, high surface area, high chemical stability, non-toxicity and so on. Investigation of the surface properties of functionalized carbon materials is essential and important, since the catalytic reactions take place on the catalyst's surface. XPS, being a surface-sensitive technique, is able to provide detailed information on the chemical state and composition of carbon materials. In this chapter, we described how the XPS spectra obtained from various forms of carbon materials can be interpreted. Analysis strategies and the essential elements of the XPS spectra, such as peak shape, binding energies, *etc.*, are provided. A brief overview of previous work on metal-free functionalized carbon materials showed that XPS is a powerful tool to investigate their surface properties, which helps us to understand the surface-active centres to further develop highly active catalysts.

References

1. H. Ba, Y. Liu, L. Truong-Phuoc, C. Duong-Viet, X. Mu and W. H. Doh, *et al.*, A highly N-doped carbon phase "dressing" of macroscopic supports for catalytic applications, *Chem. Commun.*, 2015, **51**(76), 14393–14396.
2. M. Zhou, H.-L. Wang and S. Guo, Towards high-efficiency nanoelectrocatalysts for oxygen reduction through engineering advanced carbon nanomaterials, *Chem. Soc. Rev.*, 2016, **45**(5), 1273–1307.

3. H. P. Boehm, Some aspects of the surface chemistry of carbon blacks and other carbons, *Carbon*, 1994, **32**(5), 759–769.
4. K. Siegbahn and K. Edvarson, β-ray spectroscopy in the precision range of 1 : 105, *Nucl. Phys.*, 1956, **1**(3), 137–159.
5. J. Qi, X. Hu, H. Zhu and M. Zheng, First-principles studies on the structural and spectral properties of C72 isomers and the chlorinated derivative C72Cl4, *Phys. Chem. Chem. Phys.*, 2016, **18**(11), 8049–8058.
6. A. Takashima, Y. Izumi, E. Ikenaga, T. Ohkochi, M. Kotsugi and T. Matsushita, *et al.*, Low-temperature catalyst activator: mechanism of dense carbon nanotube forest growth studied using synchrotron radiation, *IUCrJ*, 2014, **1**, 221–227.
7. A. Knop-Gericke, E. Kleimenov, M. Hävecker, R. Blume, D. Teschner and S. Zafeiratos, *et al.*, Chapter 4 X-ray photoelectron spectroscopy for investigation of heterogeneous catalytic processes, in *Advances in Catalysis*, 2009, pp. 213–272.
8. M. Smith, L. Scudiero, J. Espinal, J. S. McEwen and M. Garcia-Perez, Improving the deconvolution and interpretation of XPS spectra from chars by *ab initio* calculations, *Carbon*, 2016, **110**, 155–171.
9. S. Takabayashi, K. Okamoto, K. Motomitsu, A. Terayama, T. Nakatani and H. Sakaue, *et al.*, New method to calibrate binding energy using Au nanocolloids in X-ray photoelectron analysis of diamondlike carbon films with different electrical resistivities, *Appl. Surf. Sci.*, 2008, **254**(9), 2666–2670.
10. D. Briggs and M. P. Seah, *Practical Surface Analysis in Auger and X-ray Photoelectron Spectroscopy*, John Wiley & Sons, New York, 1983, vol. 1, pp. 397–427.
11. J. J. Yeh and I. Lindau, Atomic subshell photoionization cross sections and asymmetry parameters: 1 < *Z* < 103, *At. Data Nucl. Data Tables*, 1985, 1–155.
12. S. Tanuma, C. J. Powell and D. R. Penn, Calculations of electron inelastic mean free paths. V. Data for 14 organic compounds over the 50–2000 eV range, *Surf. Interface Anal.*, 1994, **21**(3), 165–176.
13. S. V. Merzlikin, N. N. Tolkachev, T. Strunskus, G. Witte, T. Glogowski and C. Woell, *et al.*, Resolving the depth coordinate in photoelectron spectroscopy - comparison of excitation energy variation *vs.* angular-re solved XPS for the analysis of a self-assembled monolayer model system, *Surf. Sci.*, 2008, **602**(3), 755–767.
14. C. J. Powell and A. Jablonski, Surface sensitivity of X-ray photoelectron spectroscopy, *Nucl. Instrum. Methods Phys. Res., Sect. A*, 2009, **601**(1–2), 54–65.
15. D. Briggs and J. Grant, *Surface Analysis by Auger and X-Ray Photoelectron Spectroscopy*, Surface Sp. UK: SurfaceSpectra, 2003, p. 840.
16. W. H. Doh, V. Papaefthimiou, T. Dintzer, V. Dupuis and S. Zafeiratos, Synchrotron radiation X-ray photoelectron spectroscopy as a tool to resolve the dimensions of spherical core/shell nanoparticles, *J. Phys. Chem. C*, 2014, **118**(46), 26621–26628.

17. R. Blume, D. Rosenthal, J. P. Tessonnier, H. Li, A. Knop-Gericke and R. Schlogl, Characterizing graphitic carbon with X-ray photoelectron spectroscopy: a step-by-step approach, *ChemCatChem*, 2015, **7**(18), 2871–2881.
18. J.-O. Müller, D. S. Su, U. Wild and R. Schlögl, Bulk and surface structural investigations of diesel engine soot and carbon black, *Phys. Chem. Chem. Phys.*, 2007, **9**(30), 4018–4025.
19. T. Susi, T. Pichler and P. Ayala, X-ray photoelectron spectroscopy of graphitic carbon nanomaterials doped with heteroatoms, *Beilstein J. Nanotechnol.*, 2015, **6**, 177–192.
20. K. C. Prince, I. Ulrych, M. Peloi, B. Ressel, V. Chab and C. Crotti, *et al.*, Core-level photoemission from graphite, *Phys. Rev. B: Condens. Matter Mater. Phys.*, 2000, **62**(11), 6866–6868.
21. K. V. Emtsev, F. Speck, T. Seyller, L. Ley and J. D. Riley, Interaction, growth, and ordering of epitaxial graphene on SiC{0001} surfaces: a comparative photoelectron spectroscopy study, *Phys. Rev. B: Condens. Matter Mater. Phys.*, 2008, **77**(15), 155303.
22. J. Avila, I. Razado, S. Lorcy, R. Fleurier, E. Pichonat and D. Vignaud, *et al.*, Exploring electronic structure of one-atom thick polycrystalline graphene films: a nano angle resolved photoemission study, *Sci. Rep.*, 2013, **3**(1), 2439.
23. A. B. Preobrajenski, M. L. Ng, A. S. Vinogradov and N. Mårtensson, Controlling graphene corrugation on lattice-mismatched substrates, *Phys. Rev. B: Condens. Matter Mater. Phys.*, 2008, **78**(7), 73401.
24. H. Hibino, H. Kageshima, M. Kotsugi, F. Maeda, F.-Z. Guo and Y. Watanabe, Dependence of electronic properties of epitaxial few-layer graphene on the number of layers investigated by photoelectron emission microscopy, *Phys. Rev. B: Condens. Matter Mater. Phys.*, 2009, **79**(12), 125437.
25. P. Merino, M. Švec, J. I. Martinez, P. Jelinek, P. Lacovig and M. Dalmiglio, *et al.*, Graphene etching on SiC grains as a path to interstellar polycyclic aromatic hydrocarbons formation, *Nat. Commun.*, 2014, **5**, 1–9.
26. M. Scardamaglia, B. Aleman, M. Amati, C. Ewels, P. Pochet and N. Reckinger, *et al.*, Nitrogen implantation of suspended graphene flakes: annealing effects and selectivity of sp^2 nitrogen species, *Carbon*, 2014, **73**, 371–381.
27. I. Janowska, K. Chizari, O. Ersen, S. Zafeiratos, D. Soubane and V. Da Costa, *et al.*, Microwave synthesis of large few-layer graphene sheets in aqueous solution of ammonia, *Nano Res.*, 2010, **3**(2), 126–137.
28. J. Schiessling, L. Kjeldgaard, F. Rohmund, L. K. L. Falk, E. E. B. Campbell and J. Nordgren, *et al.*, Synchrotron radiation study of the electronic structure of multiwalled carbon nanotubes, *J. Phys.: Condens. Matter*, 2003, **15**(38), 6563–6579.
29. P. Ayala, Y. Miyata, K. De Blauwe, H. Shiozawa, Y. Feng and K. Yanagi, *et al.*, Disentanglement of the electronic properties of metallicity-selected single-walled carbon nanotubes, *Phys. Rev. B: Condens. Matter Mater. Phys.*, 2009, **80**(20), 205427.

30. J. A. Leiro, M. H. Heinonen, T. Laiho and I. G. Batirev, Core-level XPS spectra of fullerene, highly oriented pyrolitic graphite, and glassy carbon, *J. Electron Spectrosc. Relat. Phenom.*, 2003, **128**(2), 205–213.
31. H. A. Girard, T. Petit, S. Perruchas, T. Gacoin, C. Gesset and J. C. Arnault, *et al.*, Surface properties of hydrogenated nanodiamonds: a chemical investigation, *Phys. Chem. Chem. Phys.*, 2011, **13**(24), 11517–11523.
32. H. Ba, S. Podila, Y. Liu, X. Mu, J.-M. Nhut and V. Papaefthimiou, *et al.*, Nanodiamond decorated few-layer graphene composite as an efficient metal-free dehydrogenation catalyst for styrene production, *Catal. Today*, 2015, **249**, 167–175.
33. A. Fujimoto, Y. Yamada, M. Koinuma and S. Sato, Origins of sp^3C peaks in C_{1s} X-ray photoelectron spectra of carbon materials, *Anal. Chem.*, 2016, **88**(12), 6110–6114.
34. K. A. Wepasnick, B. A. Smith, J. L. Bitter and D. Howard Fairbrother, Chemical and structural characterization of carbon nanotube surfaces, *Anal. Bioanal. Chem.*, 2010, **396**, 1003–1014.
35. W. H. Lee, S. J. Kim, W. J. Lee, J. G. Lee, R. C. Haddon and P. J. Reucroft, X-ray photoelectron spectroscopic studies of surface modified single-walled carbon nanotube material, *Appl. Surf. Sci.*, 2001, **181**(1–2), 121–127.
36. R. Blume, P. R. Kidambi, B. C. Bayer, R. S. Weatherup, Z.-J. Wang and G. Weinberg, *et al.*, The influence of intercalated oxygen on the properties of graphene on polycrystalline Cu under various environmental conditions, *Phys. Chem. Chem. Phys.*, 2014, **16**(47), 25989–26003.
37. G. Speranza and N. Laidani, Measurement of the relative abundance of sp^2 and sp^3 hybridised atoms in carbon-based materials by XPS: a critical approach. Part I, *Diamond Relat. Mater.*, 2004, **13**(3), 445–450.
38. S. Hajati and S. Tougaard, XPS for non-destructive depth profiling and 3D imaging of surface nanostructures, *Anal. Bioanal. Chem.*, 2010, **396**, 2741–2755.
39. A. Proctor and P. M. A. Sherwood, X-ray photoelectron spectroscopic studies of carbon fibre surfaces. I. carbon fibre spectra and the effects of heat treatment, *J. Electron Spectrosc. Relat. Phenom.*, 1982, **27**(1), 39–56.
40. D. M. Poirier, J. H. Weaver, K. Kikuchi and Y. Achiba, Electronic structure of C84 and KxC84: Comparison to C60 and graphite, *Z. Phys. D: At., Mol. Clusters*, 1993, **26**(1), 79–83.
41. S. Pylypenko, K. Artyushkova and J. E. Fulghum, Application of XPS spectral subtraction and multivariate analysis for the characterization of Ar+ ion beam modified polyimide surfaces, *Appl. Surf. Sci.*, 2010, **256**(10), 3204–3210.
42. H. Estrade-Szwarckopf, XPS photoemission in carbonaceous materials: a "defect" peak beside the graphitic asymmetric peak, *Carbon*, 2004, **42**(8–9), 1713–1721.
43. I. Retzko and W. E. S. Unger, Analysis of carbon materials by X-ray photoelectron spectroscopy and X-ray absorption spectroscopy, *Adv. Eng. Mater.*, 2003, **5**(7), 519–522.

44. S. Lizzit, L. Petaccia, A. Goldoni, R. Larciprete, P. Hofmann and G. C. Zampieri, 1s photoemission spectrum in graphite(0001), *Phys. Rev. B: Condens. Matter Mater. Phys.*, 2007, **76**(15), 153408.
45. A. Mezzi and S. Kaciulis, Surface investigation of carbon films: from diamond to graphite, *Surf. Interface Anal.*, 2010, **42**(6–7), 1082–1084.
46. M. Soria-Sánchez, A. Maroto-Valiente, J. Álvarez-Rodríguez, V. Muñoz-Andrés, I. Rodríguez-Ramos and A. Guerrero-Ruíz, Carbon nanostrutured materials as direct catalysts for phenol oxidation in aqueous phase, *Appl. Catal., B*, 2011, **104**(1–2), 101–109.
47. D. Pantea, H. Darmstadt, S. Kaliaguine, L. Sümmchen and C. Roy, Electrical conductivity of thermal carbon blacks: Influence of surface chemistry, *Carbon*, 2001, **39**(8), 1147–1158.
48. T. I. T. Okpalugo, P. Papakonstantinou, H. Murphy, J. McLaughlin and N. M. D. Brown, High resolution XPS characterization of chemical functionalised MWCNTs and SWCNTs, *Carbon*, 2005, **43**(1), 153–161.
49. S. Doniach and M. Sunjic, Many-electron singularity in X-ray photoemission and X-ray line spectra from metals, *J. Phys. C: Solid State Phys.*, 1970, **3**(2), 285–291.
50. J. Castle and A. Salvi, The characterization and use of intrinsic background in Xps, *Anal. Sci.*, 2001, **17**, i147–i150.
51. T. T. P. Cheung, X-ray photoemission of carbon: lineshape analysis and application to studies of coals, *J. Appl. Phys.*, 1982, **53**(10), 6857–6862.
52. J. Díaz, G. Paolicelli, S. Ferrer and F. Comin, Separation of the sp^3 and sp^2 components in the C1s photoemission spectra of amorphous carbon films, *Phys. Rev. B: Condens. Matter Mater. Phys.*, 1996, **54**(11), 8064–8069.
53. A. Barinov, L. Gregoratti, P. Dudin, S. La Rosa and M. Kiskinova, Imaging and spectroscopy of multiwalled carbon nanotubes during oxidation: defects and oxygen bonding, *Adv. Mater.*, 2009, **21**(19), 1916–1920.
54. D. Q. Yang and E. Sacher, s-p hybridization in highly oriented pyrolytic graphite and its change on surface modification, as studied by X-ray photoelectron and Raman spectroscopies, *Surf. Sci.*, 2002, **504**, 125–137.
55. A. Barinov, H. Ustunel, S. Fabris, L. Gregoratti, L. Aballe and P. Dudin, *et al.*, Defect-controlled transport properties of metallic atoms along carbon nanotube surfaces, *Phys. Rev. Lett.*, 2007, **99**(4), 46803.
56. S. Dieckhoff, D. Ochs, J. Günster and V. Kempter, Metastable impact electron spectroscopy (MIES) study of chemical vapour deposited (CVD) diamond films, *Surf. Sci.*, 1999, **423**(1), 53–60.
57. W. M. Lau, L. J. Huang, I. Bello, S.-T. Yiu, Y. M. Lee, I. Belle and Y. M. Yiu, *et al.*, Modification of surface band bending of diamond by low energy argon and carbon ion bombardment, *J. Appl. Phys.*, 1994, **75**, 3385.
58. P. Reinke, G. Francz, P. Oelhafen and J. Ullmann, Structural changes in diamond and amorphous carbon induced by low-energy ion irradiation, *Phys. Rev. B: Condens. Matter Mater. Phys.*, 1996, **54**(10), 7067–7073.
59. P. K. Chu and L. Li, Characterization of amorphous and nanocrystalline carbon films, *Mater. Chem. Phys.*, 2006, **96**(2–3), 253–277.

60. P. M. T. M. Van Attekum and G. K. Wertheim, Excitonic effects in core-hole screening, *Phys. Rev. Lett.*, 1979, **43**(25), 1896–1898.
61. A. Barinov, L. Gregoratti, P. Dudin, R. S. La and M. Kiskinova, Imaging and spectroscopy of multiwalled carbon nanotubes during oxidation: defects and oxygen bonding, *Adv. Mater.*, 2009, **21**(19), 1916–1920.
62. J. Hahn, H. Kang, S. Song and I. Jeon, Observation of charge enhancement induced by graphite atomic vacancy: a comparative STM and AFM study, *Phys. Rev. B: Condens. Matter Mater. Phys.*, 1996, **53**(4), R1725–R1728.
63. V. Datsyuk, M. Kalyva, K. Papagelis, J. Parthenios, D. Tasis and A. Siokou, *et al.*, Chemical oxidation of multiwalled carbon nanotubes, *Carbon*, 2008, **46**(6), 833–840.
64. M. Grujicic, G. Cao, A. M. Rao, T. M. Tritt and S. Nayak, UV-light enhanced oxidation of carbon nanotubes, *Appl. Surf. Sci.*, 2003, **214**(1–4), 289–303.
65. A. Felten, C. Bittencourt and J. J. Pireaux, Gold clusters on oxygen plasma functionalized carbon nanotubes: XPS and TEM studies, *Nanotechnology*, 2006, **17**(8), 1954–1959.
66. S. C. S. Tsang, P. J. F. Harris and M. L. H. Green, Thinning and opening of carbon nanotubes by oxidation using carbon dioxide, *Nature*, 1993, **362**(6420), 520–522.
67. W. Baaziz, X. Liu, I. Florea, S. Begin-Colin, B. P. Pichon and C. Ulhaq, *et al.*, Carbon nanotube channels selectively filled with monodispersed $Fe_{3-x}O_4$ nanoparticles, *J. Mater. Chem. A*, 2013, **1**(44), 13853–13861.
68. H. Gong, S. T. Kim, J. D. Lee and S. Yim, Simple quantification of surface carboxylic acids on chemically oxidized multi-walled carbon nanotubes, *Appl. Surf. Sci.*, 2013, **266**, 219–224.
69. C. Chen, B. Liang, A. Ogino, X. Wang and M. Nagatsu, Oxygen functionalization of multiwall carbon nanotubes by microwave-excited surface-wave plasma treatment, *J. Phys. Chem. C*, 2009, **113**(18), 7659–7665.
70. J. Zhang, X. Liu, R. Blume, A. Zhang, R. Schlögl and D. S. Su, Surface-modified carbon nanotubes catalyze oxidative dehydrogenation of n-butane, *Science*, 2008, **322**(5898), 73–77.
71. M. F. R. Pereira, J. L. Figueiredo, J. J. M. Órfao, P. Serp, P. Kalck and Y. Kihn, Catalytic activity of carbon nanotubes in the oxidative dehydrogenation of ethylbenzene, *Carbon*, 2004, **42**(14), 2807–2813.
72. S. Wu, G. Wen, R. Schlögl and D. S. Su, Carbon nanotubes oxidized by a green method as efficient metal-free catalysts for nitroarene reduction, *Phys. Chem. Chem. Phys.*, 2015, **17**(3), 1567–1571.
73. Y. Shao, G. Yin, J. Zhang and Y. Gao, Comparative investigation of the resistance to electrochemical oxidation of carbon black and carbon nanotubes in aqueous sulfuric acid solution, *Electrochim. Acta*, 2006, **51**(26), 5853–5857.
74. S. Kundu, Y. Wang, W. Xia and M. Muhler, Thermal stability and reducibility of oxygen-containing functional groups on multiwalled carbon nanotube surfaces: A quantitative high-resolution xps and TPD/TPR study, *J. Phys. Chem. C*, 2008, **112**(43), 16869–16878.

75. S. Reiche, R. Blume, X. C. Zhao, D. Su, E. Kunkes and M. Behrens, *et al.*, Reactivity of mesoporous carbon against water – An *in situ* XPS study, *Carbon*, 2014, **77**, 175–183.
76. D. S. Su, J. Zhang, B. Frank, A. Thomas, X. Wang and J. Paraknowitsch, *et al.*, Metal-free heterogeneous catalysis for sustainable chemistry, *ChemSusChem*, 2010, **3**(2), 169–180.
77. D. Yu, E. Nagelli, F. Du and L. Dai, Metal-free carbon nanomaterials become more active than metal catalysts and last longer, *J. Phys. Chem. Lett.*, 2010, **1**, 2165–2173.
78. K. Chizari, A. Deneuve, O. Ersen, I. Florea, Y. Liu and D. Edouard, *et al.*, Nitrogen-doped carbon nanotubes as a highly active metal-free catalyst for selective oxidation, *ChemSusChem*, 2012, **5**(1), 102–108.
79. C. Duong-Viet, H. Ba, Y. Liu, L. Truong-Phuoc, J.-M. Nhut and C. Pham-Huu, Nitrogen-doped carbon nanotubes on silicon carbide as a metal-free catalyst, *Chin. J. Catal.*, 2014, **35**(6), 906–913.
80. D. Guo, R. Shibuya, C. Akiba, S. Saji, T. Kondo and J. Nakamura, Active sites of nitrogen-doped carbon materials for oxygen reduction reaction clarified using model catalysts, *Science*, 2016, **351**(6271), 361–365.
81. Y. Shao, J. Sui, G. Yin and Y. Gao, Nitrogen-doped carbon nanostructures and their composites as catalytic materials for proton exchange membrane fuel cell, *Appl. Catal., B*, 2008, **79**(1), 89–99.
82. H. Wang, T. Maiyalagan and X. Wang, Review on recent progress in nitrogen-doped graphene: synthesis, characterization, and its potential applications, *ACS Catal.*, 2012, **2**, 781–794.
83. K. Gong, F. Du, Z. Xia, M. Durstock and L. Dai, Nitrogen-doped carbon nanotube arrays with high electrocatalytic activity for oxygen reduction, *Science*, 2009, **323**(5915), 760–764.
84. L. Qu, Y. Liu, J. B. Baek and L. Dai, Nitrogen-doped graphene as efficient metal-free electrocatalyst for oxygen reduction in fuel cells, *ACS Nano*, 2010, **4**(3), 1321–1326.
85. Z. H. Sheng, L. Shao, J. J. Chen, W. J. Bao, W. F. Bin and X. H. Xia, Catalyst-free synthesis of nitrogen-doped graphene *via* thermal annealing graphite oxide with melamine and its excellent electrocatalysis, *ACS Nano*, 2011, **5**(6), 4350–4358.
86. C. P. Ewels and M. Glerup, Nitrogen doping in carbon nanotubes, *J. Nanosci. Nanotechnol.*, 2005, **5**(9), 1345–1363.
87. C. Zhang, L. Fu, N. Liu, M. Liu, Y. Wang and Z. Liu, Synthesis of nitrogen-doped graphene using embedded carbon and nitrogen sources, *Adv. Mater.*, 2011, **23**(8), 1020–1024.
88. C. Zhang, R. Hao, H. Liao and Y. Hou, Synthesis of amino-functionalized graphene as metal-free catalyst and exploration of the roles of various nitrogen states in oxygen reduction reaction, *Nano Energy*, 2013, **2**(1), 88–97.
89. S. Stankovich, D. A. Dikin, R. D. Piner, K. A. Kohlhaas, A. Kleinhammes and Y. Jia, *et al.*, Synthesis of graphene-based nanosheets *via* chemical reduction of exfoliated graphite oxide, *Carbon*, 2007, **45**(7), 1558–1565.

90. Z.-H. Sheng, H.-L. Gao, W.-J. Bao, F.-B. Wang and X.-H. Xia, Synthesis of boron doped graphene for oxygen reduction reaction in fuel cells, *J. Mater. Chem.*, 2012, **22**(2), 390–395.
91. L. Yang, S. Jiang, Y. Zhao, L. Zhu, S. Chen and X. Wang, *et al.*, Boron-doped carbon nanotubes as metal-free electrocatalysts for the oxygen reduction reaction, *Angew. Chem., Int. Ed.*, 2011, **50**(31), 7132–7135.
92. Z. Yang, Z. Yao, G. Li, G. Fang, H. Nie and Z. Liu, *et al.*, Sulfur-doped graphene as an efficient metal-free cathode catalyst for oxygen reduction, *ACS Nano*, 2012, **6**(1), 205–211.
93. Y. Su, Y. Zhang, X. Zhuang, S. Li, D. Wu and F. Zhang, *et al.*, Low-temperature synthesis of nitrogen/sulfur co-doped three-dimensional graphene frameworks as efficient metal-free electrocatalyst for oxygen reduction reaction, *Carbon*, 2013, **62**, 296–301.
94. C. H. Choi, S. H. Park and S. I. Woo, Binary and ternary doping of nitrogen, boron, and phosphorus into carbon for enhancing electrochemical oxygen reduction activity, *ACS Nano*, 2012, **6**(8), 7084–7091.
95. D.-S. Yang, D. Bhattacharjya, S. Inamdar, J. Park and J.-S. Yu, Phosphorus-doped ordered mesoporous carbons with different lengths as efficient metal-free electrocatalysts for oxygen reduction reaction in alkaline media, *J. Am. Chem. Soc.*, 2012, **134**, 16127–16130.
96. Y. Wang, J. Zhang, X. Wang, M. Antonietti and H. Li, Boron- and fluorine-containing mesoporous carbon nitride polymers: metal-free catalysts for cyclohexane oxidation, *Angew. Chem., Int. Ed.*, 2010, **49**(19), 3356–3359.
97. X. Sun, Y. Zhang, P. Song, J. Pan, L. Zhuang and W. Xu, *et al.*, Fluorine-doped carbon blacks: highly efficient metal-free electrocatalysts for oxygen reduction reaction, *ACS Catal.*, 2013, **3**(8), 1726–1729.
98. I.-Y. Jeon, H.-J. Choi, M. Choi, J.-M. Seo, S.-M. Jung and M.-J. Kim, *et al.*, Facile, scalable synthesis of edge-halogenated graphene nanoplatelets as efficient metal-free eletrocatalysts for oxygen reduction reaction, *Sci. Rep.*, 2013, **3**(1), 1810.
99. L. G. Jacobsohn, R. K. Schulze, M. E. H. Maia Da Costa and M. Nastasi, X-ray photoelectron spectroscopy investigation of boron carbide films deposited by sputtering, *Surf. Sci.*, 2004, **572**(2–3), 418–424.
100. Y. Zhao, L. Yang, S. Chen, X. Wang, Y. Ma and Q. Wu, *et al.*, Can boron and nitrogen co-doping improve oxygen reduction reaction activity of carbon nanotubes?, *J. Am. Chem. Soc.*, 2013, **135**(4), 1201–1204.
101. J. Liang, Y. Jiao, M. Jaroniec and S. Z. Qiao, Sulfur and N. dual-doped mesoporous graphene electrocatalyst for oxygen reduction with synergistically enhanced performance, *Angew. Chem., Int. Ed.*, 2012, **51**(46), 11496–11500.
102. S.-A. Wohlgemuth, R. J. White, M.-G. Willinger, M.-M. Titirici and M. Antonietti, A one-pot hydrothermal synthesis of sulfur and nitrogen doped carbon aerogels with enhanced electrocatalytic activity in the oxygen reduction reaction, *Green Chem.*, 2012, **14**, 1515–1523.

103. C. H. Choi, S. H. Park and S. I. Woo, Heteroatom doped carbons prepared by the pyrolysis of bio-derived amino acids as highly active catalysts for oxygen electro-reduction reactions, *Green Chem.*, 2011, **13**(2), 406–412.
104. S. Yang, L. Zhi, K. Tang, X. Feng, J. Maier and K. Müllen, Efficient synthesis of heteroatom (N or S)-doped graphene based on ultrathin graphene oxide-porous silica sheets for oxygen reduction reactions, *Adv. Funct. Mater.*, 2012, **22**(17), 3634–3640.
105. Y. Su, Y. Zhang, X. Zhuang, S. Li, D. Wu and F. Zhang, *et al.*, Low-temperature synthesis of nitrogen/sulfur co-doped three-dimensional graphene frameworks as efficient metal-free electrocatalyst for oxygen reduction reaction, *Carbon*, 2013, **62**, 296–301.
106. J. P. Paraknowitsch and A. Thomas, Doping carbons beyond nitrogen: an overview of advanced heteroatom doped carbons with boron, sulphur and phosphorus for energy applications, *Energy Environ. Sci.*, 2013, **6**(10), 2839.
107. J. Wu, C. Jin, Z. Yang, J. Tian and R. Yang, Synthesis of phosphorus-doped carbon hollow spheres as efficient metal-free electrocatalysts for oxygen reduction, *Carbon*, 2015, **82**(C), 562–571.
108. C. Zhang, N. Mahmood, H. Yin, F. Liu and Y. Hou, Synthesis of phosphorus-doped graphene and its multifunctional applications for oxygen reduction reaction and lithium ion batteries, *Adv. Mater.*, 2013, **25**(35), 4932–4937.
109. C. H. Choi, S. H. Park and S. I. Woo, Phosphorus–nitrogen dual doped carbon as an effective catalyst for oxygen reduction reaction in acidic media: effects of the amount of P-doping on the physical and electrochemical properties of carbon, *J. Mater. Chem.*, 2012, **22**(24), 12107–12115.
110. N. Daems, X. Sheng, I. F. J. Vankelecom and P. P. Pescarmona, Metal-free doped carbon materials as electrocatalysts for the oxygen reduction reaction, *J. Mater. Chem. A*, 2014, **2**(12), 4085–4110.
111. T. Ishizaki, Y. Wada, S. Chiba, S. Kumagai, H. Lee and A. Serizawa, *et al.*, Effects of halogen doping on nanocarbon catalysts synthesized by a solution plasma process for the oxygen reduction reaction, *Phys. Chem. Chem. Phys.*, 2016, **18**(31), 21843–21851.
112. Z. Yao, H. Nie, Z. Yang, X. Zhou, Z. Liu and S. Huang, Catalyst-free synthesis of iodine-doped graphene *via* a facile thermal annealing process and its use for electrocatalytic oxygen reduction in an alkaline medium, *Chem. Commun.*, 2012, **48**(7), 1027.
113. G. Nanse, E. Papirer, P. Fioux, F. Moguet and A. Tressaud, Fluorination of carbon blacks: an X-ray photoelectron spectroscopy study: iii. Fluorination of different carbon blacks with gaseous fluorine at temperatures below 100 °C influence of the morphology, structure and physico-chemical characteristics of t, *Carbon*, 1997, **35**(4), 515–528.
114. E. Papirer, R. Lacroix, J.-B. Donnet, G. Nanse and P. Fioux, XPS study of the halogenation of carbon black – part 1. Bromination, *Carbon*, 1994, **32**(7), 1341–1358.

115. E. Papirer, R. Lacroix, J.-B. Donnet, G. Nanse and P. Fioux, XPS study of the halogenation of carbon black – part 2. Chlorination, *Carbon*, 1994, **32**(1), 63–72.
116. J. Zheng, H.-T. Liu, B. Wu, C.-A. Di, Y.-L. Guo and T. Wu, *et al.*, Production of graphite chloride and bromide using microwave sparks, *Sci. Rep.*, 2012, **2**, 662.
117. G. Panomsuwan, N. Saito and T. Ishizaki, Electrochemistry communications electrocatalytic oxygen reduction activity of boron-doped carbon nanoparticles synthesized *via* solution plasma process, *Electrochem. Commun.*, 2015, **59**, 81–85.
118. J. Wu, Z. Yang, X. Li, Q. Sun, C. Jin and P. Strasser, *et al.*, Phosphorus-doped porous carbons as efficient electrocatalysts for oxygen reduction,*J. Mater. Chem. A*, 2013, **1**(34), 9889.
119. D. S. Yang, D. Bhattacharjya, M. Y. Song and J. S. Yu, Highly efficient metal-free phosphorus-doped platelet ordered mesoporous carbon for electrocatalytic oxygen reduction, *Carbon*, 2014, **67**, 736–743.
120. Y. Zhang, M. Chu, L. Yang, W. Deng, Y. Tan and M. Ma, *et al.*, Synthesis and oxygen reduction properties of three-dimensional sulfur-doped graphene networks, *Chem. Commun.*, 2014, **50**(48), 6382.
121. G. Panomsuwan, N. Saito and T. Ishizaki, Simple one-step synthesis of fluorine-doped carbon nanoparticles as potential alternative metal-free electrocatalysts for oxygen reduction reaction, *J. Mater. Chem. A*, 2015, **3**(18), 9972–9981.
122. A. Liu, W. Li, H. Jin, X. Yu, Y. Bu and Y. He, *et al.*, The enhanced electrocatalytic activity of graphene co-doped with chlorine and fluorine atoms, *Electrochim. Acta*, 2015, **177**, 36–42.
123. Q. Han, C. Hu, F. Zhao, Z. Zhang, N. Chen and L. Qu, One-step preparation of iodine-doped graphitic carbon nitride nanosheets as efficient photocatalysts for visible light water splitting, *J. Mater. Chem. A*, 2015, **3**(8), 4612–4619.

Section III

Applications

CHAPTER 6

Metal-free Functionalized Carbon in Liquid Phase Reactions

CARINE EDITH CHAN-THAW

[a]Department of Chemistry, Università degli Studi di Milano, via Golgi 19, 20133 Milan, Italy; [b]Institut pour la Maitrise de l'Energie, Université d' Antananarivo, B.P. 566 - 101 Antananarivo, Madagascar
*E-mail: carine.chanthaw@gmail.com

6.1 Introduction

At present, a tremendous amount of attention is being focused on finding green alternatives to liquid- or gas-phase catalytic transformations. Ideally, the use of biomass as a feedstock for the chemical industry would be of benefit, and it would be even better if such catalytic transformations could occur on metal-free catalysts. Indeed, sustainable chemistry researchers have been demonstrating a high interest in metal-free catalysis.[1]

Both the academic and the industrial world are dedicating more attention to finding and improving alternatives to fossil-based carbon sources. The most straightforward candidate, due to its availability, low cost and abundance on Earth, is carbon.[2–4] Carbon-based materials are widely used either as supports for metal-based catalysts, or as catalysts themselves.[5] Metal-based

Catalysis Series No. 31
Metal-free Functionalized Carbons in Catalysis: Synthesis, Characterization and Applications
Edited by Alberto Villa and Nikolaos Dimitratos

Published by the Royal Society of Chemistry, www.rsc.org

carbon catalysts are active in the aerobic oxidation of alcohols,[6–9] the selective reduction of nitroarene and the oxidative dehydrogenation of ethylbenze,[10] among other examples. Returning to carbon as a catalyst, carbon is the ideal candidate because of its position in the periodic table. Diverse bonding states lead to different structural arrangements: sp^1-hybridization results in a carbon chain structure, whilst sp^2-hybridization leads to a planar carbon structure, and, finally, sp^3-hybridization results in a tetrahedral carbon structure. For the target liquid phase reaction, tailoring the physical (surface area and porosity) and chemical (surface functional groups) properties of carbon catalysts is possible through careful and precise synthesis. Sustainable and proficient metal-free carbon catalysts are, therefore, excellent candidates. Corrosion resistance, no heavy metal pollution and environmental friendliness are some of the desired advantages.

Metal-free carbon catalysts, functionalized or not, would, therefore, be able to replace metals or metal-oxide-based catalysts in several reactions, such as hydrogenation and dehydrogenation reactions.[11,12] Recent reviews[13,14] have reported the use of metal-free carbon materials as catalysts (also called carbocatalysis) for liquid phase reactions. The catalyst design often determines the catalytic performance of these carbon materials. The design of the catalyst is tailored to the target liquid phase reaction (among them dehydrogenation, alcohol oxidation, transesterification, electrocatalysis and photocatalysis). Indeed, the expected catalytic properties of the catalyst suggest how the material surface should be prepared. Thus, the synthesis of these metal-free carbonaceous materials constitutes one of the key elements of a successful reaction. One prerequisite of such a catalyst preparation lies in the total absence of any trace metal elements inside the materials. Only when this condition is satisfied can the catalyst be called a "metal-free carbonaceous material". The completion of such a condition is difficult, as almost all preparation methods for carbon materials include the use of metal. As an example, carbon nanotubes (CNTs) are obtained by chemical vapor deposition.[15] Furthermore, to make these materials active for any reaction, active sites are of importance. The addition of functional groups (heteroatoms) results in a modification of the electronic properties. The insertion of functional groups inside or outside these carbon materials alters the nature (basic or acidic), as well as the physical properties, of the final metal-free carbonaceous materials and, thus, their catalytic performance.

The aim of this chapter is to summarize the main liquid phase reactions carried out with metal-free carbon materials.

6.2 Hydrocarbon Oxidation

The oxidation of hydrocarbons into oxygen-containing compounds to fine chemicals relies upon efficient C–H bond activation. In heterogeneous metal-based catalysis, breaking the C–H bond is possible in the presence of metals.

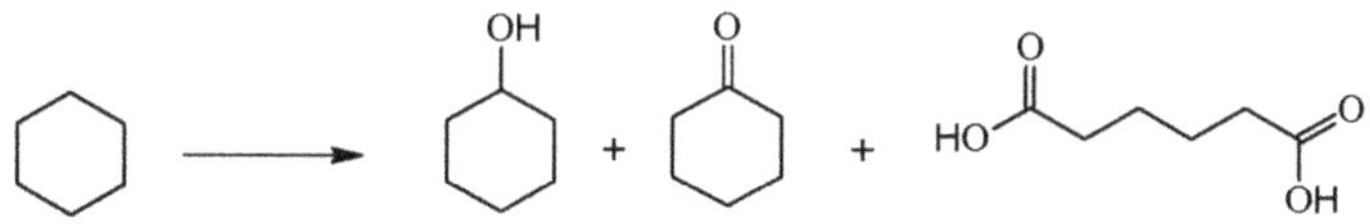

Scheme 6.1 Cyclohexane oxidation to cyclohexanone, cyclohexanol and adipic acid.

6.2.1 Cyclohexane Oxidation

The oxidation of cyclohexane (see Scheme 6.1) is an important industrial reaction, which takes place on a metal catalyst[16] or on a bimetallic catalyst.[17] The oxidation of cyclohexane results in the production of cyclohexanone, cyclohexanol and adipic acid, which play a significance role in producing nylon polymers.[18] The production of cyclohexanone and cyclohexanol is a major process in industrial chemistry, since these two products are chemical precursors for the manufacture of nylon-6 and nylon-6,6 fibres *via* oxidation to adipic acid.[19] Jointly, however, with a cyclohexane conversion and a low selectivity to the target molecule, toxic organic acid waste is produced. Industrially, low cyclohexane conversion is maintained (4–12%) to raise a selectivity over 70% to alcohol and ketones.[20]

It is, therefore, of interest to find a substitute catalyst to fulfil this lack of conversion and selectivity. Some authors have investigated several different catalysts and suggested a series of carbon materials.[21] Graphene (sp^2-hybridized carbon) oxides, micrometer-sized diamond and nanosized diamond (sp^3 bonding configuration) catalyzed the oxidation of the inert cyclohexane to the oxygenated chemicals at a pressure of 1.5 MPa oxygen. Because they contain various structural defects, such as vacancies, the authors also found that non-treated sp^2-hybridized carbon materials are more active in the reaction mentioned above. Indeed, sp^2 carbon could easily decompose the intermediate molecule $C_6H_{11}COOH$.

With the aim to improve the cyclohexane conversion, some research groups explored the idea of activating the carbon-based materials by doping with N atoms, for example. It is in this respect that Ma *et al.* utilized N-doped graphene as a catalyst and the cyclohexane oxidation proceeded with 24.7% yield.[22] In order to achieve this result, they used oxygen and a small (3.0 mmol) amount of TBHP (*tert*-butyl hydroperoxide) as oxidants. Moreover, they demonstrated that the amount of N improves the catalytic performance significantly. Wang and co-authors also attributed the improved activity to the presence of graphitic sp^2 nitrogen. Although the graphitic sp^2 nitrogen is not directly correlated to the catalytic mechanism, it participates in the change of the electronic structure of the adjacent carbon atoms, as well as in the formation of reactive oxygen species.

The selective oxidation of cyclohexane to the organic acid, namely adipic acid, also faces challenges due to low activity and safety problems. Zheng *et al.* performed this reaction over nitrogen-doped CNTs with a 59.7% selectivity at 45.3% conversion.[23] In addition to being active and selective, the catalyst was also stable over five runs.

Scheme 6.2 Selective oxidation of ethylbenzene to acetophenone.

6.2.2 Ethylbenzene Oxidation

Acetophenone (AcPO) is an intermediate material in the production of perfumes, pharmaceuticals, resins, alcohol, esters, and aldehydes. It can be derived from the selective oxidation of ethylbenzene (EB) (see Scheme 6.2).

Several catalysts are active for this reaction, such as MCM-41 molecular sieves with Mn and Co,[24] NiAl hydrocalcite,[25] and manganese nanocatalysts.[26] These three catalysts require the addition of a small amount of TBHP as an oxidant.

However, the main drawbacks of using metallic species or an acidic solvent lie in the costs and the environmental impact. This has aroused more interest in finding a catalyst that can assuage these worries, one that is metal-free, cheap, environmentally friendly, recyclable, efficient and selective in the oxidation of C–H bonds under mild conditions. The candidate that has shown impressive improved results is carbon nanotubes, CNTs, as reported by Zheng *et al.*[27] With a 38.2% conversion of EB, they obtained 60.9% selectivity to AcPO at 155 °C and at 1.5 MPa oxygen for four hours. The group suggested that the oxidation of EB proceeds by a free-radical mechanism, as already proposed elsewhere.[28] The surface carboxylic groups on the CNTs do not facilitate EB oxidation, because they could obstruct the adsorption of radical species and peroxides on CNTs.[29] By increasing the carboxylic group content on the CNTs' surface, a severe decrease of conversion was found. The CNTs played an important role in the decomposition of 1-phenylethyl hydroperoxide (PEHP), and contributed to the production of AcPO through π–π interactions between the radical species, peroxide, and the graphene sheets of the CNTs. The catalyst was stable under recycling and the theoretical calculations performed were consistent with the experimental results.

Theoretical calculations of the reaction mentioned above were also carried out by Sicilia *et al.*[30] They investigated the mechanism in which the OOH radical is the active peroxide species that drives the reaction, by using the cluster approach. They concluded that it is thermodynamically favorable. Moreover, the formation of the hydroperoxyl radical and its decomposition in other active oxygen species are associated with small activation energy barriers, which are compatible with the high catalytic activity of nitrogen-doped graphene.

Ma and coworkers also performed the selective oxidation of EB to AcPO over N-doped graphene materials using TBHP and observed that the N-doped catalyst could activate the C–H bond of EB. After a 24 h reaction time, a 91.3% yield of AcPO was reported.[22] There was no doubt as to the role played by the N atoms in the catalytic performance according to the results of X-ray absorption spectroscopy (XAS) on the graphitic N atoms. The XAS results showed that the N atoms maintained their original chemical status during the catalytic run, whilst the EB and the oxidant changed the chemical status of the carbon atoms. As in the case of cyclohexane oxidation, the same authors underlined the importance of doping the carbon nanotubes with N atoms to improve the EB conversion. Thanks to DFT calculations, they observed that the graphitic sp^2 N atoms strongly modify the neighboring carbon atoms by fluctuating the electronic partial density of states of these carbon atoms. Such alterations in the charge distribution will successively facilitate the adsorption of active oxygen groups and, thus, the oxidation of EB to AcPO.

6.2.3 Selective Oxidation of Toluene

Benzaldehyde, commonly used for its strong almond flavor, is derived from the selective oxidation of toluene (see Scheme 6.3). Because of the easy over-oxidation of benzaldehyde, benzoic acid is the side product. Benzaldehyde is the preferred product of toluene oxidation, because it is a precursor for dyes, perfumes and pharmaceuticals.

The liquid phase oxidation of toluene with homogeneous metal salt catalysts was industrially realized using oxygen or peroxides as oxidants. Commercial benzaldehyde is produced mainly by the chlorination of toluene followed by saponification.[31] In addition to low conversion, halogen ions and acidic solvents are indispensable in this reaction and they can easily cause the erosion of the facility. Hutchings *et al.* proposed bimetallic Au–Pd alloy heterogeneous catalysts for toluene oxidation. They were successful in the formation of benzyl benzoate and overcame the over-oxidation of the as-formed benzaldehyde with increased reactivity relative to toluene.[32]

Antonietti *et al.* proposed mesoporous g-C_3N_4 (mpg-C_3N_4) using O_2 under solvent-free conditions at 160 °C as an alternative catalyst for the selective oxidation of toluene and derivates to aldehydes.[33] For comparison, they tuned the specific surface area of mpg-C_3N_4 from 8 to 301 $m^2\ g^{-1}$. For an aldehyde selectivity above 99%, they reached less than 3% conversion. With respect to the Au or Pd catalysts,[32] conversion and benzaldehyde selectivity are comparable under similar conditions, calculated based on the total mass of catalysts involved. They explained the suppression of benzaldehyde over-oxidation by the mpg-C_3N_4 nanopores. The catalyst with the higher specific surface area (the more porous) exhibited the highest benzaldehyde selectivity, while the less porous catalyst (surface area of 62 $m^2\ g^{-1}$) failed in suppressing the over-oxidation of benzaldehyde. They suggest that the enlarged surface area enhanced the selectivity by providing more active sites.

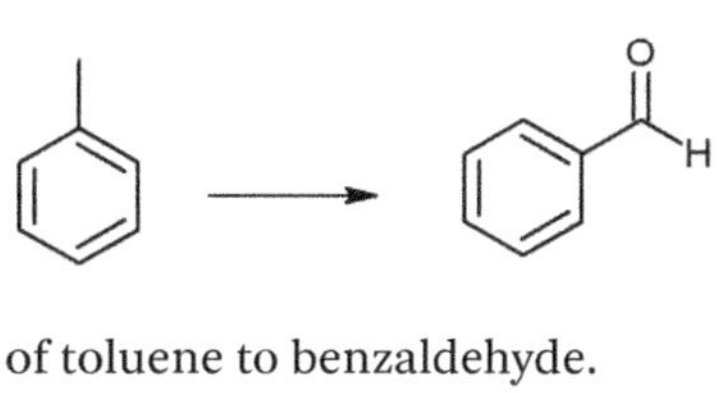

Scheme 6.3 Oxidation of toluene to benzaldehyde.

We have covered the main examples of the oxidation of the hydrocarbonic C–H bond. Oxidations of methylbenzene to aldehydes and diarylmethanes to ketones were possible over graphene oxides.[34]

6.3 Alcohol Oxidation

Selective oxidation of alcohols to the corresponding aldehydes with oxygen represents a well-studied reaction because of its wide application in chemical transformations. In heterogeneous catalysis, the reaction often takes place over metal nanoparticles (NPs) supported on carbon materials. One drawback of these catalysts lies in the over-oxidation of the products, which leads to low activity and poor selectivity. The introduction of heteroatoms into supports can be beneficial for the catalytic activity of metal nanoparticles.[7] Chan-Thaw *et al.* reported a 66% selectivity to benzaldehyde at 90% BA conversion over Pd NPs supported on a covalent triazine framework catalyst at 80 °C and 2 atm oxygen.[8] The introduction of electron-rich nitrogen atoms into the graphitic carbon domain allowed the enhancement of the electron transfer and, consecutively, the catalytic activity.

Bielawski reported for the first time the oxidation of benzyl alcohol (BA) to benzaldehyde over graphene oxide by air in 2010.[35] Thanks to a high catalyst loading (200 wt% graphene oxide), a 72% selectivity of benzaldehyde at 98% BA conversion was reported (at a temperature of 100 °C and a 24 h reaction time). Catalyst recovery was found to be both convenient and effective using simple filtration techniques. It is, however, worth mentioning that ICP-MS analysis performed by the authors demonstrated the presence of Mn, Co, Pd and Ni at a percentage inferior to 36 ppb.

Two years later, DFT work by Bielawski and co-authors suggested that BA oxidation over graphene oxide starts first with the hydrogen transfer of the $-CH_2$ of BA to the catalyst surface.[36] Carboxylic acid groups decorate the edge of graphene oxides and chemically reactive epoxy groups occupy their basal planes. The epoxy groups can be easily modified through ring-opening reactions and result in the formation of diols that subsequently undertake dehydration. Molecular oxygen in the media could revive the partially reduced catalyst and this, therefore, implies a catalyst turnover. It is in this respect, with a graphene oxide catalyst, that Bielawski has suggested a mechanism involving radicals.

The n-type dopants, N or P functionalities, favor electron transfer between graphene sheets of carbon nanotubes and reactive radicals. Indeed, several research groups have focused their attention on investigating P-doped or N-doped carbocatalysts. Patel *et al.* demonstrated that P-doped carbon materials can be used as selective metal-free catalysts for the aerobic oxidation of benzylic alcohols to benzaldehydes with a selectivity over 98%.[37] They were successful in oxidizing 4-methoxybenzyl alcohol, 4-methylbenzyl alcohol and 4-nitrobenzyl alcohol over heteroatom-doped carbon materials. They connected the reactivity to the electron-donating or -withdrawing properties of the functional groups attached to the aromatic ring of the benzylic alcohol. The P bond configuration influences the work function of P-doped carbon (PG) materials. Different to Bielawski, Patel and coworkers proposed a non-radical mechanism with P-doped carbocatalysts. P=O and P–OH functionalities cohabit along with the graphitic regions on PG and play an important role in BA oxidation. In the first step of catalysis, an alcoholate intermediate is formed as a result of the condensation between the alcohol and P=O moieties on PG. Both the interaction of the alcohol with the PG surface by π–π interactions with the graphitic domains and hydrogen bonding with the polar groups (such as P–OH) enable the condensation. In the second step of the reaction, H transfer takes place, and results in the release of the aldehyde product and a water molecule. The presence of oxygen is critical for regenerating the active –P=O sites on the phosphorus-doped graphitic carbon catalysts. This was also observed utilizing N-doped carbon materials.

Watanabe *et al.* performed BA oxidation over a nitrogen-doped metal-free activated carbon (AC) catalyst.[38] A maximum BA conversion of 23% was achieved with the AC catalyst prepared through a H_2O_2 treatment at 130 °C and a subsequent treatment with NH_3 at 800 °C. Through their results, the authors highlighted the importance of catalyst preparation. Moreover, they underlined the non-negligible correlation between the amount of graphite-type nitrogen species and BA conversion. They assume a key role of these nitrogen species in the creation of active sites on the surface of AC. Their results are coherent with those reported by Long *et al.*[39] Among the nitrogen species doped into the graphene lattice—pyridinic N, pyrrolic N, and graphitic N—Long *et al.* reported the role played by the graphitic sp^2 N species in the aerobic alcohol oxidation reaction. Indeed, they suggested the possible formation of a sp^2 N–O_2 adduct transition state, which can oxidize alcohols directly to aldehydes (99%) without any by-product. However, only 15% BA conversion was achieved with about 200 wt% loading of the catalyst. They also reported a linear correlation between the initial reaction rate and the amount of graphitic N atoms.

These N-doped graphene materials could also catalyze the oxidative degradation of phenol using peroxymonosulfate as the oxidant.[40] They highlight the improved catalytic performance of the N-doped materials (80 times more) compared to the undoped ones. DFT calculations underlined that N-doped materials favor the decomposition of peroxymonosulfate, leading to the formation of active radicals necessary for the degradation of phenols.

Cinnamyl alcohol, 5-(hydroxymethyl)furfural (HMF) and BA can be oxidized easily by nitrogen-doped metal-free carbon catalysts. Indeed, the carbon bonding to the hydroxyl group may be activated by the adjacent conjugated system. Because of this lack of proximity between a conjugated system and the hydroxyl group, cyclohexylmethanol, 3-phenyl-1-propanol and 1-heptanol were not activated by nitrogen-doped activated carbon catalysts.[38]

Later, Nguyen *et al.* reported the successful aerobic oxidation of HMF to 2,5-furandicarboxylic acids (FDCA) over a metal-free high nitrogen-doped nanoporous graphitic carbon catalyst at 80 °C.[41] The results obtained were similar to those obtained with metal-containing catalysts. The catalyst was stable under recycling reactions.

In a recent article, Gupta *et al.* successfully oxidized glycerol over metal-free nitrogen-containing carbon nanotubes with TBHP as an oxidant.[42] The authors assigned the activity of the catalyst to the pyridinic groups present on its surface. They assume the pyridinic groups are converted into N–O groups formed by the edge oxidation of N-doped carbon nanotubes by TBHP.

6.4 Acid or Base Catalysis

We have previously discussed the role of n-type dopants, N or P functionalities, in alcohol oxidations. The acidity or basicity of the carbon materials can be altered by the addition of doped heteroatoms or surface functional groups.[43] Acidity is easily altered, as the carbon material itself has a lower acidity than its solid acid counterparts.

Sulfonated or amino functionalized carbon were reported as an environmentally friendly solid acid or base catalyst.[44,45]

The sulfonation of carbon materials often leads to the addition of $-SO_3H$ groups that are more acidic than carboxylic acid. Wen *et al.*, using Boehm titration, scaled the acidity of different oxygenated groups as follows: carboxylic acid > lactone > phenol > carbonyl.[43] However, sulfonated carbocatalysts suffer from deactivation due to leaching in a solution of $-SO_3H$ species.[46] In the case of graphite oxide as a carbon material, we cannot exclude the presence of $-SO_3H$ groups. Graphite oxide is commonly derived from graphite by the Hummers method, where sulfuric acid is added together with sodium nitrate and potassium permanganate.[47]

The introduction of basic groups within a carbon structure—particularly N-containing groups—or grafting amine groups onto carbon, modify the acid/base properties of carbon materials. Chan-Thaw *et al.* introduced N functionalities into activated carbon supports in two different ways.[9] The reference activated carbon, AC, was modified either by physisorption of pyridine (ACpyr) or N functionality was introduced (N-AC) by treating pristine AC with HNO_3. The autogenerated pH values of the three support materials (AC, N-AC and ACpyr) were 6.4, 8.3 and 8.6, respectively. The pH value increased concomitantly with the introduced basicity.

6.4.1 Knoevenagel Condensation Reaction

Small organic fragments are united to form larger molecules through a nucleophilic addition between an aldehyde or ketone and an active hydrogen compound in the presence of a basic catalyst (see Scheme 6.4). Such a reaction results in the formation of a C–C bond. The active hydrogen compound contains a C–H bond, which can be deprotonated by the basic catalyst. The reaction is usually followed by spontaneous dehydration resulting in an unsaturated product.

Graphene oxides that derive from the exfoliation of graphite oxide to one or several layers, are generally rich in defects and contain a substantial amount of oxygen groups. They are, therefore, a better candidate for

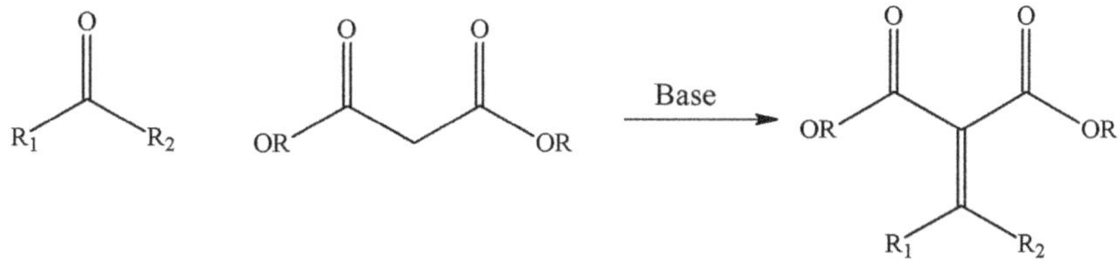

Scheme 6.4 Knoevenagel condensation.

catalysis reactions. Graphene oxides are weak acid catalysts thanks to the oxygen groups. These oxygen groups allow the grafting of some functional groups to increase either the acidity[48] or the basicity of the catalyst.[49,50] Xue and coworkers modified graphene oxide with ethylenediamine at room temperature.[51] The catalyst was active and selective in the Knoevenagel condensation of benzaldehyde with malononitrile. The number of basic active sites generated by ethylenediamine has a significant effect on the catalytic performance. The amine grafting occurs through a nucleophilic addition of ethylenediamine on the epoxide carbon and carboxylic groups of graphene oxides. The catalyst was active and selective, 96% and 97%, respectively, at 40 °C for four hours' reaction time.

Xu *et al.* investigated mesostructured graphitic carbon nitride materials in the Knoevenagel reaction.[52] Through the use of XPS characterization, they explained that the presence of two types of N species, namely N atoms trigonally bonded (399.6 eV) and sp^2-bonded (397.6 eV) to C atoms, was influencing the catalytic performance positively. Reacting benzaldehyde and malononitrile to produce benzylidene malononitrile, 84% conversion was obtained. Moreover, the catalyst was stable under recycling runs. Ansari *et al.* also used metal-free mesoporous carbon nitride materials for the Knoevenagel condensation of ethyl cyanoacetate with aromatic aldehydes under microwave irradiation.[53] They reported 90% conversion and over 99% selectivity to the Knoevenagel product when reacting equimolar (1 mmol) benzaldehyde and ethyl cyanoacetate and 20 mg of catalyst at 120 °C. They assume that the Lewis base sites of carbon nitride materials lead to the abstraction of an acidic proton from the active methylene group of ethyl cyanoacetate.

Because it is possible to enhance the carbon material's basicity, N-doped carbon nanotubes (N-CNTs) were also envisaged as catalysts for the same condensation reaction. Bitter *et al.* reported the preparation of basic nitrogen-containing CNTs by chemical vapor deposition.[54] Good activity in the Knoevenagel condensation of ethyl-α-cyanoacetate with benzaldehyde is associated with the amount of pyridinic nitrogen incorporated into the CNT structure.

6.4.2 Transesterification

Both acid and base catalysts promote transesterification reactions. This reaction takes place when a carboxylic ester is placed in a large excess of alcohol and leads to the formation of a different carboxylic ester.

Jin *et al.* reported the efficiency of mesoporous carbon nitride materials in the transesterification reaction of β-keto esters with different primary alcohols.[55] Because the basicity of carbon materials is known to be weak, they suggest the basicity and reactivity of the catalyst is given by the nitrogen functional groups. These functional groups can be NH, NH_2, and/or N bonded with the carbon matrix in the wall structure of mesoporous carbon nitride. The β-keto esters were tested with aryl, aliphatic, and cyclic primary alcohols. High reactivity and high selectivity to the corresponding ester at a short reaction time were reported.

Later, Ananda and coworkers performed the transesterification of ethyl acetoacetate with various alcohols, such as 1-butanol, 1-octanol, cyclohexanol, benzyl alcohol and furfuryl alcohol, over metal-free mesoporous carbon nitride.[56] They assumed that the activity of their solid base catalyst in the transesterification of ethyl acetoacetate is promoted by the presence of NH or NH_2 groups that offer strong Lewis basic sites.

Villa *et al.* carried out the transesterification of glyceryl tributyrate with methanol, varying the basicity of the support by adding triethylamine, ethylamine or pyrrolidine (Et_3N-CNT, $EtNH_2$-CNT and C_4H_9N-CNTs, respectively).[49,50] Basic properties, as well as the amount of accessible basic sites, were characterized by acid–base titration with HCl as the titrant. Basicity densities of 1.00, 0.99 and 0.54 mmol g^{-1}, respectively, for grafted $-Et_3N$, $-EtNH_2$ and $-C_4H_9N$ functionalized CNTs was given. The corresponding pH values were 10.30, 10.10 and 9.5, respectively. The catalyst with the highest performance (77% conversion after eight hours) is the most basic and has the highest basicity density: Et_3N-CNTs. One should, therefore, pay attention to the basicity of the amine-functionalized CNTs and not to the initial pK_a of the corresponding amine. C_4H_9N-CNT, which exhibits the highest pK_a (11.31) value, showed a lower basicity in comparison to the other catalysts.

Nitrogen-doped nanoporous carbon (NNC) materials were utilized by Yuan *et al.* for the transesterification of dimethyl oxalate with phenol. The catalyst was active and selective to methylphenyl oxalate and diphenyl oxalate. The catalyst contains both oxygen-containing functional groups and a lot of nitrogen on its surface. However, the catalytic performance was justified by the presence of stable basic sites like pyridinic-N on the surfaces of the NNC materials.[57]

6.4.3 Hydrolysis Reactions

The catalytic hydrolysis of ethyl acetate over a solid acid sulfonated graphene, SO_3H-graphene, catalyst was reported by Ji *et al.*[58] They quantified to 6.4% the amount of elemental sulfur grafted homogeneously on the entire surface of the graphene sheets by energy-dispersive X-ray spectroscopy (EDS). The density of the SO_3H groups was measured to be about 2 mmol g^{-1}. With such a quantity of accessible active sites, the SO_3H graphene catalyst was more active and stable than the reference material. After five runs, regardless of the presence of water, the group stated a negligible loss of SO_3H active sites. The catalyst was compared to the commercial Nafion NR50 catalyst.

The same reaction was also performed over a solid basic amino-grafted catalyst, NEt_3-graphene, by Yuan *et al.*[59] For a nitrogen content of 4.8 at%, the catalyst exhibited 67% conversion, almost five times more than that (15%) obtained with the reference material, which was chemically and thermally reduced graphene oxide. Over five recycling runs, a 15% loss of nitrogen was characterized with a final nitrogen content of 4.08 at%.

It would be very desirable to apply this cleavage of chemical bonds in the presence of water to biomass. Indeed, cellulose hydrolysis is usually performed by a strongly homogeneous acid catalyst, which results in separation and environmental concerns. Zhao and coauthors were successful in the hydrolysis of cellulose over graphene oxide.[60] After a 24 h reaction time at 150 °C, they obtained a 21% yield of glucose at 30.2% conversion of cellulose (0.04 g catalyst, 0.27 g cellulose and 27 mL water), far better than the one obtained with graphite oxide or conventional carbon solid acid catalysts (sulfonated carbon). They ascribed the amazing results of graphene oxide to the synergy of its layered, soft structure and the abundance of its hydroxyl/carboxyl functionalities. However, the catalyst suffers from the loss of grafted S species and the decomposition of labile O functionalities (O_2 species, including anhydride, carboxylic, and hydroxylic groups). The essential role of the sulfur in the graphene oxide for this reaction was highlighted. They pretreated the graphene oxide under hydrothermal conditions for 12 hours at 150 °C to make sure that any traces of S left during the preparation of the catalyst were removed. The resulting S-free graphene oxide was inactive in the cellulose hydrolysis, less than 3% of conversion for 1.1% glucose yield.

6.5 Coupling Reactions

Graphite oxides, as well as graphene oxides, were reported to be active in coupling reactions thanks to the oxygen groups they contain. When their acidity favors acid-catalyzed reactions, their nucleophilic aptitudes activate some cations in the reactions. In their study, Hu and coauthors reported the use of graphene oxide to catalyze the direct Friedel–Crafts alkylation reaction between styrenes and alcohols.[61] They produced high yields of the diarylmethane product (up to 99%) with admirable regioselectivity. In their catalytic conditions, excellent performance was achieved when coupling 1,3-dimethoxybenzene with benzyl alcohols. In order to explain these results, they underline first the fundamental role played by the activation of the olefin coupling partners by hydration generated *in situ* from the reduction of graphene. Subsequently, the location of the arene nucleophile for a concerted C–C bond formation by π-stacking interactions with the carbon-based material constitutes the second step of the reaction. When these two conditions are fulfilled, the overall reaction leads to the formation of an alkylated product and to the regeneration of the catalyst after the release of water.[62] Such a mechanism was made possible by the presence of polar oxygen groups and holes in the graphene layers (see Figure 6.1).

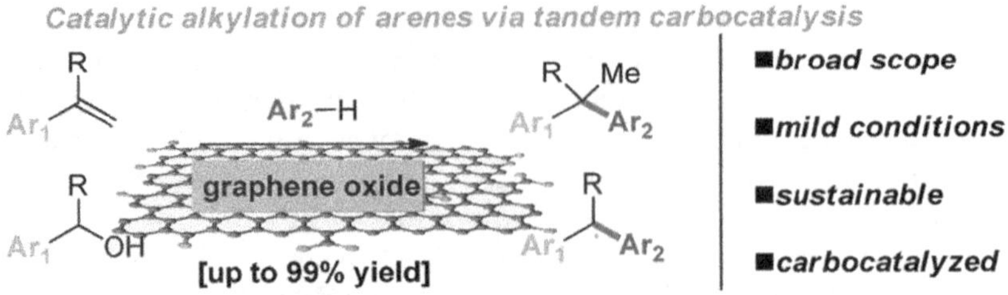

Figure 6.1 Catalytic alkylation of arenes over graphene oxide (reprinted with permission from F. Hu, M. Patel, F. Luo, C. Flach, R. Mendelsohn, E. Garfunkel, H. He and M. Szostak, *J. Am. Chem. Soc.*, 2015, **137**, 14473. Copyright (2016) American Chemical Society).[61]

The key role of oxygen functional groups was also underlined by Gao *et al.* in the direct alkylation reaction between iodobenzene and benzene over graphene oxides.[63] A strong base, KO*t*Bu, was necessary to help the catalyst to perform the reaction, as when it was replaced by another base, the coupling did not take place.

The significant role of oxygen groups in this reaction was confirmed by the observations made. Indeed, a linear correlation between the biaryl compound yield and the oxygen amount was highlighted. Additionally, the group carried out DFT calculations on the arylation of benzene on graphene oxide. Without hydroxy groups, the interaction between KO*t*Bu and graphene is very weak. Potassium cations, K^+, can strongly interact with oxygen species at the edge of a graphene nanopore. The negative charge of the oxygen groups that contain graphene oxide stabilizes the positive charge of the potassium cations. It follows that K^+ activates the C–I bond of the iodobenzene, which leads both to the separation of I radicals and the formation of C_6H_5 radicals. At the same time, the positively charged benzylic hydrogen atoms facilitate the stabilization of C_6H_5 radicals, aside from the π–π interaction stabilization of the graphene.

6.6 Reduction Reactions

In the field of carbocatalysis, fewer reduction reactions compared to oxidation reactions were reported. Graphene-based structures, prepared by chemical reduction of exfoliated graphite oxide to a monolayer or a few layers, have a natural π-conjugated structure and are rich in defects and oxygen-containing functional groups. These oxygen functional groups are mainly carboxyl (–COOH), carbonyl (–C=O), hydroxyl (–OH), ketone (–C=O), and epoxy (–C–O–C–) groups.

Nitroarenes' selective reduction (see Scheme 6.5) is one example of a reduction reaction with carbocatalysts. Traditional active carbon and natural graphite were reported to catalyze the conversion of nitrobenzene to aniline by hydrazine, a strong reducing agent.[64,65] Besides the electron transfer between nitrobenzene and hydrazine, adsorption of the substrate was also possible thanks to the carbon catalyst.[64]

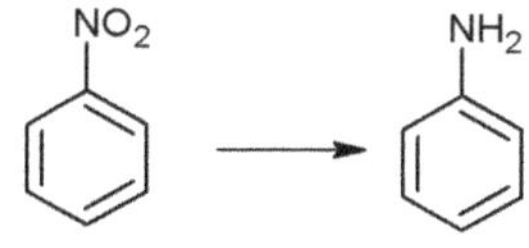

Scheme 6.5 Arene reduction to anilines.

The use of carbon-based materials and, more specifically, graphene oxides for nitrobenzene reduction to aniline, even at room temperature, was first reported by Gao *et al.*[66] Hydrazine hydrate was the reductant. They achieved a 97.4% yield after 48 hours' reaction time. The method of preparing graphene oxide from graphite oxide plays a central role due to its electronic states, as the graphite itself did not show any activity under the selected reaction conditions.

Carbon nanotubes (CNTs) oxidized by hydrogen peroxide were reported as valid metal-free catalysts in the reduction of nitroarenes. Wu and coworkers oxidized commercial CNTs with concentrated H_2O_2 at 30% or HNO_3, labelled CNTs-HP and oCNTs.[67] Both catalysts were tested in the reduction of nitrobenzene at 95 °C using hydrazine monohydrate as the reducing agent. CNTs-HP were revealed to be more active than oCNTs and reached almost full conversion after four hours. The authors mentioned a linear relationship between the initial conversion and the CNTs-HP loading. The temperature of oxidation was optimized to obtain a more performant catalyst. 60 °C was found to be the optimal temperature for the best catalyst, CNT-HP60. The catalyst obtained by this method was tested for the reduction of different nitroarenes such as 1,3-dinitrobenzene, 4-nitroanisole and 2-nitrofluorene. Almost full conversion was obtained (over 99.5%) with a selectivity to the corresponding amines above 86.7%. As expected, the amount of oxygen functional groups was different on both catalysts, oCNTs and CNT-HP60. The total content of surface oxygen was 3.2 at% and 2.7 at% for oCNTs and CNT-HP60, respectively. Through XPS investigation, the carbonyl group (–C=O) was calculated to be 0.3 at% after deconvolution for CNT-HP60 and 0.5 at% for oCNTs. CNT-HP60 contains fewer carboxyl groups (0.7 at%) than oCNTs (1.6 at%).

6.7 Discussion

(a) One should keep in mind that during the production of graphene-based carbon by Hummer's method, strong acids and oxidants ($NaNO_3$ and $KMnO_4$ in concentrated H_2SO_4) are needed, leading to environmental impacts. In addition to this environmental concern, when speaking about metal-free carbocatalysis, care should be taken to ensure the total absence of any metallic impurities. Metal elements, even in traces or at a level of a few ppm, can interfere in any catalytic reaction. Graphite materials, both natural and synthetic, contain intrinsic metallic impurities.[68] During the preparation of most carbonaceous materials, metals in several forms are involved (particles, salts or reagents). Dreyer and coauthors reported the presence of Mn, Co, Pb and Ni at ≤36 ppb by ICP-MS analysis on graphite oxide, which was believed to be metal-free.[69] To avoid any misunderstanding as regards the promising catalytic performance of metal-free carbonaceous materials, careful analysis should be performed. A combination of both combustion and ICP-MS analysis could be a recommendation. Excluding the role of residual metal is, therefore, a requirement for carbocatalysis research. Several research groups are using XPS, Raman spectroscopy as well as TPD. XPS is sensitive to the local electronic structure and is a quantitative method. However, in the field of nanocarbon, its application is hindered by a lack of surface sensitivity. Raman is a bulk sensitive technique, so it could be difficult to evaluate the surface modification of nanocarbons. Finally, TPD experiments would be difficult to interpret because of the transport limitation for heat.

An alternative to preclude any participation of the suspected metal is to increase or decrease its quantity during the production process of the carbon material. During the synthesis of carbon materials, metal impurities could be encapsulated and not removed by washing. These impurities could be released by oxidation or hydrogenation into solution, and alter the catalytic results depending on their chemical state. An incredible step forward has been made in the synthesis and application of highly efficient metal-free catalysts. To continue to optimize the use of these promising metal-free carbon catalysts, as well as get a fundamental understanding of their potentialities, suitable characterization is a critical issue.

(b) Deliberately modifying the surface chemistry of these carbonaceous materials led to an enhanced catalytic performance in several reactions. The addition of heteroatoms such as N, S, P or B into the carbon network positively alters the electronic properties of carbons for the target application. Mesoporous carbons and graphene were revealed to be excellent solid acid or basic catalysts thanks to the functionalization of the carbon surface with sulfonic or amino groups. Sulfonic- or amino-functionalized carbon have demonstrated a respectable performance in esterification and condensation or transesterification reactions. Indeed, modification of the carbon surface through heteroatom variation implies an alteration of the defects and the structure, but also the surface chemistry.

(c) From a green (environmental) and sustainable point of view, the design of such carbocatalysts with specific functional groups is still a challenge. From an industrial perspective, being able to intensively design and produce these carbonaceous materials in a simple way, at low cost, would represent a financial advantage.

References

1. D. M. Alonso, S. G. Wettstein and J. A. Dumesic, *Chem. Soc. Rev.*, 2012, **41**, 8075.
2. Y. Wang, X. C. Wang and M. Antonietti, *Angew. Chem., Int. Ed.*, 2012, **51**, 68.
3. M. Hayashi, *Chem. Rec.*, 2008, **8**, 252.
4. D. S. Su, J. Zhang, B. Frank, A. Thomas, X. Wang, J. Paraknowitsch and R. Schlögl, *Chemsuschem*, 2010, **3**, 169.
5. P. Serp, M. Corrias and P. Kalck, *Appl. Catal., A*, 2003, **253**, 337.
6. C. E. Chan-Thaw, A. Villa, P. Katekomol, D. Su, A. Thomas and L. Prati, *Nano Lett.*, 2010, **10**, 537.
7. C. E. Chan-Thaw, A. Villa, G. M. Veith, K. Kailasam, L. A. Adamczyk, R. R. Unocic, L. Prati and A. Thomas, *Chem.–Asian J.*, 2012, **7**, 387.
8. C. E. Chan-Thaw, A. Villa, L. Prati and A. Thomas, *Chem.–Eur. J.*, 2011, **17**, 1052.
9. C. E. Chan-Thaw, A. Villa, G. M. Veith and L. Prati, *ChemCatChem*, 2015, **7**, 1338.
10. Y. Iwasawa, H. Nobe and S. Ogasawara, *J. Catal.*, 1973, **31**, 444.
11. F. Cavani, F. Trifirò and A. Vaccari, *Appl. Catal., A*, 1995, **133**, 219.
12. N. A. Beckers, S. Huynh, X. Zhang, E. J. Luber and J. M. Buriak, *ACS Catal.*, 2012, **2**, 1524.
13. D. S. Su, G. Wen, S. Wu, F. Peng and R. Schlögl, *Angew. Chem., Int. Ed. Engl.*, 2017, **56**, 936.
14. L. Liu, Y.-P. Zhu, M. Su and Z.-Y. Yuan, *Catalysts*, 2015, **7**, 2765.
15. J.-P. Tessonnier and D. S. Su, *ChemSusChem*, 2011, **4**, 824.
16. M. Conte, X. Liu, D. M. Murphy, K. Whiston and G. J. Hutchings, *Phys. Chem. Chem. Phys.*, 2012, **14**, 16279.
17. X. Liu, M. Conte, M. Sankar, Q. He, D. M. Murphy, D. Morgan, R. L. Jenkins, D. Knight, K. Whiston, C. J. Kiely and G. J. Hutchings, *Appl. Catal., A*, 2015, **504**, 373.
18. R. Zhao, D. Ji, G. Lv, G. Qian, L. Yan, X. Wang and J. Suo, *Chem. Commun.*, 2004, 904.
19. J. G. Speight, *Chemical and Process Design Handbook*, McGraw-Hill, New York, 2002, p. 2.30.
20. U. Schuchardt, D. Cardoso, R. Sercheli, R. Pereira, R. S. de Cruz, M. C. Guerreiro, D. Mandelli, E. V. Spinace and E. L. Fires, *Appl. Catal., A*, 2001, **211**, 1.
21. Y. Cao, X. Luo, H. Yu, F. Peng, H. Wang and G. Ning, *Catal. Sci. Technol.*, 2013, **3**, 2654.
22. Y. Gao, G. Hu, J. Zhong, Z. Shi, Y. Zhu, D. S. Su, J. Wang, X. Bao and D. Ma, *Angew. Chem., Int. Ed.*, 2013, **52**, 2109.
23. H. Yu, F. Peng, J. Tan, X. Hu, H. Wang, J. Yang and W. Zheng, *Angew. Chem.*, 2011, **123**, 4064.
24. C. Guo, Q. Peng, Q. Liu and G. Jiang, *J. Mol. Catal. A: Chem.*, 2003, **192**, 295.

25. S. K. Jana, P. Wu and T. Tatsumi, *J. Catal.*, 2006, **240**, 268.
26. M. Arshadi, M. Ghiaci, A. Rahmanian, H. Ghaziaskar and A. Gil, *Appl. Catal., B*, 2012, **119–120**, 81.
27. J. Luo, F. Peng, H. Yu, H. Wang and W. Zheng, *ChemCatChem*, 2013, **5**, 1578.
28. I. Hermans, J. Peeters and P. A. Jacobs, *J. Org. Chem.*, 2007, **72**, 3057.
29. H. Ma, J. Xu, C. Chen, Q. Zhang, J. Ning, H. Miao, L. Zhou and X. Li, *Catal. Lett.*, 2007, **113**, 104.
30. C. Ricca, F. Labat, N. Russo, C. Adamo and E. Sicilia, *J. Phys. Chem. C*, 2014, **118**, 12275.
31. W. Partenheimer, *Catal. Today*, 1995, **23**, 69.
32. L. Kesavan, R. Tiruvalam, M. H. A. Rahim, M. I. bin Saiman, D. I. Enache, R. L. Jenkins, N. Dimitratos, J. A. Lopez-Sanchez, S. H. Taylor, D. W. Knight, C. J. Kiely and G. J. Hutchings, *Science*, 2011, **331**, 195.
33. X.-H. Li, X. Wang and M. Antonietti, *ACS Catal.*, 2012, **2**, 2082.
34. H.-P. Jia, D. R. Dreyer and C. W. Bielawski, *Tetrahedron*, 2011, **67**, 4431.
35. D. R. Dreyer, H.-P. Jia and C. W. Bielawski, *Angew. Chem., Int. Ed.*, 2010, **49**, 6813.
36. D. W. Boukhvalov, D. R. Dreyer, C. W. Bielawski and Y.-W. Son, *ChemCatChem*, 2012, **4**, 1844.
37. M. A. Patel, F. Luo, M. R. Khoshi, E. Rabie, Q. Zhang, C. R. Flach, R. Mendelsohn, E. Garfunkel, M. Szostak and H. He, *ACS Nano*, 2016, **10**, 2305.
38. H. Watanabe, S. Asano, S.-i. Fujita, H. Yoshida and M. Arai, *ACS Catal.*, 2015, **5**, 2886.
39. J. Long, X. Xie, J. Xu, Q. Gu, L. Chen and X. Wang, *ACS Catal.*, 2012, **2**, 622.
40. X. Duan, Z. Ao, H. Sun, S. Indrawirawan, Y. Wang, J. Kang, F. Liang, Z. H. Zhu and S. Wang, *ACS Appl. Mater. Interfaces*, 2015, **7**, 4169.
41. C. V. Nguyen, Y.-T. Liao, T.-C. Kang, J. E. Chen, T. Yoshikawa, Y. Nakasaka, T. Masuda and K. C.-W. Wu, *Green Chem.*, 2016, **18**, 5957.
42. N. Gupta, O. Khavryuchenko, A. Villa and D. Su, *ChemSusChem*, 2017, **10**, 3030.
43. G. Wen, J. Diao, S. Wu, W. Yang, R. Schlögl and D. S. Su, *ACS Catal.*, 2015, **5**, 3600.
44. R. Xing, Y. Liu, Y. Wang, L. Chen, H. Wu, Y. Jiang, M. He and P. Wu, *Microporous Mesoporous Mater.*, 2007, **105**, 41.
45. L. Peng, A. Philippaerts, X. Ke, J. V. Noyen, F. D. Clippel, G. V. Tendeloo, P. A. Jacobs and B. F. Sels, *Catal. Today*, 2010, **150**, 140.
46. D. Lee, *Molecules*, 2013, **18**, 8168.
47. W. S. Hummers and R. E. Offeman, *J. Am. Chem. Soc.*, 1958, **80**, 1339.
48. S. H. Shuit, K. F. Yee, K. T. Lee, B. Subhash and S. H. Tan, *RSC Adv.*, 2013, **3**, 9070.
49. A. Villa, J.-P. Tessonnier, O. Majoulet, D. S. Su and R. Schlögl, *Chem. Commun.*, 2009, 4405.
50. A. Villa, J.-P. Tessonnier, O. Majoulet, D. S. Su and R. Schlögl, *ChemSusChem*, 2010, **3**, 241.
51. B. Xue, J. G. Zhu, N. Liu and Y. X. Li, *Catal. Commun.*, 2015, **64**, 105.

52. J. Xu, K. Shen, B. Xue, Y.-X. Li and Y. Cao, *Catal. Lett.*, 2013, **143**, 600.
53. M. B. Ansari, H. Jin, M. N. Parvin and S.-E. Park, *Catal. Today*, 2012, **185**, 211.
54. S. van Dommele, K. P. de Jong and J. H. Bitter, *Chem. Commun.*, 2006, 4859.
55. X. Jin, V. V. Balasubramanian, S. T. Selvan, D. P. Sawant, M. A. Chari, G. Q. Lu and A. Vinu, *Angew. Chem., Int. Ed.*, 2009, **48**, 7884.
56. C. Ananda, S. Vishnu Priya, G. Lawrence, G. P. Mane, D. S. Dhawale, K. S. Prasad, V. V. Balasubramanian, M. A. Wahab and A. Vinu, *Catal. Today*, 2013, **204**, 164.
57. X. Yuan, M. Zhang, X. Chen, N. An, G. Liu, Y. Liu, W. Zhang, W. Yan and M. Jia, *Appl. Catal., A*, 2012, **439–440**, 149.
58. J. Ji, G. Zhang, H. Chen, S. Wang, G. Zhang, F. Zhang and X. Fan, *Chem. Sci.*, 2011, **2**, 484.
59. C. Yuan, W. Chen and L. Yan, *J. Mater. Chem.*, 2012, **22**, 7456.
60. X. Zhao, J. Wang, C. Chen, Y. Huang, A. Wang and T. Zhang, *Chem. Commun.*, 2014, **50**, 3439.
61. F. Hu, M. Patel, F. Luo, C. Flach, R. Mendelsohn, E. Garfunkel, H. He and M. Szostak, *J. Am. Chem. Soc.*, 2015, **137**, 14473.
62. P. Rys, P. Skrabal and H. Zollinger, *Angew. Chem., Int. Ed. Engl.*, 1972, **11**, 874.
63. Y. Gao, P. Tang, H. Zhou, W. Zhang, H. Yang, N. Yan, G. Hu, D. Mei, J. Wang and D. Ma, *Angew. Chem., Int. Ed.*, 2016, **55**, 3124.
64. J. W. Larsen, M. Freund, K. Y. Kim, M. Sidovar and J. L. Stuart, *Carbon*, 2000, **38**, 655.
65. H. H. Byung, H. S. Dae and Y. C. Sung, *Tetrahedron Lett.*, 1985, **26**, 6233.
66. Y. J. Gao, D. Ma, C. L. Wang, J. Guan and X. H. Bao, *Chem. Commun.*, 2011, **47**, 2432.
67. S. Wu, G. Wen, R. Schlögl and D. S. Su, *Phys. Chem. Chem. Phys.*, 2015, **17**, 1567.
68. A. Ambrosi, C. K. Chua, B. Khezri, Z. Sofer, R. D. Webster and M. Pumera, *Proc. Natl. Acad. Sci. U. S. A.*, 2012, **109**, 12899.
69. D. R. Dreyer, H.-P. Jia and C. W. Bielawski, *Angew. Chem.*, 2010, **122**, 6965.

CHAPTER 7

Carbon-based Metal-free Catalysts for Dehydrogenation of Hydrocarbons

XI LIU*, AJIN CHERUVATHUR AND RYAN SHARPE

Synfuels China Co. Ltd, Beijing, China
*E-mail: liuxi@synfuelschina.com.cn

7.1 Introduction

Since Jöns Jakob Berzelius coined the term "catalysis" in 1835 to describe the acceleration of reactions by substances that remain unchanged after the reaction, heterogeneous catalysis has played an increasingly more crucial role in the chemical industry and in domestic life. The Fe-catalyzed ammonia synthesis, a key step in the mass production of fertilizers from chemically inert N_2, led to one of the greatest changes in our world since the 1960s and 1970s, which was the so-called "Green Revolution".[1] It is generally considered that the extra production of food significantly diminished food shortages and famine in Asian and Latin America. An estimated 40% of the world's population relies on chemical fertilizers to develop agricultural activities.[2] Another successful example is the catalytic converter for fuel vehicles and stationary engines, which is used to reduce toxic exhaust gas pollutants such as CO, NO_x, hydrocarbons and small particles. In 1975, the catalytic converter was first introduced in the U.S. due to the strict regulation of exhaust emissions

Catalysis Series No. 31
Metal-free Functionalized Carbons in Catalysis: Synthesis, Characterization and Applications
Edited by Alberto Villa and Nikolaos Dimitratos

Published by the Royal Society of Chemistry, www.rsc.org

required by the U.S. Environmental Protection Agency. This has reduced the emission of pollutants by more than 10 billion tons in the U.S.[3] Automotive catalysts have been regarded as one of the greatest engineering achievements of the 20th century.[3] Clearly, any technical breakthrough in catalysis has a tremendous impact on the chemical industry, medicine, material science, and sustainable energy, which affects our society profoundly. During work spanning 100 years, scientists have found that the most reactive candidates for industrial catalysts are transition metals with variable oxidation states, like Fe, Co, Cu, V, Mo *etc*. What makes these transition metals serve as active catalysts is the ability to facilely lend out and store electrons once reactants adsorb and desorb from the surface of the catalysts, respectively. It suggests that non-metal elements, like the carbon group elements (C and Si), nitrogen group elements (N and P) and oxygen group elements (O and S), cannot act as efficient heterogeneous catalysts by themselves. During the last few decades, non-metal nanomaterials, mainly carbonaceous materials or carbon hybrids, have been intensively studied with respect to their unique structural complexities and resulting novel physicochemical properties, and have shown great potential in many fields like green energy, electronics, functional materials, medicine, and heterogeneous catalysis. It has been proved that the catalytic performance of supported metal active species can be greatly influenced by the chemical characteristics of the catalyst supports in reference to their surface functionalities. However, what was overlooked was that the flexibility of controlling electronic properties by modulating surface functionalities likely allowed reversible electron transfer between absorbents and the specific surface functionalities with the desired redox potential, which has been supported by both experimental data and DFT calculations.[4] This means that non-metal inorganic materials can not only serve as unique catalyst supports, but also as catalysts themselves for various catalytic reactions. The potential application of these nanomaterials as metal-free catalysts will lead to a fundamental breakthrough in heterogeneous catalysis as basic concepts related with catalyst activation, surface chemisorption and other issues could be significantly expanded with respect to the novel catalytic functions of the nanomaterials. The mechanistic understanding of metal-free catalysis will subsequently facilitate industrial practice in the design and synthesis of fine-tunable catalysts with high efficiency and selectivity. Other advantages of metal-free catalysts include: (i) nanomaterials typically present a large surface area and high porosity with an abundance of surface functional groups; (ii) the microstructure can be adjusted to meet the requirements of different reactions; (iii) there are enormous physical forms with relative structural stability; (iv) surface functional groups can be easily regenerated; (v) the nanomaterials are environmentally benign, and disposal of the used catalysts will have a less negative impact on the environment; (vi) there is no subsurface involved in catalytic reactions, which will confine the scientific interest to quasi two-dimensional systems. However, the disadvantages of metal-free catalysts, in particular carbonaceous catalysts, also seriously limit their industrial applications with regards to their thermal instability and reproducibility. During the last few decades, the catalytic activities of carbon-based metal-free materials have attracted increasing attention in various fields, including catalytic oxidation, selective reduction, dehydration, electrocatalysis and so on.[5–9] In this chapter, we will summarize those contributions dedicated to exploring carbon-based materials as novel carbonaceous catalysts for the oxidative dehydrogenation (ODH) and direct dehydrogenation (DDH) of ethylbenzene and alkanes, as well as highlight the understanding of the chemical nature of metal-free catalysis. We expect the discussions will shed light on this new catalysis with great potential for green and sustainable synthetic chemistry.

7.2 Oxidative Dehydrogenation of Ethylbenzene

Industry has a long history of the utilization of activated carbon as a catalyst in dechlorination and desulfation of waste gases.[10] It has also been industrially used to produce SO_2Cl_2 by the reaction of chlorine with SO_2 catalyzed by a carbon catalyst. Aside from these well-established processes, other industrial applications of activated carbon catalysts include fuel gas cleaning, removal of halogens, production of glyphosate and so on (see Table 7.1).[11]

It was expected that the surface area of activated carbon catalysts should be a primary factor in determining their catalytic activities. However, in most cases, this correlation cannot be established when using activated carbons with different precursors, preparation history, and structural features for a

Table 7.1 Reactions catalyzed by carbon catalysts (Reprinted with permission from R. W. Coughlin, Carbon as Adsorbent and Catalyst, *Ind. Eng. Chem. Prod. Res. Dev.*, 1969, **8**, 12. Copyright 1969 American Chemical Society).[11]

Oxidation–reduction	$SO_2 + 1/2O_2 \rightarrow SO_3$
	$NO + 1/2O_2 \rightarrow NO_2$
	$2H_2S + O_2 \rightarrow S_2 + 2H_2O$
	$C_6H_5C_2H_5 + 1/2O_2 \rightarrow C_6H_5C_2H_3 + H_2O$
	Toxin oxidation (creatinine)
	Oxidation of industrial effluents (oxalic acid)
Hydrogenation–dehydrogenation	$H_2 + D_2 \rightarrow 2HD$
	$RX + H_2 \rightarrow RH + HX$ (X = Cl, Br)
	$HCOOH \rightarrow CO_2 + H_2$
	$CH_3CHOHCH_3 \rightarrow CH_3COCH_3 + H_2$
Combination with halogens	$H_2 + Br_2 \rightarrow {}_2HBr$
	$CO + Cl_2 \rightarrow COCl_2$ (phosgene)
	$C_2H_4 + 5Cl_2 \rightarrow C_2Cl_6 + 4HCl$
	$SO_2 + Cl_2 \rightarrow SO_2Cl_2$
	$C_6H_5CH_3 + Cl_2 \rightarrow C_6H_5CH_2Cl + HCl$
Decomposition	$2H_2O_2 \rightarrow 2H_2O + O_2$
	$CH_4 \rightarrow C + 2H_2$
Dehydration, isomerization, and polymerization	$HCOOH \rightarrow H_2O + CO$
	$3C_2H_2 \rightarrow C_6H_6$
	α-Olefins → poly(α-olefins)
	α-Oxime → β-oxime

given catalytic reaction.[5] This suggests that the surface chemical characteristics, rather than the surface area of the carbon catalysts, strongly modulates their catalytic activities. Investigation of the relationship between the physicochemical properties of carbon materials and their catalytic activities has been conducted for several decades. Two kinds of approaches have been widely applied in the surface chemistry of carbonaceous materials; one is the "organic surface groups" approach and the other is the "solid state chemistry" approach. The former studies the chemical characteristics of surface groups, whilst the latter studies the crystalline properties of carbon materials. The "organic surface groups" approach deals with the features and functionalities of the surface complexes of oxygen and other compounds chemisorbed at surface defects (see Scheme 7.1; both oxygen and nitrogen functional groups on a graphitic carbon surface are summarized).[12]

The "solid state chemistry" approach also considers defects on the surface of carbon materials, which arise from the limited crystallinity of the

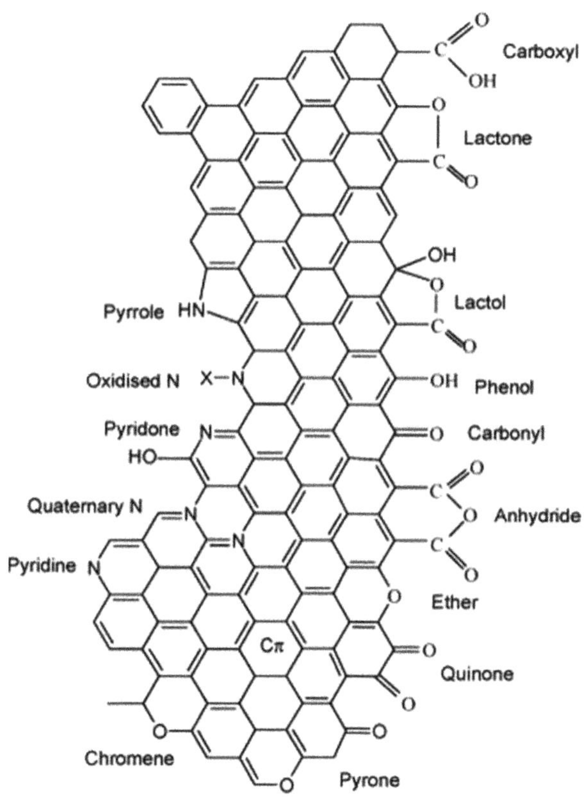

Scheme 7.1 Nitrogen and oxygen surface groups on carbon (Reprinted from *Catalysis Today*, **150**, J. L. Figueiredo and M. F. R. Pereira, The role of surface chemistry in catalysis with carbons, 2–7, Copyright 2010, with permission from Elsevier).[18]

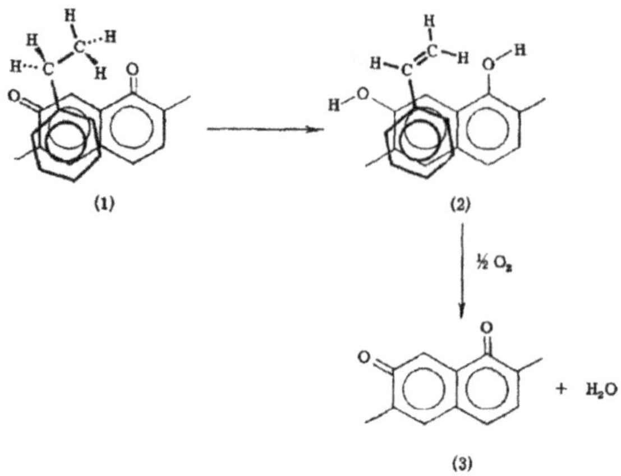

Scheme 7.2 Transfer of hydrogen between diquinoinic groups and ethylbenzene, leading to the formation of styrene and water (Reprinted from *Catalysis*, **31**, Y. Iwasawa, H. Nobe and S. Ogasawara, Reaction mechanism for styrene synthesis over polynaphthoquinone, 444–449, Copyright 1973, with permission from Elsevier).[14]

carbon materials, as the main active sites, since the edge-side carbon atoms are more chemically reactive. Since the 1960s, polymers, including pyrolyzed polyacrylonitrile and polynaphthoquinone, have attracted considerable interest in connection to their interesting physical and electronic properties, which exhibit encouraging catalytic activities for the decomposition of peroxide, autoxidation, dehydrogenation, dehydration and catalytic degradation of hydrazine.[13] These works rationalized that surface functionalities could act as strong hydrogen acceptors, making them suitable for catalytic dehydrogenation. Afterwards, Iwasawa *et al.* examined the catalytic activity of polynaphthoquinone for the oxidative dehydrogenation (ODH) of ethylbenzene to styrene and addressed a key role of quinonic groups as active centers evolved in the transfer of hydrogen (see Scheme 7.2).[14]

On the other hand, in 1969, Robert W. Coughlin discussed the correlation between the catalytic activity of carbon material in a heterogeneous system with its physicochemical properties.[11] The author assumed that, with regard to their conductivity, conductive carbon (graphite), semi-conductive carbon (polycrystalline graphite) and insulated carbon (coke) likely exhibited analogous activities as metal, metal oxide and salt/acid catalysts in selected catalytic reactions, respectively (Scheme 7.3).

However, he realized that it was critically difficult to build up the correlation between the electronic properties of carbon materials and their catalytic activity; thereby, the author attributed the dilemma to a broad variety of polycrystalline properties of graphite materials and inhomogeneity in their microstructure and composition. Meanwhile, studies on the catalytic behavior of acidic catalysts for the ODH of ethylbenzene disclosed a correlation between the promoted catalytic performance and formation of a

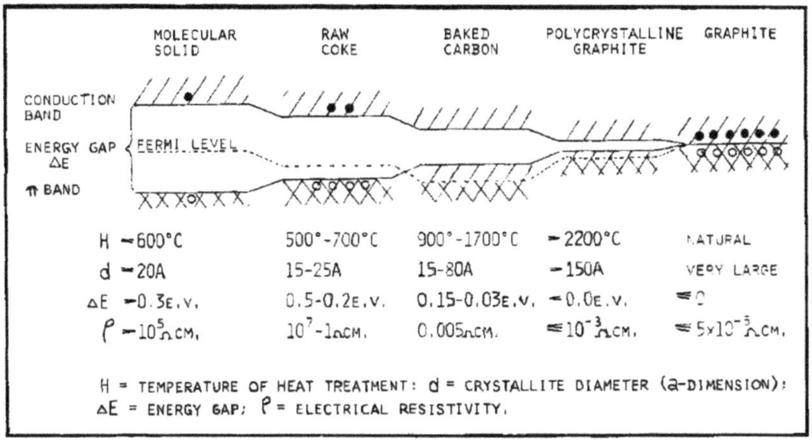

Scheme 7.3 Band model of the electronic structure of different carbons (Reprinted with permission from R. W. Coughlin, Carbon as Adsorbent and Catalyst, *Ind. Eng. Chem. Prod. Res. Dev.*, 1969, **8**, 12. Copyright 1969 American Chemical Society).[11]

carbon deposit, which implied that the real catalytic species should be the newly-formed carbonaceous material rather than the metal oxides themselves.[15–17] Cadus further confirmed the redox nature of the ODH of ethylbenzene to styrene over oxygen-rich coke species and also suggested that electronegative functional groups, most likely quinonic groups, served as active sites for the dehydrogenation reaction.[15] These exploratory works form a solid basis for understanding the chemical reactivity of the surface functionalities for the ODH of hydrocarbons.

Figueiredo *et al.* first employed activated carbon as a catalyst for the ODH of ethylbenzene and developed systematic studies to qualitatively and quantitatively distinguish contributions of different surface functionalities to the catalytic performance.[12,18] Non-treated activated carbon displayed a moderate initial catalytic activity of about 20% yield of styrene. After the oxidation treatment, the initial styrene productivity of the treated activated carbon increased up to 50%, which suggested that oxygen surface functionalities took part in the ODH reaction (Figure 7.1(a)).

However, serious deactivation of the treated activated carbons was also observed, which was attributed to the thermal instability of surface functionalities and coke deposition.[9] Both advantages and disadvantages of activated carbon as a catalyst for catalytic ODH were discussed in these works: they were cheap and could be feasibly produced in large quantities; they had a high surface area with a porous microstructure; but they also displayed a vast variety in physicochemical properties related with their fabrication processing, reproducibility of preparation and existence of impurities were not satisfactory, the microporous structure could not stand severe reaction conditions, and regeneration of deactivated activated carbon

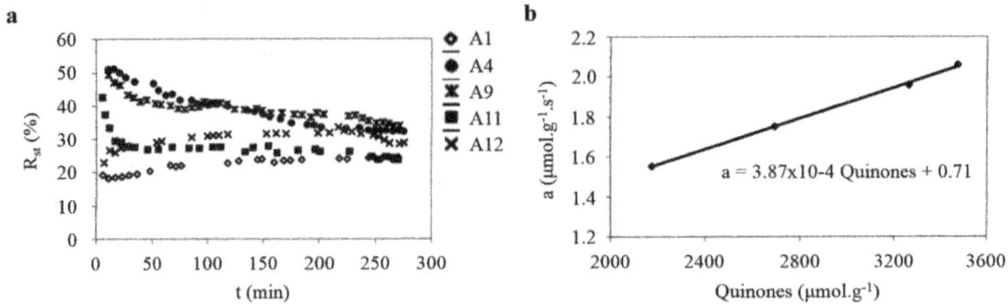

Figure 7.1 (a) Styrene yield obtained as a function of reaction time for activated carbon catalysts prepared by different oxidation treatments: A1: pristine carbon; A4: oxidized with O_2 (weight loss during oxidation of 21.5%); A9: oxidized with N_2O (weight loss during oxidation of 21%); A11: oxidized with HNO_3; A12: oxidized with H_2O_2; (Reproduced with permission from ref. 12). (b) Relation between the activity in the ODH of ethylbenzene and the amount of ketone/quinone groups (Reprinted from *Applied Catalysis A: General*, **184**, M. F. R. Pereira, J. J. M. Orfao and J. L. Figueiredo, Oxidative dehydrogenation of ethylbenzene on activated carbon catalysts. I. Influence of surface chemical groups, 153–160, Copyright 1999, with permission from Elsevier.)[12]

catalysts was also likely impossible since blocked micropores could not be restored. Since the 1990s, extensive experimental and theoretical studies on low-dimensional graphitic materials, like fullerene, carbon nanotubes (CNTs), and graphene, have stimulated enormous interest throughout the catalysis community in exploring their potential applications in heterogeneous catalysis instead of activated carbons. Their uniform microstructure, surface reactivity and structural stability would allow a better catalytic performance as well as chemical tunability in comparison to traditional carbon materials. The catalytic activities of different forms of graphitic nanomaterials (see Figure 7.2), like fullerene, carbon nanotubes/nanofilaments, honeycomb-like carbon, mesoporous carbon, onion-like carbon, or even porous graphene oxide, for the ODH of ethylbenzene have been tested (see Table 7.2).[19–32]

The carbonaceous catalysts exhibited catalytic activities that were comparable, or in some cases even better, than those of metal oxide catalysts at relatively lower temperatures.[33–35] CNT catalysts also displayed remarkable catalytic stability even after 200 hours of catalytic testing; no coke formation was identified.[22] It is worth mentioning that the apparent difference between the catalytic stability of the distinct carbonaceous catalysts with and without long-range order structure can be attributed to two factors: (1) the ordered graphitic framework capable of anchoring particular oxygen functionalities to surface defects even at elevated temperatures, and (2) higher structural resistance to coke formation. The disadvantages of coke deposition can be attributed to two reasons: (1) fatal blockage of micropores and (2) inactivity of the formed coke for the ODH of ethylbenzene even if it contains various oxygen functionalities.[12]

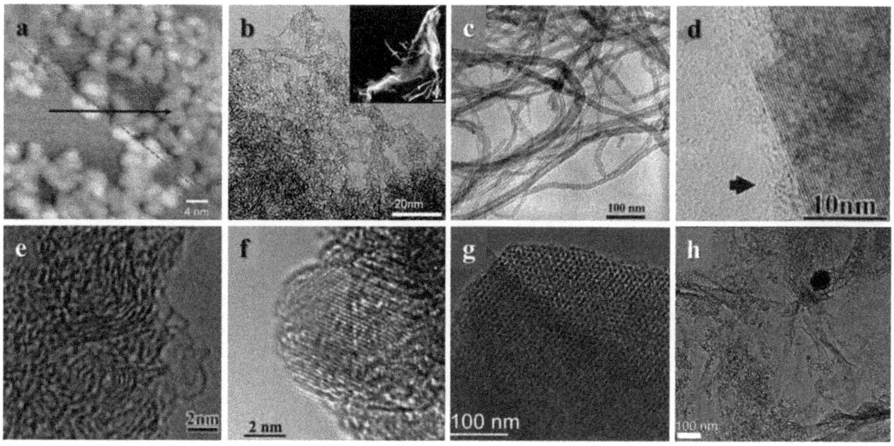

Figure 7.2 Various forms of carbon catalysts for catalyst dehydrogenation reactions: (a) fullerene (Reproduced with permission from ref. 27 with permission from The Royal Society of Chemistry), (b) porous graphene (Reproduced with permission from ref. 26 with permission from The Royal Society of Chemistry), (c) multi-walled carbon nanotubes (Reprinted from *Carbon*, **42**, M. F. R. Pereira, J. L. Figueiredo, J. J. M. Orfao, P. Serp, P. Kalc and Y. Kihn, Catalytic activity of carbon nanotubes in the oxidative dehydrogenation of ethylbenzene, 2807–2813, Copyright 2004, with permission from Elsevier),[22] (d) carbon filaments (Reprinted from *Catalysis Today*, **102**, D. S. Su, N. Maksimova, J. J. Delgado, N. Keller, G. Mestl, M. J. Ledoux and R. Schlögl, Nanocarbons in selective oxidative dehydrogenation reaction, 110–114, Copyright 2005, with permission from Elsevier),[28] (e) onion-like carbon (Reprinted from *Catalysis Today*, **102**, D. S. Su, N. Maksimova, J. J. Delgado, N. Keller, G. Mestl, M. J. Ledoux and R. Schlögl, Nanocarbons in selective oxidative dehydrogenation reaction, 110–114, Copyright 2005, with permission from Elsevier),[28] (f) nanodiamond (Reprinted from *Carbon*, **45**, D. Su, N. I. Maksimova, G. Mestl, V. L. Kuznetsov, V. Keller, R. Schlögl and N. Keller, 2145–2151),[30] (g) mesoporous carbon (Reprinted with permission from Z. Zhao, Y. Dai, J. Lin and G. Wang, *Chem. Mater.*, 2014, **26**, 3151. Copyright 2014 American Chemical Society),[31] (h) nanodiamond–graphene hybrids (Reproduced with permission from ref. 25 with permission from The Royal Society of Chemistry).

Despite the variable dimensions and curvature of nanocarbons, the basic structural element in terms of planar graphene sheets can be summarized in Scheme 7.1, whose edge-sites and defects are decorated with various surface functionalities. Non-defective graphene sheets exhibit significant chemical stability to many reactions. For instance, chemisorption of O_2 on defect-free graphitic nanocarbons is highly endothermic and thermodynamically disfavored, which makes them unable to activate the gaseous reactant for catalytic reactions.[36] In contrast to the inertness of the carbon in a hexagonal ring structure, the carbon in a pentagonal ring structure strongly reacts with O_2 exothermically to form either ether or carbonyl (Figure 7.3). The oxygen-functionalized defects are less reactive, but still more active than the defect-free graphene sheets.[36]

Table 7.2 Catalytic performance of carbon catalysts for the ODH of ethylbenzene.

	Temperature °C	Conversion %	Selectivity %	Yield %	Reaction rate (mmol-ST/g h)
Polynaphthoquinone[14]	270		95	9.1	2.2
Carbon black[28]	517	—	—	~12	0.6
Activated carbon[12,18]	350	48	73	35	7.9
Graphite[28]	517	—	—	44	2.1
CNTs[22]	450	28	67	19	2.7
CNFs[28]	517			54	2.6
Graphene[26]	400	65	90	59	9.27
Onion-like carbon[28,32]	517	90	72	62	3.0
Honeycomb-like carbon[24]	400	48	86	41	6.2
Nanodiamond[30]	515	—	—	38	1
Al_2O_3[15]	445	~35	~35	~12	0.9
20%V–Mg–O[33]	530	32	96	31	11
6%V/CeZr/SiO_2[34]	400	20	92		4.0
$FePO_4$-A-450[35]	440	44	99	44	0.35

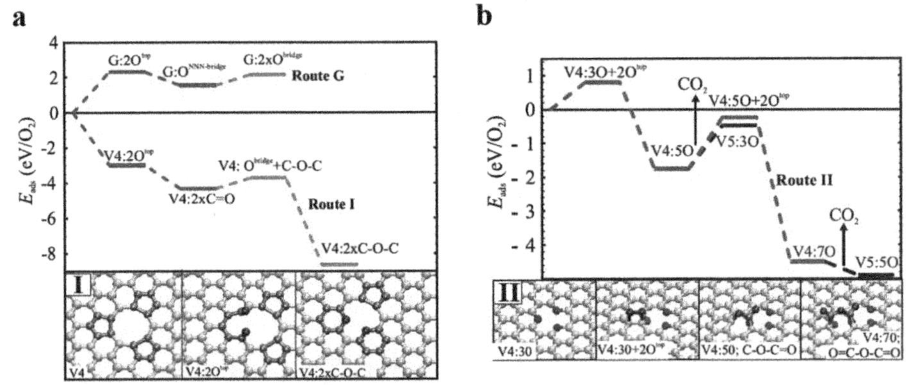

Figure 7.3 (a) Activation of molecular oxygen, O_2, on defect-free (upper, route G) and defective (lower, route I) basal plane graphene. Red lines indicate O_2 adsorption and blue lines O-atom diffusion. The corresponding geometries of the stable intermediate adsorption structures along route I are shown in the lower panel; (b) the adsorption energy for dissociative chemisorption of O_2 at an oxygen-saturated vacancy. The corresponding geometries of the stable intermediate adsorption structures are shown in the lower panel (Reprinted with permission from J. M. Carlsson, F. Hanke, S. Linic and M. Scheffler, *Phys. Rev. Lett.*, **102**, 2009, 166104, Copyright 2009 American Physical Society).[36]

On the other hand, the curved microstructure of CNTs will slightly favor the chemisorption of O_2 on the surface of the CNTs, but the defects and tips of CNTs are still the most reactive to gas reactants.[37] The catalytic activities of various functionalities in regard to their physicochemical properties have been widely discussed in the literature (see Table 7.3).[18,38–48]

Table 7.3 The type of surface chemistry required or the nature of the active sites for carbon-catalyzing reactions.

Reactions	Surface chemistry/Active sites
Oxidative dehydrogenation	Quinones[18]
Dehydration of alcohols	Carboxylic acids[38]
Dehydrogenation of alcohols	Lewis acids and basic sites[39]
NO_x reduction	Acidic oxides + basic sites[40,41]
NO oxidation	Basic sites, vacancies[42]
SO_2 oxidation	Basic sites[43]
H_2S oxidation	Basic sites[44]
Hydrogen peroxide reactions	Basic sites[45]
Catalytic ozonation	Basic sites[46]
Catalytic wet air oxidation	Basic sites[47]

For redox reactions, basic functionalities, like quinone/ketone groups, have been considered as active sites.[12] To the best of our knowledge, the earliest model for the catalytic oxidation of ethylbenzene was proposed by Iwasawa *et al.* in 1973 (see Scheme 7.2) with the help of a model compound, which demonstrated the production of styrene *via* hydrogen transfer from ethylbenzene to diquinone groups associated with the formation of hydroquinone groups.[14] The hydrogenated catalysts could be regenerated after exposure to O_2. In Iwasawa's work, diquinone groups were regarded as the sole active site for the ODH reaction due to the uniform chemical properties of polynaphthoquinone. In 1988, Cadus *et al.* first identified the dependence of styrene productivity on the coke content formed on acidic oxide and quantitatively analyzed the function of surface oxygen functionalities.[15] They also suspected that electrophilic (electronegative) oxygen functional groups like quinone-type oxygen served as the active sites for the ODH of ethylbenzene. Afterwards, the assumption was strongly evidenced by Figueiredo *et al.*'s work, which disclosed a linear correlation between amounts of quinonic groups and its catalytic performance for the ODH of ethylbenzene (see Figure 7.1(b)).[12] To date, it has been widely acknowledged that basic oxygen functionalities with respect to their nucleophilicity act as active sites for C–H homocleavage. The identity and quantity of surface oxygen functionalities in reference to their thermal stability, electronegativity, vibrational function, and acidity, can be determined by temperature-programmed desorption (TPD) (Figure 7.4(a)),[18,19] X-ray photoelectron spectroscopy (XPS) (Table 7.4),[18] IR (Table 7.5),[19] and Boehm titration,[49,50] respectively.

The thermal decomposition temperatures of various oxygen functionalities determined by TPD are summarized in Figure 7.4(a), corresponding to the release of CO and CO_2 as a function of temperature.[18] The determination of distinct oxygen functional groups with respect to their electronegativity is listed in Table 7.4, which is evidence by both experimental and theoretical results.[18] The underlying principle of the Boehm titration is that oxygen functionalities on carbon surfaces have different acidities and can be neutralized by bases of different strengths.[49,50] Linear relationships between the number

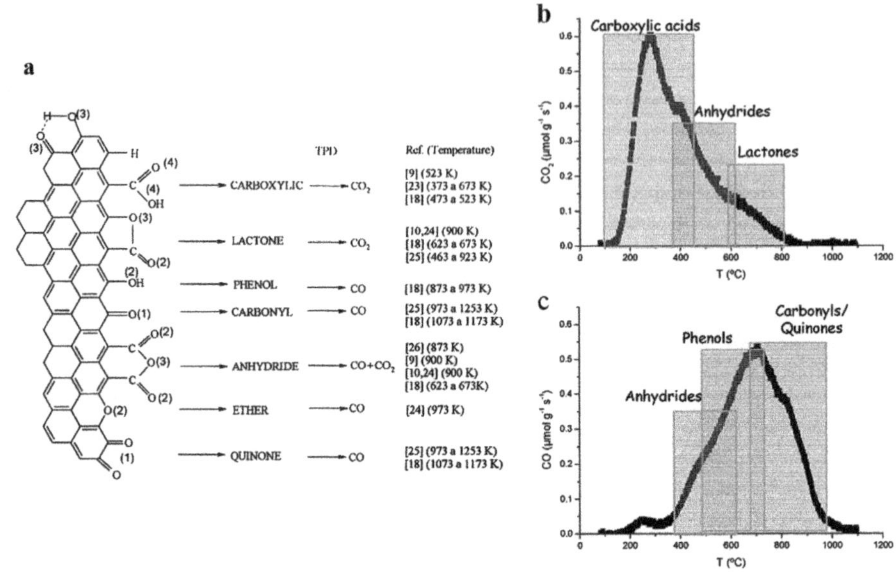

Figure 7.4 (a) Thermal decomposition temperature of oxygen functionalities on activated carbon (Reprinted from *Carbon*, 37, J. L. Figueiredo, M. F. R. Pereira, M. M. A. Freitas and J. J. M. Orfao, Modification of the surface chemistry of activated carbons, 1379–1389, Copyright 1999, with permission from Elsevier)[19] and evolution of (b) CO_2 and (c) CO upon decomposition of the oxygen functionalities as a function of temperature (Reprinted from *Catalysis Today*, **150**, J. L. Figueiredo and M. F. R. Pereira, The role of surface chemistry in catalysis with carbons, 2–7, Copyright 2010, with permission from Elsevier).[18]

Table 7.4 XPS binding energies of the O 1s regions for oxygen functional groups (Reprinted from *Catalysis Today*, **150**, J. L. Figueiredo and M. F. R. Pereira, The role of surface chemistry in catalysis with carbons, 2–7, Copyright 2010, with permission from Elsevier).[18]

Group	XPS Binding energy (eV)
C=O	531.1 ± 0.3
Hydroxyls, ethers and C=O in esters, amides, anhydrides	532.3 ± 0.3
C–O in esters and anhydrides	533.3 ± 0.3
Carboxylic groups	534.2 ± 0.3
Water	536.1 ± 0.3

of groups determined by using Boehm titration and the number of groups determined with TPD or XPS could be found (Figure 7.5), indicating the consistency and reliability of the results obtained from the different methods.[50,51]

However, it is worth mentioning that the thermal decomposition temperature of specific surface functionality can change up to a maximum extent of 200 °C.[19] It is not easy to identify particular oxygen functionalities with regard

Table 7.5 Characteristic vibrations of oxygen functionalities (Reprinted from *Carbon*, 37, J. L. Figueiredo, M. F. R. Pereira, M. M. A. Freitas and J. J. M. Orfao, Modification of the surface chemistry of activated carbons, 1379–1389, Copyright 1999, with permission from Elsevier).[19]

Group	IR assignment regions (cm^{-1})		
	1000–1500	1500–2050	2050–3700
C–O stretch of ethers	1000–1300		
Ether bridge between rings	1230–1250		
Cyclic ethers containing COCOC groups	1025–1141		
Alcohols	1049–1276		3200–3640
Phenolic groups:			2500–3620
C–OH stretch	1000–1220		
O–H bend/stretch	1160–1200		
Carbonates: Carboxyl-carbonates	1100–1500	1590–1600	
Aromatic C=C stretching		1585–1600	
Quinones		1550–1680	
Carboxylic acids (COOH)	1120–1200	1665–1760	2500–3300
Lactones	1160–1370	1675–1790	
Anhydrides	980–1300	1740–1880	
Ketenes (C=C=O)			2080–2200
C–H stretch			2600–3000

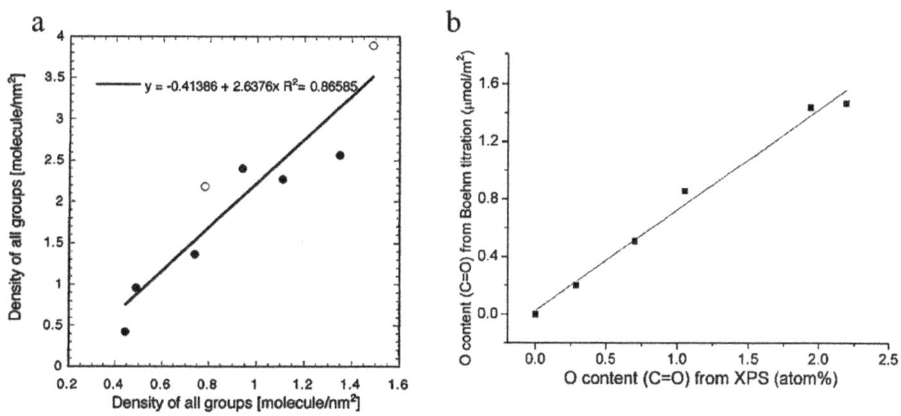

Figure 7.5 (a) Relationship between the number of groups determined from Boehm titration (*x*-axis) and TPD (*y*-axis) (Reprinted from *Journal of Colloid and Interface Science*, **240**, I. I. Salame and T. J. Bandoszz, Surface chemistry of activated carbons: combining the results of temperature-programmed desorption, Boehm and potentiometric titrations, 252–258, Copyright 2001, with permission from Elsevier),[50] (b) Relationship between the number of carbonyl groups determined from Boehm titration and XPS (Reprinted with permission from G. Wen, J. Diao, S. Wu, W. Yang, R. Schlögl and D. S. Su, *ACS Catal.*, 2015, **5**, 3600. Copyright 2015 American Chemical Society).[51]

to their thermal decomposition temperature or binding energy, which is evidenced by the overlap between the evolution of CO/CO_2 species upon decomposition of the oxygen functionalities. This ambiguity arises because the chemical properties of the functionalities can be strongly modulated by their chemical environment,[52,53] which suggests that reactivity of quinone/ketone on different sites might not be identical. Therefore, we speculate that the microstructure of carbonaceous catalysts should make great contributions to overall catalytic activity since we find that it is difficult to quantitatively compare catalytic activities of different carbonaceous catalysts in terms of styrene productivity normalized by the number of quinone/ketone groups. The dependence of the chemical reactivity of surface quinone groups and of the graphene surface itself on the local atomic structure has been discussed in the literature, indicating dramatically enhanced chemical reactivity at the curved or bent area of CNTs in comparison to the planar graphene surface (Figure 7.6).[52–55]

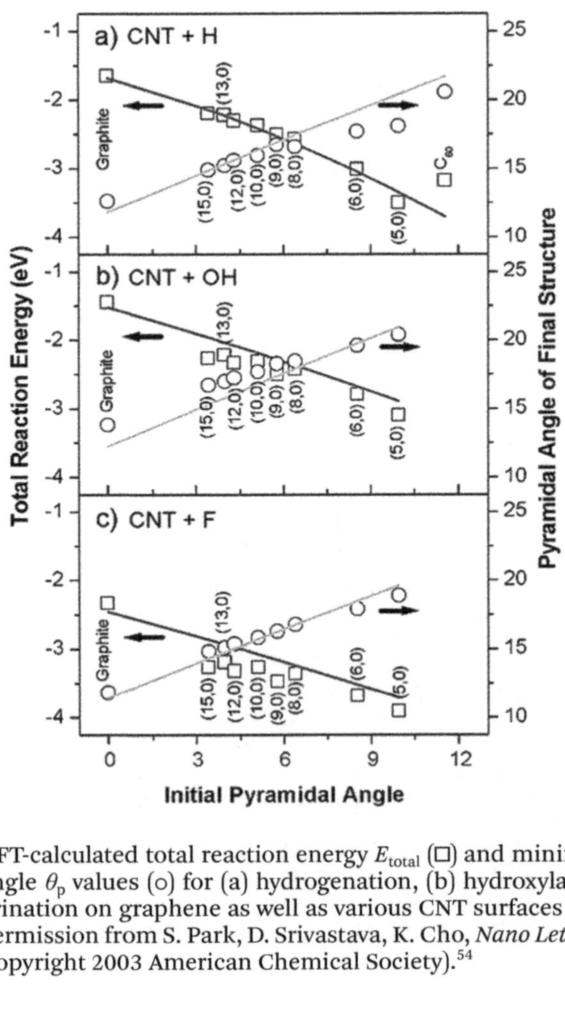

Figure 7.6 DFT-calculated total reaction energy E_{total} (□) and minimum pyramidal angle θ_p values (o) for (a) hydrogenation, (b) hydroxylation and (c) fluorination on graphene as well as various CNT surfaces (Reprinted with permission from S. Park, D. Srivastava, K. Cho, *Nano Lett.*, 2003, 3, 1273. Copyright 2003 American Chemical Society).[54]

Obviously, a combination of the aforementioned "organic surface groups" approach and the "solid state chemistry" approach could lead to a deeper insight into the realistic catalytic reaction occurring upon the surface of carbon materials. Moreover, the reaction kinetics related to the hydrogen transfer has not been well studied and the proposed mechanism of molecular oxygen activation is also disputed.[14] The proposed oxygen evolution process involving activation of O_2 on a surface of graphene sheets and consecutive regeneration of quinone group remains mostly speculative. Soot combustion and relevant oxygen activation have been extensively studied over decades, which demonstrated that the active sites capable of O_2 activation were also potential points of attack for O_2, leading to degradation of the graphitic structure.[56] DFT calculations suggested that the molecular oxygen activation should be more complicated than what we expected, and the oxygen-saturated carbon defects could alternatively serve as the active sites for molecular oxygen activation.[57,58]

In order to investigate the reaction mechanism and quantitatively identify the function of distinguished oxygen functionalities under reaction conditions, Su *et al.* developed a novel *in situ* chemical titration method to determine the dependence of the catalytic performance of carbonaceous catalysts on the number of quinoin functionalities.[59,60] Their studies demonstrated that the ethylbenzene conversion rate decreased immediately as phenyl hydrazine was introduced to neutralize carbonyl groups, and the conversion dropped to almost zero when the surface carbonyl groups were saturated by phenyl hydrazine. The linear dependence of ethylbenzene conversion on the uptake of phenyl hydrazine (see Figure 7.7) confirmed the key role of carbonyl groups as active sites for the ODH reaction.

Moreover, the productivity of styrene could be successfully normalized per amounts of carbonyl groups with a TOF value of 10×10^{-4} molecules of EB converted per C=O group per second. A good agreement between the TOF value obtained with the *in situ* chemical titration method (10×10^{-4} s^{-1}) and

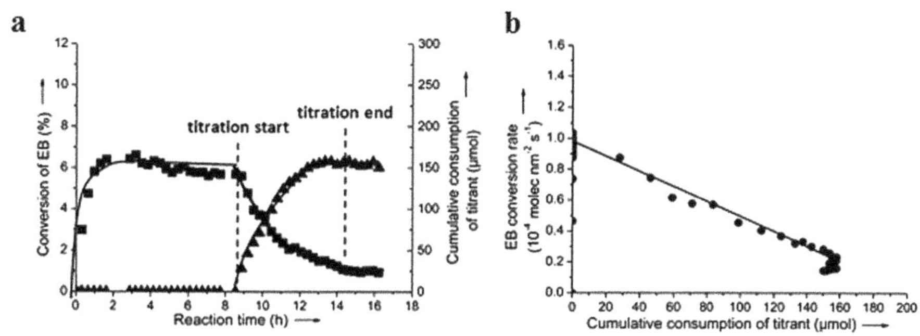

Figure 7.7 (a) EB conversion (■) and consumption of titrant (▲) as a function of time on stream, (b) EB conversion rate as a function of the cumulative consumption of the titrant pH (Reproduced with permission from ref. 59, © 2015 Wiley-VCH Verlag GmbH &Co. KGaA, Weinheim).

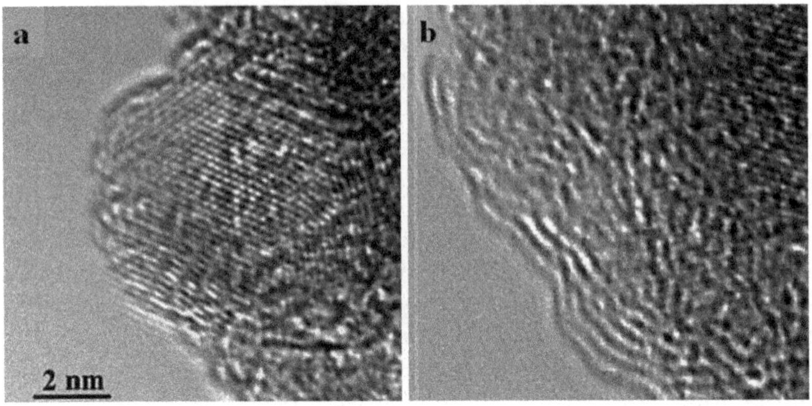

Figure 7.8 HRTEM images of a nanodiamond catalyst (a) before and (b) after oxidative dehydrogenation of ethylbenzene (Reprinted from *Carbon*, **45**, D. Su, N. I. Maksimova, G. Mestl, V. L. Kuznetsov, V. Keller, R. Schlögl, N. Keller, Oxidative dehydrogenation of ethylbenzene to styrene over ultra-dispersed diamond and onion-like carbon, 2145–2151, Copyright 2007, with permission from Elsevier).[30]

the value obtained with the *ex situ* TPD (3.87×10^{-4} s^{-1} Figure 7.1(b)) suggests a similar chemical nature of active centers in different forms of nanocarbon catalysts,[12,59,60] which is less relevant to their microstructure.

Apart from the sp^2-hybrid carbonaceous materials, an sp^3-hybrid carbon, like nanodiamond, was also tested as a catalyst for the ODH of hydrocarbons (see Figure 7.2(f) and Table 7.3).[30,61,62] The nanodiamond initially displayed low activity for the ODH of ethylbenzene, but its catalytic activity gradually increased as a function of reaction time. However, HRTEM images of the used catalyst showed formation of graphitized outer layers in the nanodiamond catalyst (Figure 7.8). This suggests that the reactive graphitic layers might make a major contribution to the catalytic performance. This assumption was supported by further experimental results that pre-graphized nanodiamond with a diamond–graphite core–shell structure displayed a better catalytic performance than non-treated nanodiamond.[63,64]

As we discussed above, the basicity of the surface functional groups governs the reactivity of the carbonaceous catalysts for the ODH reaction. It means that we can introduce heteroatoms to tailor the electronic and chemical properties of carbonaceous material as well as their catalytic activity.[65] For instance, the introduction of nitrogen functional groups into the surface of activated carbon can promote basicity of the carbon surface, resulting in a better capacity to adsorb acidic chemicals, like HCl, SO_2, CO_2, phenol, valeric acid and so on.[66–68] Formation of basic nitrogen functional groups (see Scheme 7.1) can be achieved by either reaction with nitrogen-containing reagents like ammonia and amines, or pyrolysis of carbon materials with melamine, pyridine, and urea.[69] Stohr's work disclosed that activated carbons treated by NH_3 or HCN displayed enhanced catalytic activities to catalytic oxidation reactions.[70] Later on, Zhao, *et al.* prepared nitrogen-doped carbon nanotubes *via* a pyrolysis method as an efficient catalyst for the direct dehydrogenation of ethylbenzene; they reported a production of 4.0 mmol g^{-1} h^{-1} of styrene in comparison with the 2.6 mmol g^{-1} h^{-1} of styrene obtained by using non-treated CNTs.[36] However, we cannot conclude that the formed nitrogen functionalities act as active sites for the dehydrogenation reaction since the modified CNTs also contained a certain amount of oxygen functional groups, including quinone/ketone groups. DFT calculations revealed that C atoms adjacent to N dopants, not the N atoms themselves, acted as active sites for chemisorption and activation of molecular O_2.[71] Accordingly, Su *et al.* ascribed the promoted catalytic activity to increased surface quinone groups and a higher content of surface defects,[72] while the catalytic role of the nitrogen functional groups was not clear at all. Meanwhile, the surface chemical nature of ammoniated CNTs have been widely investigated, demonstrating a thermally-induced reaction pathway between nitrogen and oxygen functional groups.[73] The evolution of a relative concentration of nitrogen functional groups as a function of ammoniation temperature indicated complex surface characteristics with respect to the different thermal stabilities of the nitrogen functional groups and resulting acid–base bifunctional catalysis (Table 7.6, Figure 7.9).[74]

The synthesis of amine-functionalized carbonaceous materials attracts increasing attention due to their potential applications in biomedical and material science.[65,73] However, as we mentioned above, amine functional groups are not thermally stable and can undergo rearrangement, resulting in the formation of N-functionalities with higher thermal stabilities, like pyridine

Table 7.6 Thermal decomposition temperatures of various nitrogen functional groups (Reprinted with permission from K. F. Ortega, R. Arrigo, B. Frank, R. Schlögl, A. Trunschke, *Chem. Mater.*, 2016, **28**, 6826. Copyright 2016 American Chemical Society).[74]

Name	Structure	TPD temperature peak maxima (T_{max} °C)
Ammonium carboxylate	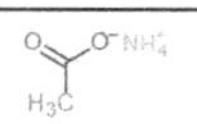	200
Nitrile	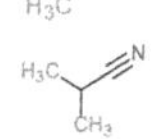	250
Imide	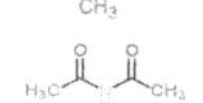	500
Lactam (pyridone)	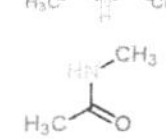	680
Pyrrole		785
Pyridine	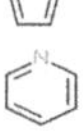	905
Quaternary N	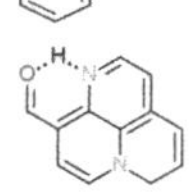	

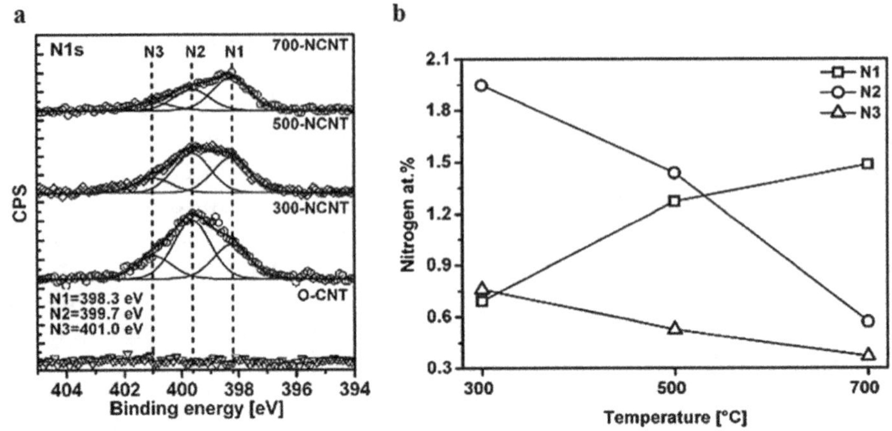

Figure 7.9 (a) XPS spectra of surface nitrogen species on CNTs with different ammoniation temperatures, (b) relative concentration of different nitrogen functional groups as a function of ammoniation temperature based on XPS. N1: pyridine, N2: pyrrole, imide, lactam, N3: quaternary N35 (Reprinted with permission from K. F. Ortega, R. Arrigo, B. Frank, R. Schlögl and A. Trunschke, *Chem. Mater.*, 2016, **28**, 6826. Copyright 2016 American Chemical Society).[74]

and pyrrole. This suggests that the amine-functionalized carbonaceous materials might not be good candidates for heterogeneous catalysis under severe conditions. To date, the enhanced catalytic activities of other kinds of N-doped nanocarbons or carbon hybrids have been reported in the literature, indicating the potential applications of the novel materials in heterogeneous catalysis.[75–78]

7.3 Catalytic Oxidation of Alkenes and Alkanes

Production of unsaturated hydrocarbons *via* catalytic oxidation of light alkanes and alkenes is very challenging but is very attractive from an industrial point of view. It can alternatively synthesize monomers instead of the catalytic dehydrogenation process as well as steam cracking of naphtha or liquefied petroleum gas. Considerable efforts have been carried out for the oxidative transformation of light alkanes using different transition metal oxides, like vanadium and molybdenum in particular.[79–85] These studies addressed a crucial role of oxygen species for C–H cleavage, which was considered as the rate-determining step for the formation of alkenes. However, the dehydrogenated products always displayed higher reactivity than the reactants, leading to further oxidation and low selectivity.[81] This suggests that a catalytic performance with both high conversion and high selectivity for the ODH of light alkanes could not be achieved from a thermodynamic point of view (see Figure 7.10), which makes the catalytic oxidation reaction less economically attractive to the chemical industry.[81]

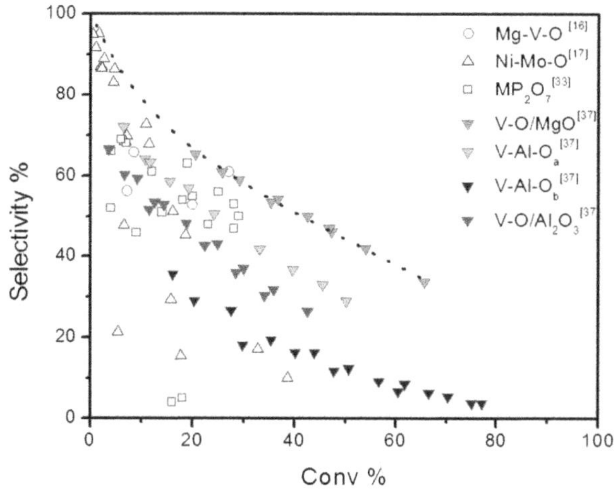

Figure 7.10 Selectivity-conversion plot for *n*-butane ODH alkanes.[41,82,85]

Despite decades of effort on catalyst design and optimization, none of these ODH catalysts have been commercialized on a large scale due to serious technical defaults. The promising catalytic performance of the nanocarbon catalysts in the ODH of ethylbenzene motivated explorative studies of carbonaceous materials as metal-free catalysts for the selective oxidation of alkanes/alkenes (Table 7.7).[14,86–89]

Iwasawa first employed polynaphthoquinone as a metal-free catalyst for the ODH of cyclohexane and cyclohexene and assigned quinonic groups as active sites for alkyl C–H bond cleavage.[14] Later on, coal was used as a catalyst for the oxidative dehydrogenation of butane.[86] However, the experimental result suggests that ash content, not graphitic carbon, should be the active species, since coal with a higher carbon content showed a poorer catalytic performance. Sui *et al.* used carbon nanofibers as catalysts for the selective oxidation of propane, which exhibited considerable catalytic activity in comparison to metal oxides (Figure 7.11).[87]

However, the catalysts also underwent serious gasification under the reaction conditions. In order to improve thermal stability, phosphoric acid was introduced to modify the carbon catalysts and, thereby, a catalytic performance of 16% propene yield was obtained after six hours' reaction.[87] Carbonyl groups were still considered as the active sites for the selective oxidation of propane based on TPSR (temperature-programmed surface reaction) and TPD results.[87] In 2006, Figueiredo and coauthors evaluated the catalytic performance of different kinds of activated carbon for the selective dehydrogenation of isobutene.[90] In this work, correlation between initial isobutene productivity and amounts of quinone groups was discussed, demonstrating the catalytic capabilities of the surface functionalities for the activation of the alkane/alkene. Su and coauthors reported the promising

Table 7.7 Catalytic performance of carbon catalysts in the ODH of light hydrocarbons.

Catalysts	Reactants	Products	Temperature K	Conv. %	Selec. %	Yield %
Coal[86]	Butane	Butene, butadiene	973	~40	~17	7
Polynaphthoquinone[14]	Cyclohexane	Cyclohexene Benzene	673	3.3	100%	1.8 (cyclohexene) 1.5 (benzene)
Polynaphthoquinone[14]	Cyclohexene	Cyclohexadiene Benzene	200	2.5	100	0.4 (cyclohexadiene) 2.15 (benzene)
CNTs[87]	Propane	Propene	500	42	40	17
CNTs[88]	Butene	Butadiene	400	72	60	43
Glassy carbon[89]	Butane	Butane	600	72	62	
Activated carbon[89]	Butene	Butane	600	85	62	
Graphite[89]		Butane	600	48	72	
Phosphorus modified graphite[89]		Butane	600	50	92	
Graphene[89]		Butane	600	70	68	
SWCNTs[89]			600	42	70	
Activated carbon[90]	Isobutane	Isobutene	648	25	60	15

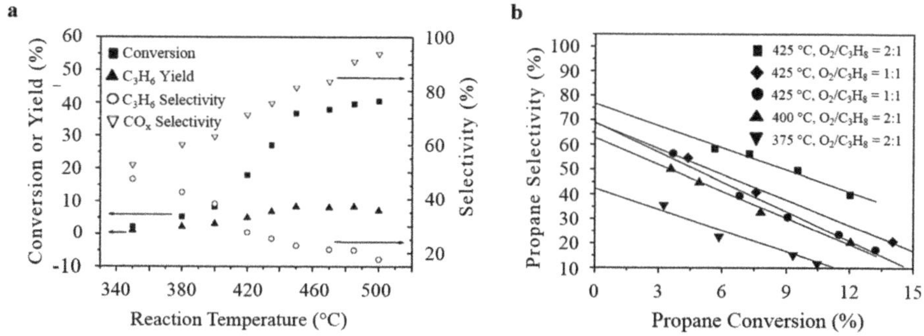

Figure 7.11 (a) Initial catalytic performance of carbon nanofibers as a function of reaction temperature, and (b) propene selectivity *versus* conversion under different reaction conditions (Reprinted from *Catalysis Today*, **106**, Z.-J. Sui, J.-H. Z. Y.-C. Dai and W.-K. Yuan, Oxidative dehydrogenation of propane over catalysts based on carbon nanofibers, 90–94, Copyright 2005, with permission from Elsevier).[87]

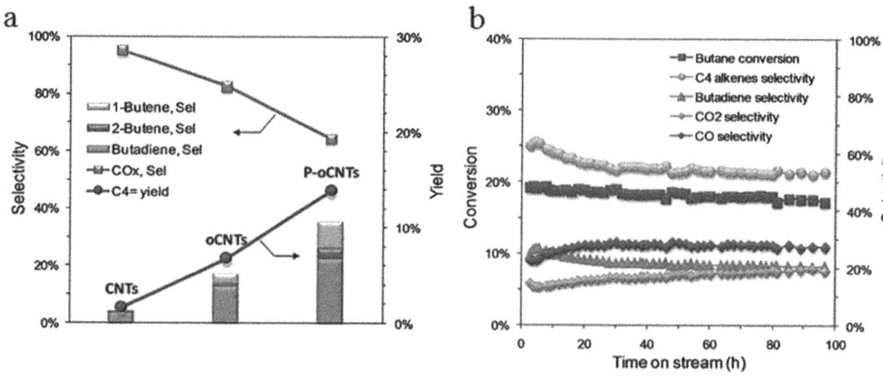

Figure 7.12 (a) Catalytic performance of pristine CNTs and CNTs treated in different ways, (b) catalytic performance of phosphoric-modified CNTs over 100 hours' testing (From J. Zhang, X. Liu, R. Blume, A. Zhang, R. Schlçgl and D. S. Su, *Science*, 2008, **322**, 73. Reprinted with permission from AAAS).[91]

catalytic performance of phosphoric-acid-modified CNTs for the selective oxidation of butane (Figure 7.12).[91]

The chemical properties of the surface oxygen functional groups of the modified CNTs were measured with near-ambient XPS under reaction conditions, which were compared with those of non-treated CNTs under identical conditions. The authors ascribe the promoted catalytic performance of the modified CNTs to the relatively higher content of quinono/ketone groups (Figure 7.13).

Afterwards, the surface chemistry of carbonaceous catalysts and the influence of heteroatoms, like B and P, on the ODH of light alkanes were systematically investigated by Su *et al.*[92,93] They found that activation of molecular

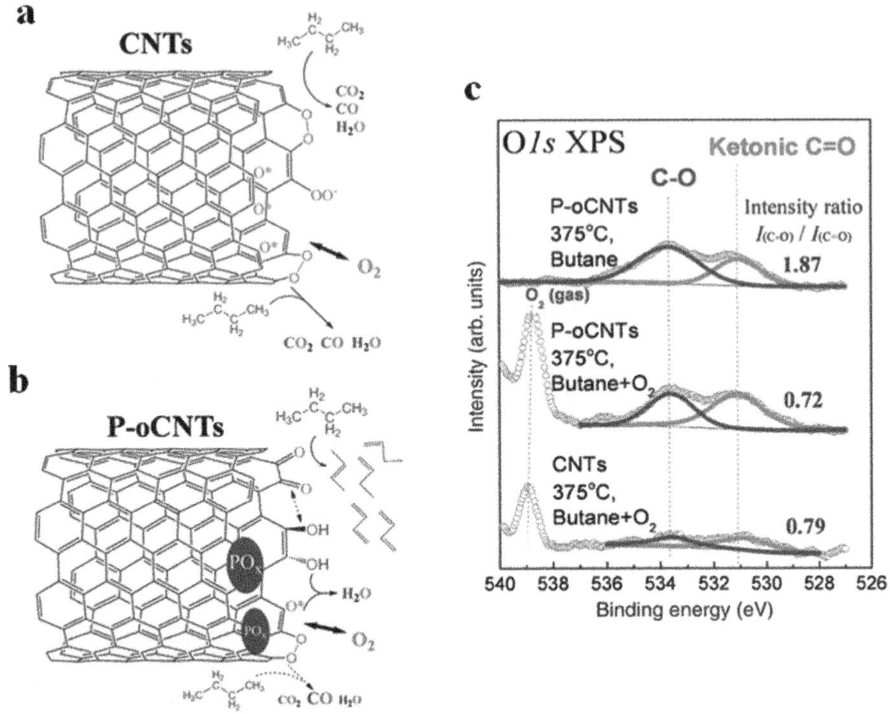

Figure 7.13 (a),(b) Schematic illustration of the oxidative dehydrogenation of butane upon oxidized CNTs and phosphoric-modified CNTs, respectively, (c) XPS O1s profiles of CNTs obtained under reaction conditions (From J. Zhang, X. Liu, R. Blume, A. Zhang, R. Schlçgl and D. S. Su, *Science*, 2008, **322**, 73. Reprinted with permission from AAAS).[91]

oxygen upon the surface of non-treated CNT catalysts led to the formation of both nucleophilic and electrophilic peroxide or superoxide intermediates, of which the latter were responsible for an uncontrollable complete oxidation of hydrocarbons as well as gasification of the carbon catalysts themselves. The authors assumed that there were two kinds of active sites for the catalytic oxidation of propane, which were responsible for the selective dehydrogenation and total oxidation, respectively. Addition of boron oxide or phosphorus oxide could quench the active sites for the non-selective oxidation, but did not affect the chemical nature of the active sites for the selective oxidation. This assumption was further supported by DFT studies on the chemisorption of O_2 on defected graphitic vacancies with and without boron oxide, which disclosed that dopants of B could weaken the chemisorption of O_2 and also inhibit combustion of graphitic carbon.[94–96] Jang *et al.* systematically tested various forms of graphitic carbon, including activated carbon, glassy carbon, graphite, graphene and single-walled carbon nanotubes, for the selective oxidation of butane under different reaction conditions (see Figure 7.14).[89]

They reported that a higher selectivity for alkenes can be obtained with a higher oxygen partial pressure or by using oxidized graphitic catalysts.

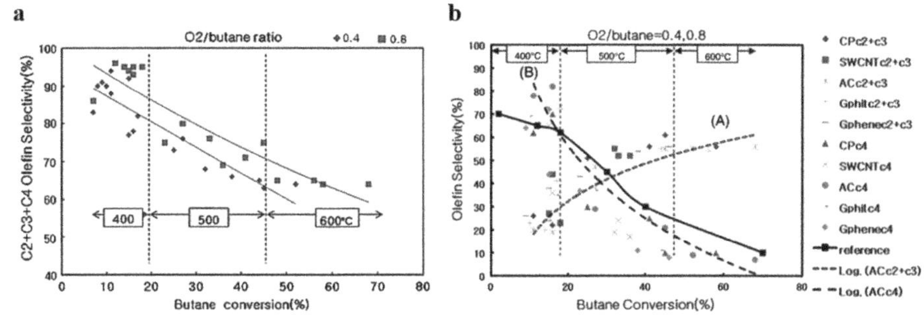

Figure 7.14 a) Plot of conversion *vs.* total olefin selectivity in the ODH of butane over carbon catalysts, and b) plot of conversion *vs.* butenes (B) and ethane + propene selectivity (A) in the ODH of butane over different carbon catalysts. (Reproduced from *Research on Chemical Intermediates*, Oxidative dehydrogenation of *n*-butane on nanocarbon catalysts having graphitic structures, 37, 2011, D. Y. Jang, © Springer Science + Business Media B.V. 2011. With permission of Springer).[89]

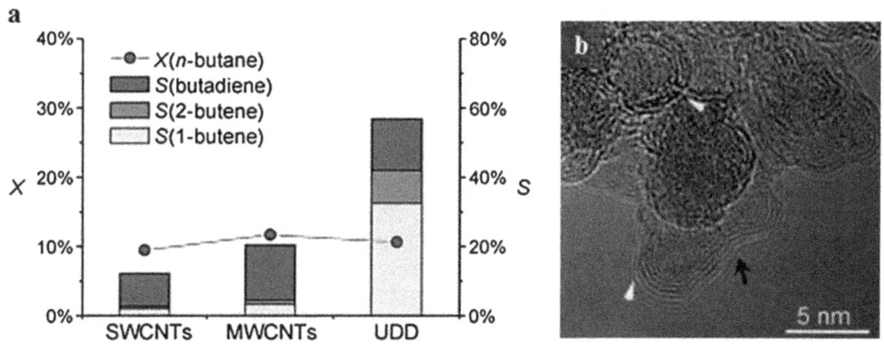

Figure 7.15 (a) Catalytic performance of various carbon catalysts for the oxidative dehydrogenation of butane (X = conversion, S = selectivity) (Reproduced with permission from ref. 100); (b) HRTEM image of used nanodiamond catalyst (Reproduced with permission from ref. 100, © 2011 Wiley-VCH Verlag GmbH & Co. KGaA, Weinheim).

The effect of phosphorus dopants on the catalytic performance was also examined in this work, which concluded that the phosphorus promotion only suppressed the combustion rather than facilitated the formation of alkenes.[97,98] Schwartz *et al.* also reported an explorative work on the ODH of isobutane using phosphorus-modified mesoporous carbon as a catalyst, which was consistent with the above-mentioned notion that phosphorus did not change the nature of the active centers for the ODH reaction, but instead inhibited formation of electrophilic oxygen species.[99]

Apart from the graphitic materials, non-treated nanodiamond also displayed a superior catalytic performance with a three-fold increase in the yield of butenes in comparison to CNTs (Figure 7.15(a)).[100] However, HRTEM

Table 7.8 Fraction of sp^2 carbon and oxygen content in annealed nanodiamond catalysts before and after reaction (Reproduced with permission from ref. 101, © 2014 Wiley-VCH Verlag GmbH & Co. KGaA, Weinheim).

	Fresh		Used	
Annealing temperature (°C)	sp^2%	Oxygen content %	sp^2%	Oxygen content %
Untreated	32	9.7	63	2.4
500	33	5.5	66	2.8
800	70	2.6	75	1.9
1100	75	2.0	79	1.9
1300	88	0.7	88	0.8

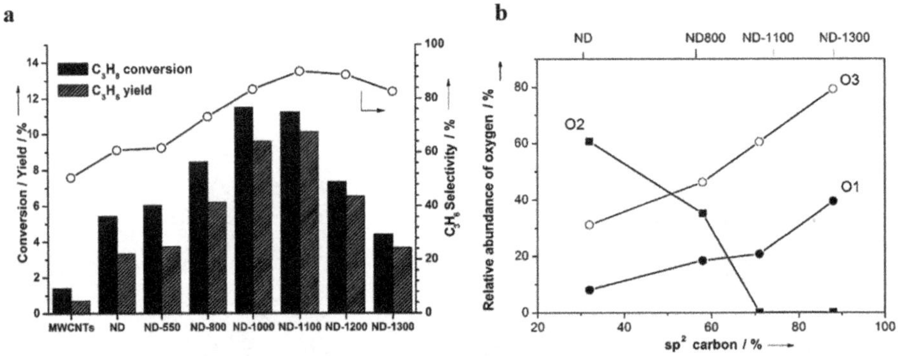

Figure 7.16 (a) Catalytic performance of nanodiamond catalysts annealed at different temperatures, and (b) concentration of different oxygen species *versus* fractions of sp^2 carbon O1: ketonic/quinonic carbonyl oxygen, O2: oxygen atoms in hydroxyls or ethers; O3: CO in esters and anhydrides (Reproduced with permission from ref. 101, © 2014 Wiley-VCH Verlag GmbH & Co. KGaA, Weinheim).

images showed that the used catalyst had a core–shell microstructure, with a diamond core covered by a concentric graphitic shell (Figure 7.15(b)).

The authors assumed that the formation of graphitic outer layers should be attributed to a lattice sp^3-to-sp^2 rearrangement induced by the chemical environment. This suggests that the graphitic defects and grafted surface functionalities should function as real active species for the selective oxidation of the light alkane. The structural characteristics related with the surface curvature and resulting partial sp^3 hybridization might enhance the catalytic performance as compared to the CNT catalysts. In order to investigate the synergistic effect arising from the core–shell microstructure and influence of the extent of surface graphization on the catalytic performance, Su *et al.* annealed nanodiamond at different temperatures to obtain a series of hybrid nanocomposites with a diamond core and graphitic outer surface with different sp^2/sp^3 hybrid contents (Table 7.8).[101,102] They found that the graphized nanodiamond catalysts with less oxygen content still exhibited a promising catalytic performance for the DDH of propane (Figure 7.16), which suggested that the bare defects/vacancies in the graphitic outer surface made their own contributions to the activation of the C–H bond in the light alkane. Theoretical calculations indicated the enhanced catalytic activity of oxygen functional groups and carbon defects on the surface of the sp^2@sp^3 core–shell nanocarbons in comparison to the counterparts in graphitic carbons.[103]

Later on, Su *et al.* discussed an alternative dehydrogenation pathway with single ketone groups as active centers, which was complementary to the well-known diquinone active centers.[104,105] The catalytic performance of nanodiamond for the ODH of propane could be further enhanced by modifying the catalyst with small amounts of borate.[106] These works confirmed the complex chemistry of the different kinds of functionalities involved in C–H activation behind the dehydrogenation of alkanes, in which single ketone groups, zig-zag edges, and diquinone/diketones made their own contributions to the formation of propene. These results give a new insight into the mechanism of the interaction between the heteroatom and carbon surface, as well as their influence on the catalytic behavior in the ODH reaction, which could provide principles for rational modification of nanocarbons.

7.4 Direct Dehydrogenation

The use of carbon-based materials as metal-free catalysts in the oxidative dehydrogenation of ethylbenzene and of light alkanes into the corresponding olefins has gained significant research attention as seen in the previous sections. However, the ODH process suffers from some disadvantages, such as the deep oxidation of reactants to carbon oxides (which will eventually reduce the selectivity), deep oxidation/combustion of the catalyst (which leads to loss of catalyst, eventually reducing the yield) and the use of flammable hydrocarbons at high temperatures in an oxygen atmosphere, which creates a highly flammable system and, therefore, potential safety issues. Direct dehydrogenation or dehydrogenation in the absence of any oxygen source can avoid problems related to deep oxidation and combustion of catalyst and reactant, which cause serious safety problems when extended to an industrial scale. However, commercial direct dehydrogenation catalyst systems—the K-promoted Fe-based catalyst (ethylbenzene to styrene)[107,108] and CrOx[109,110] Pt-based[111,112] catalyst (light alkanes to alkenes)—suffer from severe cracking and coking, which influence the product selectivity and catalyst stability, respectively. In addition, the adverse effects of Cr towards environmental and health systems necessitate a different catalyst. Recently, carbon materials, such as carbon nanotubes[75,113–116] mesoporous carbon,[31,117–121] nanodiamonds[62,64,101] and carbon nitride–nanodiamond hybrid materials,[122] were found to be active in the DDH of ethylbenzene[31,75,113–116] and lower alkenes.[101,117–121] The DDH over carbon-based materials has been performed at relatively higher temperatures than those of ODH (550 °C instead of 350 °C). The DDH of ethylbenzene to styrene over various carbon materials is summarized in Table 7.9.

Table 7.9 Catalytic performance of carbon catalysts for the DH of ethylbenzene.

	Temperature °C	Conversion %	Selectivity %	Yield %	Reaction rate (mmol-ST/g h)
CNTs[114]	550	—	98.5	~14.6	4.6
Mesoporous carbon[31,122]	550	—	96	~6.4	2
Nanodiamond[122]	550	—	99	~12.7	4
K–Fe[114,116]	550	—	94.5	~2.2	0.7

The optimum catalytic activity of these carbon materials depends on the structural defectiveness and/or the high chemical reactivity of the surface ketonic (C=O) groups. The ketonic groups have substantial electron density at the oxygen atom, which in turn can serve as the Lewis base to activate the C–H bond in the alkane. On CNTs and mesoporous carbon-based catalysts, the general approach to enhance the activity is N-doping and P-doping. As we discussed above, this doping improves the electron density of the ketonic groups and weakens the acidity of the overall system, which inhibits the phenolic or hydroxylic groups, thereby, promoting the cracking of hydrocarbons.[92,99,116,120,123] Hybridization of nanodiamonds to get a graphene-decorated outer surface is the general approach to enhance the DDH activity. This hybridization leads to an sp^3 to sp^2 lattice rearrangement with a defect-rich graphene layer, which provides carboxyl, anhydride, hydroxyl and ketonic functional groups.[91,100,101]

Su *et al.*[62] proposed that the ketonic groups, acting as electron donors, activate the alkane fragment of ethylbenzene. Styrene is released as the product and the hydroxyl group (C–OH) remains as an intermediate. The reaction cycle is completed by the thermal decomposition of C–OH to C=O and molecular hydrogen. This reaction is favorable at high temperatures. The strength of the covalent bond network (conjugated π electron cloud) in the styrene product is responsible for the significant activity in the ethylbenzene to styrene reaction compared to other alkane dehydrogenations. The unique electronic structure of graphene also allows the catalyst to reversibly store electrons with little structural change. These two effects lead to the easy desorption of styrene and an enhancement in the formation of H_2 from the hydroxyl (C–OH) intermediate. They also proposed that the initial deactivation of the catalyst could be due to hydrogen passivation. Zhao *et al.*[122] also attributed the initial decrease in activity due to the reduction of C=O by the H_2 formed during the reaction. However, none of these reports show the stoichiometric production of hydrogen. In all these cases, the activity was regained by exposing to air (an O_2 source).

The influence of trace amounts of O_2 is evident from the studies by Su *et al.*[64] The nanodiamond-catalyzed ethylbenzene DH under oxygen-free (DDH), oxygen-lean (LODH) and oxygen-rich (RODH) conditions shows that the lean-oxygen conditions (ethyl benzene–oxygen = 5) leads to 40% conversion and 92% styrene selectivity as opposed to 5.5% conversion and 98%

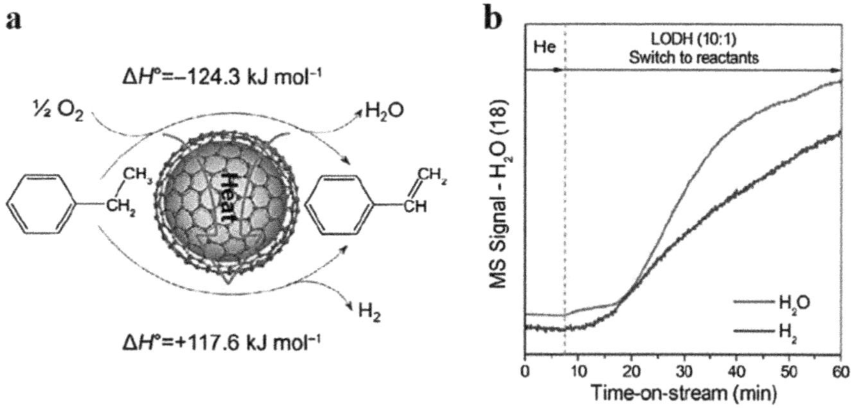

Figure 7.17 (a) Reaction process of lean oxygen dehydrogenation (LODH) of ethylbenzene, (b) online MS of water and hydrogen produced during LODH-10 (ethylbenzene–oxygen ratio 10:1) (Reproduced with permission from ref. 64, © 2016 WILEY-VCH Verlag GmbH & Co. KGaA, Weinheim).

selectivity at oxygen-free conditions. A further increase in O_2 to real ODH conditions (RODH) increases the conversion to 61% but the selectivity decreases to 89%. At LODH conditions, it is believed to be a combination of DDH and ODH. DDH is endothermic, which needs high reaction temperatures to break the C–H bond and release H_2, whilst ODH is exothermic due to the formation of water from hydrogen and oxygen. In oxygen-lean conditions (not enough oxygen to make the ethylbenzene–oxygen ratio 2:1), heat released from the ODH conditions can promptly transfer to the DDH process (as shown in Figure 7.17(a)). Mass spectroscopy signals show the evolution of H_2O and H_2; an attempt to calculate the proportion of each reaction (DDH and ODH) has also been carried out. The signal intensity of water increases faster than H_2 in the initial stage, suggesting that the DDH process benefits from ODH by saving part of the energy consumption (Figure 7.17(b)). The O_2 is also responsible for the *in situ* formation of carbonyl groups, which is active for the DH reactions, since carbonyl groups initially on the catalyst support tend to desorb or decompose during reaction conditions. The decomposition of carbonyl groups from nanodiamonds and mesoporous carbon under reaction conditions were reported.[101,102,117]

Studies on a pre-reduced ordered mesoporous carbon catalyst, which shows activity towards the DDH of isobutene, even without any oxygen-containing functional groups on the surface,[117] suggest that the carbon itself, with a lot of defects and vacancies, is responsible for the observed activity. A similar observation has been found in TPD studies before and after consecutive DDH catalytic cycles of propane over core–shell graphene/nanodiamond structures.[101] After the first TPD, where the removal of carboxyl groups has been observed, further TPD profiles in between catalytic cycles shows almost no CO evolution (apart from the carbonyl groups formed by traces of water impurity in the gas). Still, all these catalytic cycles give a similar yield, conversion and selectivity; this leads to the conclusion that the observed catalytic activity originates mainly from defects or vacancies on the surface. Normally, ketone functional groups anchored on defects can activate the C–H of alkane, but in their absence, the defects can do the same.[101] Again, none of these reports mentioned the production or fate of byproduct hydrogen.

7.5 Conclusion

The catalytic performance of functionalized nanocarbons for the oxidative or direct dehydrogenation of hydrocarbons has been briefly summarized in this chapter. As an alternative to metal oxide catalysts, novel metal-free catalysts display enhanced catalytic activities and promising stability for the production of valuable unsaturated hydrocarbons. The critical role of quinoin groups as active sites for catalytic dehydrogenation reactions has been rationalized based on experimental and theoretical results. Chemical modifications, like ammoniation and doping of phosphorus or boron atoms, can significantly enhance catalytic activity and stability by depressing non-selective oxidation and combustion of carbon catalysts themselves. The promising catalytic performance of novel metal-free catalysts with tunable architectures and surface properties indicates their potential applications as industrial catalysts.

Although promising achievements have been obtained in the synthesis and application of highly efficient metal-free catalysts, an in-depth understanding of the reaction mechanism and the chemical nature of the active sites are still limited. The activation of hydrocarbons on the quinoin groups has not been directly evidenced, and information about the intermediates formed during the catalytic cycle is still missing. Moreover, reaction pathways for oxygen evolution involving O_2 activation and generation/degradation of surface oxygen functionalities remain speculative. Therefore, detailed kinetic studies and well-defined surface characterization under realistic reaction conditions are urgently required to disclose the mechanism of carbon catalyzed reactions. The synthesis of an active carbon catalyst with a single targeted functional group could greatly favor mechanistic studies to establish a link between the chemical properties of particular functionalities to its catalytic activity. Obviously, the lack of fundamental understanding of the particular functions of metal-free catalysts and their surface functionalities will greatly hinder the development of an efficient catalyst design principle for industrial applications of the metal-free catalysis. As discussed above, there is still a lot of work to do in this new area before carbonaceous catalysts can be used for any industrial applications.

Acknowledgements

We gratefully acknowledge financial support from Syngaschem BV, Synfuels China Technology Co. Ltd and the National Natural Science Foundation of China (NSFC; grant. 21673273).

References

1. P. L. Pingali, *Proc. Natl. Acad. Sci. U. S. A.*, 2012, **109**, 12302.
2. R. M. Heck and R. J. Farrauto, *Appl. Catal., A*, 2001, **221**, 443.
3. R. M. Heck, R. J. Farrauto and S. T. Gulati, Automotive catalyst, *Catalytic Air Pollution Control*, John Wiley & Sons, Inc., Hoboken, NJ, USA, 3rd edn, 2009, vol. 6, p. 101.
4. R. Garg, N. K. Dutta and N. R. Choudhury, *Nanomaterials*, 2014, **4**, 267.
5. R. F. Rodriguez, *Carbon*, 1998, **36**, 159.
6. D. S. Su, J. Zhang, B. Frank, A. Thomas, X. Wang, J. Paraknowitsch and R. Schlögl, *ChemSusChem*, 2010, **3**, 169.
7. W. Qi and D. S. Su, *ACS Catal.*, 2014, **4**, 3212.
8. L. Liu, Y. P. Zhu, M. Su and Z. Y. Yuan, *ChemCatChem*, 2015, **7**, 2765.
9. H. Hu, J. H. Xin, H. Hu, X. Wang and Y. Kong, *Appl. Catal., A*, 2015, **492**, 1.
10. P. Serp and J. L. Figueiredo, *Carbon Materials for Catalysis*, John Wiley & Sons, Inc., Hoboken, New Jersey, 2009, p. 6.
11. R. W. Coughlin, Carbon as adsorbent and catalyst, *Ind. Eng. Chem. Prod. Res. Dev.*, 1969, **8**, 12.
12. (a) F. R. Pereira, J. J. M. Orfao and J. L. Figueiredo, *Appl. Catal., A*, 1999, **184**, 153; (b) M. F. R. Pereira, J. J. M. Orfao and J. L. Figueiredo, *Appl. Catal., A*, 2000, **196**, 43; (c) M. F. R. Pereira, J. J. M. Orfao and J. L. Figueiredo, *Appl. Catal., A*, 2001, **218**, 307.
13. (a) J. Manassen and J. Wallach, *J. Am. Chem. Soc.*, 1965, **87**, 2671; (b) J. Manassen and S. Khalif, *J. Am. Chem. Soc.*, 1966, **88**, 1943; (c) J. Manassen and S. Khalif, *J. Catal.*, 1967, **7**, 110; (d) J. Manassen and S. Khalif, *J. Catal.*, 1969, **13**, 290.
14. Y. Iwasawa, H. Nobe and S. Ogasawara, *J. Catal.*, 1973, **31**, 444.
15. L. E. Cadus, L. A. Arrua, O. F. Gorriz and J. B. Rivarola, *Ind. Eng. Chem. Res.*, 1988, **27**, 2241.
16. C. H. Collett and J. McGregor, *Catal. Sci. Technol.*, 2016, **6**, 363.
17. V. Zarubina, C. Nederlof, B. van der Linden, F. Kapteijn, H. J. Heeres, M. Makkee and I. Melián-Cabrera, *J. Mol. Catal. A: Chem.*, 2014, **381**, 179.
18. J. L. Figueiredo and M. F. R. Pereira, *Catal. Today*, 2010, **150**, 2.
19. J. L. Figueiredo, M. F. R. Pereira, M. M. A. Freitas and J. J. M. Orfao, *Carbon*, 1999, **37**, 1379.
20. K. P. De Jong and J. W. Geus, *Catal. Rev.*, 2000, **42**, 481.
21. G. Mestl, N. Maksimova, N. Keller, V. Roddatis and R. Schlögl, *Angew. Chem.*, 2001, **40**, 2066.
22. M. Pereira, J. Figueiredo, J. Orfao, P. Serp, P. Kalck and Y. Kihn, *Carbon*, 2004, **42**, 2807.
23. J. J. Delgado, R. Vieira, G. Rebmann, D. S. Su, N. Keller, M. J. Ledoux and R. Schlögl, *Carbon*, 2006, **44**, 809.
24. L. Wang, J. Zhang, D. S. Su, Y. Ji, X. Cao and F. S. Xiao, *Chem. Mater.*, 2007, **19**, 2894.
25. L. Roldan, A. M. Benito and E. García-Bordeje, *J. Mater. Chem. A*, 2015, **3**, 24379.

26. J. Diao, H. Liu, J. Wang, Z. Feng, T. Chen, C. Miao, W. Yang and D. S. Su, *Chem. Commun.*, 2015, **51**, 3423.
27. I. Soykal, H. Wang, J. Park, A. P. Li, C. Liang and V. Schwartz, *J. Mater. Chem. A*, 2015, **3**, 8667.
28. D. S. Su, N. Maksimova, J. J. Delgado, N. Keller, G. Mestl, M. J. Ledoux and R. Schlögl, *Catal. Today*, 2005, **102**, 110.
29. D. S. Su, S. Perathoner and G. Centi, *Chem. Rev.*, 2013, **113**, 5782.
30. D. S. Su, N. I. Maksimova, G. Mestl, V. L. Kuznetsov, V. Keller, R. Schlögl and N. Keller, *Carbon*, 2007, **45**, 2145.
31. Z. Zhao, Y. Dai, J. Lin and G. Wang, *Chem. Mater.*, 2014, **26**, 3151.
32. N. Keller, N. I. Maksimova, V. V. Roddatis, M. Schur, G. Mestl, Y. V. Butenko, V. L. Kuznetsov and R. Schlögl, *Angew. Chem., Int. Ed.*, 2002, **41**, 1885.
33. W. Oganowski, J. Hanuza and L. Kepinski, *Appl. Catal., A*, 1998, **171**, 145.
34. B. M. Reddy, P. Lakshmanan, S. Loridant, Y. Yamada, T. Kobayashi, C. L. Cartes, T. C. Rojas and A. J. Fernandez, *Phys. Chem. B*, 2006, **110**, 9140.
35. F. M. Bautista, J. M. Campelo, D. Luna, J. M. Marinas, R. A. Quiros and A. A. Romero, *Appl. Catal., B*, 2007, **70**, 611.
36. Z. K Zhao, W. Z. Li, Y. T. Dai G. F. Ge X. W. Guo and G. R. Wang, *ACS Sustainable Chem. Eng.*, 2015, **3**, 3355.
37. J. A. Robinson, E. S. Snow, Ş. C. Bădescu, T. L. Reinecke and F. K. Perkins, *Nano Lett.*, 2006, **6**, 1747.
38. C. M. Castilla, F. C. MarÍn, C. P. Pérez and M. V. López Ramón, *Carbon*, 2001, **39**, 869.
39. G. S. Szymański and G. Rychlicki, *Carbon*, 1993, **31**, 247.
40. H. Teng, Y. T. Tu, Y. C. Lai and C. C. Lin, *Carbon*, 2001, **39**, 575.
41. G. S. Szymański, T. Grzybek and H. Papp, *Catal. Today*, 2004, **90**, 51.
42. I. Mochida, Y. Kawabuchi, S. Kawano, Y. Matsumura and M. Yoshikawa, *Fuel*, 1997, **76**, 543.
43. E. R. Piñero, D. C. Amorós and A. L. Solano, *Carbon*, 2003, **41**, 1925.
44. F. Adib, A. Bagreev and T. J. Bandosz, *Langmuir*, 1999, **16**, 1980.
45. L. B. Khalil, B. S. Girgis and T. A. M. Tawfik, *J. Chem. Technol. Biotechnol.*, 2001, **76**, 1132.
46. M. S. Polo, U. Gunten and J. R. Utrilla, *Water Res.*, 2005, **39**, 3189.
47. W. Oganowski, J. Hanuza and L. Kepinski, *Appl. Catal., A*, 1998, **171**, 145.
48. M. Li, F. Xu, H. Li and Y. Wang, *Catal. Sci. Technol.*, 2016, **6**, 3670.
49. H. P. Boehm, *Carbon*, 1994, **32**, 759.
50. I. I. Salame and T. J. Bandosz, *J. Colloid Interface Sci.*, 2001, **240**, 252.
51. G. Wen, J. Diao, S. Wu, W. Yang, R. Schlögl and D. S. Su, *ACS Catal.*, 2015, **5**, 3600.
52. Q. C. Sun, H. L. Bai, B. K. Tay, S. Li and E. Y. Jiang, *J. Phys. Chem. B*, 2003, **107**, 7544.
53. Y. Zhang and Z. Liu, *J. Phys. Chem. B*, 2004, **108**, 11435.
54. S. Park, D. Srivastava and K. Cho, *Nano Lett.*, 2003, **3**, 1273.
55. B. Q. Pham, V. H. Nguyen and T. N. Truong, *Carbon*, 2016, **101**, 16.
56. R. Backreedy, J. M. Jones, M. Pourkashanian and A. Williams, *Faraday Discuss.*, 2002, **119**, 385.

57. A. B. S. Tapia, X. G. Carmona and L. R. Radovic, *Carbon*, 2012, **50**, 1152.
58. X. Qi, W. Song and J. Shi, *PLoS One*, 2017, **12**, e0173864.
59. W. Qi, W. Liu, X. Guo, R. Schlögl and D. S. Su, *Angew. Chem.*, 2015, **54**, 13682.
60. X. Guo, W. Qi, W. Liu, P. Yan, F. Li, C. Liang and D. S. Su, *ACS Catal.*, 2017, **7**, 1424.
61. P. E. Fanning and M. A. Vannice, *Carbon*, 1993, **31**, 721.
62. J. Zhang, D. S. Su, R. Blume, R. Schlögl, R. Wang, X. Yang and A. Gajovic, *Angew. Chem.*, 2010, **49**, 8640.
63. Y. M. Lin and D. S. Su, *ACS Nano*, 2014, **8**, 7823.
64. J. Diao, Z. Feng, R. Huang, H. Liu, S. B. A. Hamid and D. S. Su, *ChemSusChem*, 2016, **9**, 662.
65. R. Kannan and V. K. Pillai, Amine-Functionalized Carbon Nanotubes, *Surface Modification of Nanotube Fillers*, Wiley-VCH Verlag GmbH & Co. KGaA, Weinheim, Germany, 5th edn, 2011.
66. C. L. Mangun, K. R. Benak, J. Economy and K. L. Foster, *Carbon*, 2001, **39**, 1809.
67. G. Yang, H. Chen, H. Qin and Y. Feng, *Appl. Surf. Sci.*, 2014, **293**, 299.
68. Y. E. Sayeda and T. J. Bandosz, *Phys. Chem. Chem. Phys.*, 2003, **5**, 4892.
69. M. Terrones, P. Redlich, N. Grobert, S. Trasobares, W. Hsu, H. Terrones, Y. Zhu, J. P. Hare, C. L. Reeves, A. K. Cheetham, M. Rühle, H. W. Kroto and R. M. David, *Adv. Mater.*, 1999, **11**, 655.
70. B. Stohr, H. P. Boehm and R. Schlögl, *Carbon*, 1991, **29**, 707.
71. X. Hu, Y. Wu and Z. Zhang, *J. Mater. Chem.*, 2012, **22**, 15198.
72. L. Shi, W. Qi, W. Liu, P. Yan, F. Li, J. Sun, D. S. Su, *Catal. Today*, 2018, **301**, 48.
73. R. Arrigo, M. Hävecker, S. Wrabetz, R. Blume, M. Lerch, J. McGregor, E. P. J. Parrott, J. A. Zeitler, L. F. Gladden, A. Knop-Gericke, R. Schlögl and D. S. Su, *J. Am. Chem. Soc.*, 2010, **132**, 9616.
74. K. F. Ortega, R. Arrigo, B. Frank, R. Schlögl and A. Trunschke, *Chem. Mater.*, 2016, **28**, 6826.
75. Z. Zhao, Y. Dai, G. Ge, X. Guo and G. Wang, *Phys. Chem. Chem. Phys.*, 2015, **17**, 18895.
76. M. Li, F. Xu, H. Li and Y. Wang, *Catal. Sci. Technol.*, 2016, **6**, 3670.
77. D. Yu, E. Nagelli, F. Du and L. Dai, *J. Phys. Chem. Lett.*, 2010, **1**, 2165.
78. L. Liu, Y. Zhu, M. Su and Z. Yuan, *ChemCatChem*, 2015, **7**, 2765.
79. W. D. Harding, H. H. Kung, V. L. Kozhevnikov and K. R. Poeppelmeier, *J. Catal.*, 1993, **144**, 597.
80. H. H. Kung, *Adv. Catal.*, 1994, **40**, 1.
81. C. Batiot and B. K. Hodnett, *Appl. Catal., A*, 1996, **137**, 179.
82. R. M. M. Aranda, M. F. Portela, L. M. Madeira, F. Freire and M. Oliveira, *Appl. Catal., A*, 1995, **127**, 201.
83. L. M. Madeira and M. F. Portela, Catalytic oxidative dehydrogenation of *n*-butane, *Catal. Rev.*, 2002, **44**(2), 247–286.
84. I. C. Marcua, I. Sandulescu and M. J. M. Millet, *Appl. Catal., A*, 2002, **227**, 309.

85. J. M. L. Nieto, P. Concepcion, A. Dejoz, H. Knozinger, F. Melo and M. I. Vasquez, *J. Catal.*, 2000, **189**, 147.
86. F. J. M. Hodar, L. M. Madeira and M. F. Portela, *Appl. Catal., A*, 1999, **178**, 49.
87. Z. J. Sui, J. H. Zhou, Y. C. Dai and W. K. Yuan, *Catal. Today*, 2005, **106**, 90.
88. X. Liu, D. S. Su and R. Schlögl, *Carbon*, 2008, **46**, 547.
89. D. Y. Jang, H. G. Jang, G. R. Kim and G. Kim, *Res. Chem. Intermed.*, 2011, **37**, 1145.
90. J. D. D Velasquez, L. A. C. Suarez and J. L. Figueiredo, *Appl. Catal., A*, 2006, **311**, 51.
91. J. Zhang, X. Liu, R. Blume, A. Zhang, R. Schlçgl and D. S. Su, *Science*, 2008, **322**, 73.
92. B. Frank, J. Zhang, R. Blume, R. Schlögl and D. S. Su, *Angew. Chem.*, 2009, **48**, 6913.
93. B. Frank, M. Morassutto, R. Schomacker, R. Schlögl and D. S. Su, *ChemCatChem*, 2010, **2**, 644.
94. J. Liu, T. Liang, C. Wang and W. Lai, *RSC Adv.*, 2017, **7**, 3257.
95. X. K. Kong, C. L. Chen and Q. W. Chen, *Chem. Soc. Rev.*, 2014, **43**, 2841.
96. J. Albero and H. Garcia, *J. Mol. Catal. A: Chem.*, 2015, **408**, 296.
97. J. Liu, T. Liang, C. Wang and W. Lai, *RSC Adv.*, 2017, **7**, 3257.
98. X. Wu and L. R. Radovic, *Carbon*, 2006, **44**, 141.
99. V. Schwartz, H. Xie, H. M. Meyer III, S. H. Overbury and C. Liang, *Carbon*, 2011, **49**, 659.
100. X. Liu, B. Frank, W. Zhang, T. P. Cotter, R. Schlögl and D. S. Su, *Angew. Chem., Int. Ed.*, 2011, **50**, 3318.
101. R. Wang, X. Sun, B. Zhang, X. Sun and D. S. Su, *Chem.–Eur. J.*, 2014, **20**, 6324.
102. X. Sun, R. Wang, B. Zhang, R. Huang, X. Huang, D. S. Su, T. Chen, C. Miao and W. Yang, *ChemCatChem*, 2014, **6**, 2270.
103. T. Liu, S. Ali, B. Li and D. S. Su, *ACS Catal.*, 2017, **7**, 3779.
104. X. Sun, B. Li and D. S. Su, *Chem.–Asian J.*, 2016, **11**, 1668.
105. X. Sun, B. Li and D. S. Su, *Chem. Commun.*, 2014, **50**, 11016.
106. X. Sun, Y. Ding, B. Zhang, R. Huang and D. S. Su, *Chem. Commun.*, 2015, **51**, 9145.
107. B. D. Herzog and H. F. Raso, *Ind. Eng. Chem. Prod. Res. Dev.*, 1984, **23**, 187.
108. A. C. Oliveira, J. L. G. Fierro, A. Valentni, P. S. S. Norbe and M. do Carmo Rangel, *Catal. Today*, 2003, **85**, 49.
109. S. T. Korhonen, S. M. K. Airaksinen, M. A. Banares and A. O. I. Krause, *Appl. Catal., A*, 2007, **333**, 30.
110. H. Zimmermann and F. Versluis, US Pat. 5378350, 1995.
111. M. L. Casella, G. J. Siri, G. F. Santori, O. A. Ferretti and M. M. Ramirez-Corredores, *Langmuir*, 2000, **16**, 5639.
112. R. R. Herber and G. J. Thompson, US Pat. 5143886, 1992.
113. Z. Zhao, Y. Dai, G. Ge, X. Guo and G. Wang, *Green Chem.*, 2015, **17**, 3723.

114. Z. Zhao, Y. Dai, G. Ge, X. Guo and G. Wang, *RSC Adv.*, 2015, **5**, 53095.
115. Z. Zhao, Y. Dai, G. Ge and G. Wang, *AIChE J.*, 2015, **61**, 2543.
116. Z. Zhao, Y. Dai and G. Ge, *Catal. Sci. Technol.*, 2015, **5**, 1548.
117. Z. Zhang, Y. Li, J. Wang, H. Yang, N. Li, C. Ma and Z. Hao, *Catal. Sci. Technol.*, 2016, **6**, 4863.
118. L. Liu, Q. Deng, Y. Liu, T. Ren and Z. Yuan, *Catal. Commun.*, 2011, **16**, 81.
119. L. Liu, Q. Deng, B. Agula, X. Zhao, T. Ren and Z. Yuan, *Chem. Commun.*, 2011, **47**, 8334.
120. L. Li, W. Zhu, Y. Liu, L. Shi, H. Liu, Y. Ni, S. Liu, H. Zhou and Z. Liu, *RSC Adv.*, 2015, **5**, 56304.
121. Y. Li, Z. Zhang, J. Wang, C. Ma, H. Yang and Z. Hao, *Chin. J. Catal.*, 2015, **36**, 1214.
122. Z. Zhao and Y. Dai, *J. Mater. Chem. A.*, 2014, **2**, 13442.
123. Y. V. Kissin, *Catal. Rev.: Sci. Eng.*, 2001, **43**, 85.

CHAPTER 8

Electrochemistry Reactions Catalyzed by Metal-free Functionalized Carbons

CHENG TANG AND QIANG ZHANG*

Beijing Key Laboratory of Green Chemical Reaction Engineering and Technology, Department of Chemical Engineering, Tsinghua University, Beijing 100084, China
*E-mail: zhang-qiang@mails.tsinghua.edu.cn

8.1 Introduction

The exploration for new and sustainable energy sources has become one of the most pivotal global challenges. One promising alternative to current energy systems based on fossil fuels is the "hydrogen economy", which is based on splitting water *via* electrocatalysis or photocatalysis to produce clean and sustainable H_2.[1–10] However, the oxygen evolution reaction (OER, $2H_2O \rightarrow O_2 + 4H^+ + 4e^-$, in acid) at the anode is a bottleneck in water electrolyzers due to its sluggish kinetics with a high overpotential, which limits the energy efficiency and practical promotion of splitting water.[3,5] Furthermore, highly efficient and effective energy storage and conversion technologies are also required to power devices used in our daily lives based on sustainable energy sources. Fuel cells[11–14] and metal–air batteries[15–17] have been considered over the last few decades due to their high energy densities and open

Catalysis Series No. 31
Metal-free Functionalized Carbons in Catalysis: Synthesis, Characterization and Applications
Edited by Alberto Villa and Nikolaos Dimitratos

Published by the Royal Society of Chemistry, www.rsc.org

cell design. In both cases, electrocatalysts with satisfactory electrocatalytic activity for the oxygen reduction reaction (ORR, $O_2 + 4H^+ + 4e^- \rightarrow 2H_2O$, in acid) are required. Even better are bifunctional ORR/OER catalysts in oxygen electrodes for regenerative operations and to accelerate these sluggish processes.[18,19]

Noble-metal-based materials, such as Pt for ORR, and IrO_2 or RuO_2 for OER, are widely recognized as state-of-the-art electrocatalysts. However, their high cost and the scarcity of precious metal resources limit the industrial applications.[4,20–23] Therefore, tremendous efforts have been directed towards the development of noble-metal-free catalysts for ORR and OER, including perovskites,[12,24] transition metal/metal oxides,[13,25] and functionalized carbon materials.[26–32] In particular, metal-free functionalized nanocarbon materials have been intensively investigated recently, encouraged by pioneering work from Dai's group in 2009.[33] Heteroatom-doped (N, O, B, P, S, Si, *etc.*) carbon nanomaterials are well-demonstrated as excellent ORR catalysts under alkaline conditions.[34–52] Recently, some nitrogen-doped carbon materials have also shown remarkable OER activities, but limited durability.[53–55] To date, various carbon materials have been developed as promising ORR or OER catalysts, due to their high conductivity and flexibility, remarkable activity, facile preparation, tunable structure and surface chemistry, and economic viability (Figure 8.1).[36,41,56]

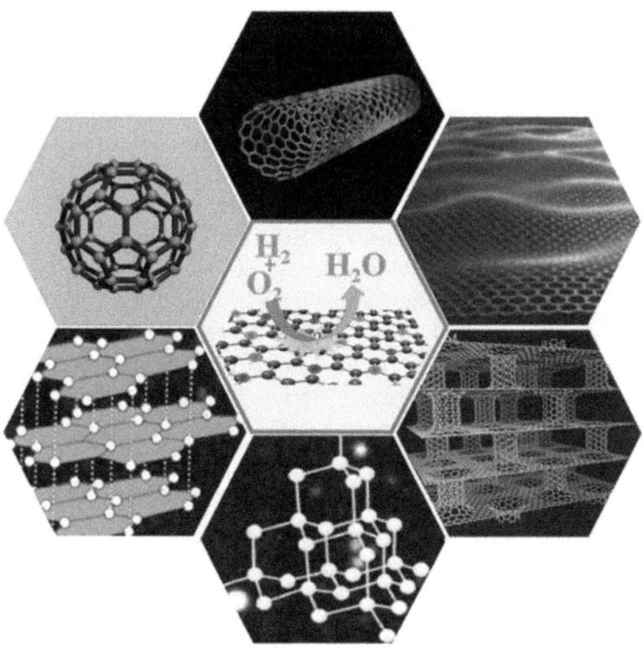

Figure 8.1 Various carbon materials for ORR catalysis. (Reprinted with permission from L. M. Dai, Y. H. Xue, L. T. Qu, H. J. Choi and J. B. Baek, *Chem. Rev.*, 2015, **115**, 4823. Copyright 2015 American Chemical Society.[41])

Along with the great advances achieved in the field of material engineering, a fundamental understanding of metal-free nanocarbon-based electrocatalysis has also been significantly enhanced. The reactivity of heteroatom-doped carbon materials is firstly ascribed to the altered charge[33,57] or spin distribution[58,59] in the sp^2 carbon matrix, facilitating oxygen adsorption or/and subsequent O–O bond breaking. Additionally, the surface functional groups, edge dangling bonds, or even defective sites in functionalized nanocarbon materials are also claimed to be beneficial for the electrocatalytic activity.[31,36,60,61] Specifically, carbon materials with higher edge exposure for nitrogen doping within the distance of the edge effect,[31,62] are expected to be more active. Most recently, the intrinsic topological defects (pentagon, zigzag edge, *etc.*) have been identified to be catalytically active sites for the ORR and OER themselves, or are able to alter the doping behaviour and improve the electrochemical activity of heteroatom-doped nanocarbon catalysts.[63–69] Despite these advances, the knowledge of critical active sites and their specific roles for the ORR and OER in metal-free functionalized carbon catalysts remains confusing and controversial, resulting in the dominant trial-and-error research for further development.[70]

As seen above, many scientific explorations have been conducted on nanocarbon materials with enhanced catalytic performances, and many excellent reviews have already appeared to afford comprehensive scenarios of this group of electrocatalysts.[3,23,36,41,56,71] In this chapter, we will focus on discussing the origins of the activity of metal-free functionalized carbon catalysts based on a comprehensive digestion of current achievements. It is not our intention to survey all the achievements in this field, but to highlight some critical and representative developments, aiming at a concise and clear summary of the underlying mechanism and also some essential insights for targeted optimization. It is fundamentally important and technically promising to promote the rational design of metal-free functionalized carbon catalysts for the ORR, OER, and other important electrochemistry reactions in sustainable chemistry. As metal-free nanocarbon catalysts are more promising for the ORR and have been intensively researched, we will primarily focus on the ORR to discuss the different activity origins, including dopants, edges, and defects. The development of functionalized carbon materials for OER catalysts, including N-doped, oxidized, and dual-doped carbons, is also included. Especially, the applications of metal-free carbons as multifunctional catalysts for the ORR, OER or hydrogen evolution reaction (HER, $2H^+ + 2e^- \rightarrow H_2$, in acid) are summarized in this chapter.

8.2 Oxygen Electrochemistry Reactions

The ORR/OER catalysis is kinetically sluggish due to complex reaction pathways, and occurs only at triple-phase boundary regions, where the oxygen gas, liquid electrolyte, and electron can meet with each other on the surface of a solid catalyst. Both the ORR and OER involve multi-step (dominant four-step) proton-coupled electron transfer paths and desorption/adsorption of various intermediates on certain surfaces.[1] Therefore, an understanding of the reaction mechanisms remains elusive due to the difficulty of

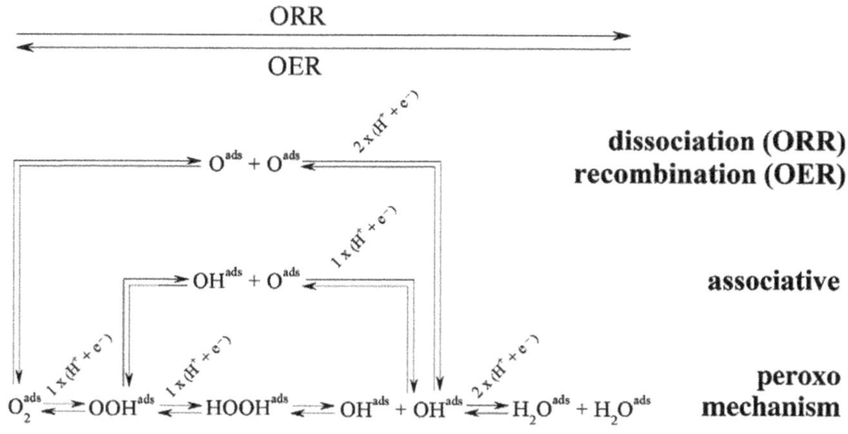

Figure 8.2 Proposed mechanism for the ORR and OER. (Reprinted with permission from ref. 1. © 2014 WILEY-VCH Verlag GmbH & Co. KGaA, Weinheim.)

unambiguously identifying the intermediates and interpreting kinetic data, which always depend on the specific systems. However, it is a prerequisite and is critical to elucidate the mechanisms for both theoretical and experimental research. Generally, the multi-step proton-coupled electron transfer paths are proposed to occur as follows and can be unified in a single scheme for both the ORR and OER (Figure 8.2).[1]

(1) Normally, the ORR proceeds through dissociation, association, or a peroxo mechanism. The first two involve four-electron transfer, while the third mechanism involves two-electron transfer with the formation of HOOH. The OER process occurs by reversing these mechanisms. Therefore, the mechanisms are only described for the ORR below.
(2) For the dissociation mechanism, the O–O bond in O_2 directly breaks upon adsorption, and then the formed O* is reduced to OH*, and to H_2O*, successively.
(3) For the association mechanism, the adsorbed O_2 transforms into OOH* without O–O bond breaking, which further breaks into O* and OH*. Then, similar to the dissociation mechanism, the formed O* is reduced to OH* and to H_2O*.
(4) For the peroxo mechanism, the adsorbed O_2 is reduced to OOH* and HOOH* successively through a two-electron transfer step. The HOOH* can be further reduced to OH* and ultimately to H_2O*.

8.3 Metal-free Functionalized Carbons for ORR Catalysis

In 2009, Dai and co-workers reported a pioneering study of metal-free nanocarbon materials as ORR catalysts.[33] The vertically aligned nitrogen-doped carbon nanotubes (VA-NCNT) exhibited outstanding electrocatalytic activity, three times higher than commercial 20% Pt/C catalysts. Besides, the nitrogen doped nanocarbon was free from methanol cross-over and CO poisoning effects in alkaline conditions, leading to better durability. Most recently, some nitrogen-doped CNTs or graphene were also demonstrated to have remarkable ORR activity and long-term stability in acidic media as well, suggesting great potential in practical applications such as acidic polymer electrolyte membrane fuel cells.[72–75] However, there has been a continuing controversy about the effect of metal impurities in so-called "metal-free" nanocarbon catalysts on ORR activity since the beginning.[76] Even a trace of metal residues is believed to significantly influence the electrocatalytic activity. Nonetheless, various research, including theoretical investigations,[57,77] non-metal precursor fabrication,[63,64] metal poisoning tests,[76] and high stability in acid,[72] *etc.*, undoubtedly confirms the intrinsic activity of metal-free functionalized carbon catalysts for the ORR.

8.3.1 Heteroatom or Molecular Doping

8.3.1.1 Nitrogen-doped Nanocarbon

The activity origin of metal-free nanocarbon catalysts is generally believed to be ascribed to doping-induced electron modulation, especially in heteroatom-doped nanocarbon materials. Nitrogen heteroatoms in carbon-based electrocatalysts are the most promising dopants. In the case of VA-NCNT electrocatalysts, the incorporation of electron-accepting nitrogen dopants into the conjugated carbon plane generates a relatively positive charge density on the adjacent carbon atoms, resulting in preferable chemisorption of oxygen molecules and lower ORR overpotential.[33] More specifically, the doping-induced charge delocalization was revealed to change the chemisorption mode of O_2 on the surface of NCNT from end-on (Pauling model) to side-on (Yeager model), thereby effectively weakening the O–O bonding and facilitating the first electron transfer (Figure 8.2).

The hierarchical nanostructure of functionalized carbon materials is also critical for the resultant catalytic performances due to the requirement of favorable triple-phase boundary regions.[78] For example, VA-NCNT with aligned NCNT and a well-defined high surface area exhibited much better performance than non-aligned counterparts. The top-end alignment of NCNT between the electrode substrate and electrolyte solution facilitates electrolyte/reactant diffusion and ensures a better demonstration of the active sites on the surface of NCNT. Therefore, it is beneficial to engineer the functionalized nanocarbon structures towards an interconnected electron/mass transfer pathway and a full exposure of active sites.[27,28,79–81] For example, Tian *et al.* proposed a CNT@NCNT coaxial nanocable to enrich the nitrogen atoms on the surface for enhanced electrocatalytic activity (Figure 8.3(a)).[27] As shown in Figure 8.3(b), N-doped turbostratic carbon layers were epitaxially grown on the outer walls of pristine CNTs *via* the chemical vapor deposition (CVD) of pyridine at 760 °C. The element mapping clarified

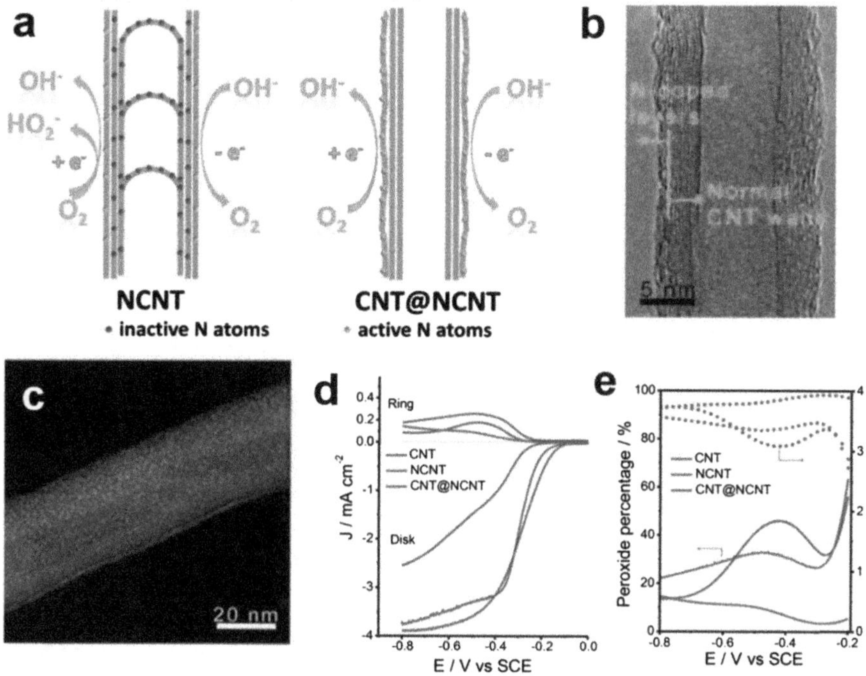

Figure 8.3 Oxygen electrocatalysis on nanostructured carbon. (a) Scheme of the CNT@NCNT structure for full exposure of active sites. (b) TEM image of the CNT@NCNT nanocable. (c) Element mapping of CNT@NCNT, showing the distribution of N, C, and O. (d) RRDE voltammograms obtained in O_2-saturated 0.10 M KOH, and (e) corresponding peroxide yield (solid line) and electron transfer number. (Reprinted with permission from ref. 27. © 2014 WILEY-VCH Verlag GmbH & Co. KGaA, Weinheim.)

the concentrated distribution of N atoms on the surface (Figure 8.3(c)). The CNT@NCNT has a higher N–C ratio on the surface than in the bulk, determined by X-ray photoelectron spectroscopy (XPS) analysis and elemental analysis. In spite of it having fewer N atoms than its bulk counterpart, CNT@NCNT affords a higher ORR current, positively shifted onset potential (~51 mV) and lower peroxide yield (below 8.68%), compared with NCNT (Figure 8.3(d) and (e)). Furthermore, the turnover frequency (TOF) and normalized current with respect to amounts of N for CNT@NCNT nanocables were both higher than those for NCNT. This substantially enhanced performance was ascribed to the unique nanostructure that contains an N-enriched surface for fully accessible and efficient utilization of active sites and an intact inner CNT for high electrical conductivity.

In addition to the apparent content of incorporated heteroatoms, the configurations and locations can notably alter the ORR activity.[41] For instance, nitrogen atoms can be doped in the forms of pyridine-like, pyrrole-like, or quaternary nitrogen (or graphitic N). The planar pyridinic N with a lone electron pair is generally claimed as the favorable configuration to improve the electron-donating capability and weaken the O–O bond.[35,71,75] Most recently, an investigation on model graphene catalysts clearly demonstrated the critical role of pyridinic N, rendering the adjacent carbon atoms' Lewis basicity as active sites for the ORR.[82] Meanwhile, some research suggested the graphitic N to be responsible for ORR activity instead of pyridinic N.[83,84] This controversy has made it challenging to reach a consensus due to the temperature-induced conversion of different doping forms, and the difficulty to precisely control and identify certain doping configurations.

During the fabrication of functionalized carbons, oxygen-containing functional groups are always generated on the surface.[85] With a higher electronegativity and more electrons compared with C atoms, the doping of oxygen can also break the electroneutrality of the sp^2 carbon lattice and thereby lead to altered ORR activities. It was demonstrated that the ORR activities of some oxygen-containing functional groups decreased in the order carboxyl > carbonyl > hydroxyl groups.[86,87] In other cases, such as S-doped graphene, the spin density redistribution induced by heteroatom doping was identified as the more favorable activity origin rather than charge modulation.[88] The charge polarization was almost negligible due to the small electronegativity difference between S (2.58) and C (2.55).[58] After the integration of S, the carbon atoms in graphene with the highest spin density promoted the four-electron transfer pathway, while the sulfur atoms with a higher charge density catalyzed the ORR following a two-electron transfer mechanism.[88]

8.3.1.2 *Activity Descriptor*

Although various heteroatom-doped nanocarbon materials have been proposed as highly efficient ORR catalysts, the research dominantly relies on trial-and-error approaches at its current stage. There is an urgency to understand and propose rational principles or intrinsic reactivity descriptors for the preferential design and optimization of most desirable electrocatalysts. The energy level of the d-band center[89] or e_g-filling[12] have been proposed as activity descriptors for metal or transition-metal-oxide perovskites. Analogously, the orbital energy theory[60] and work functions of carbon catalysts[90] have been utilized to correlate the relationship between material properties, intermediate adsorption and catalytic activities. The better match and coupling between the bonding orbitals of adsorbates and valence orbitals of nanocarbon catalysts is supposed to result in optimized adsorption and improved activity.[91]

In the case of metal-free heteroatom-doped carbon materials, both the electronegativity and electron affinity difference between dopants (E_X and A_X) and carbon (E_C and A_C) are important to achieve a superior ORR activity.[70,92] The electronegativity represents the ability to acquire electrons during the formation of a covalent bond, whereas electron affinity represents the energy released from a neutral atom or molecule when an extra electron is added to form a negative ion. Based on these critical features, Xia and co-workers proposed an activity descriptor for heteroatom-doped carbon nanomaterials to correlate the intrinsic material properties to the catalytic performance, considering the synergistic effect of electronegativity and electron affinity on charge redistribution: $\Phi = (E_X/E_C) \times (A_X/A_C)$.[70] A typical volcano relationship between the minimum overpotential *vs.* Φ can be yielded for both ORR and OER catalysis based on X-doped carbon (Figure 8.4(a)). Notably, nitrogen was again identified as the most effective dopant for the ORR. The electrocatalytic activities for doped carbon could exceed the best noble-metal-based catalysts in an optimized range of Φ.

In addition to the theoretical predictions, experimental results including onset potential and limiting current also exhibited a volcano relationship with the descriptor Φ (Figure 8.4(b) and (c)). Detailed investigations revealed that dopants with $\Phi < 1$ could result in them and their neighboring carbon atoms having a positive charge due to the electron accepting/donating difference,

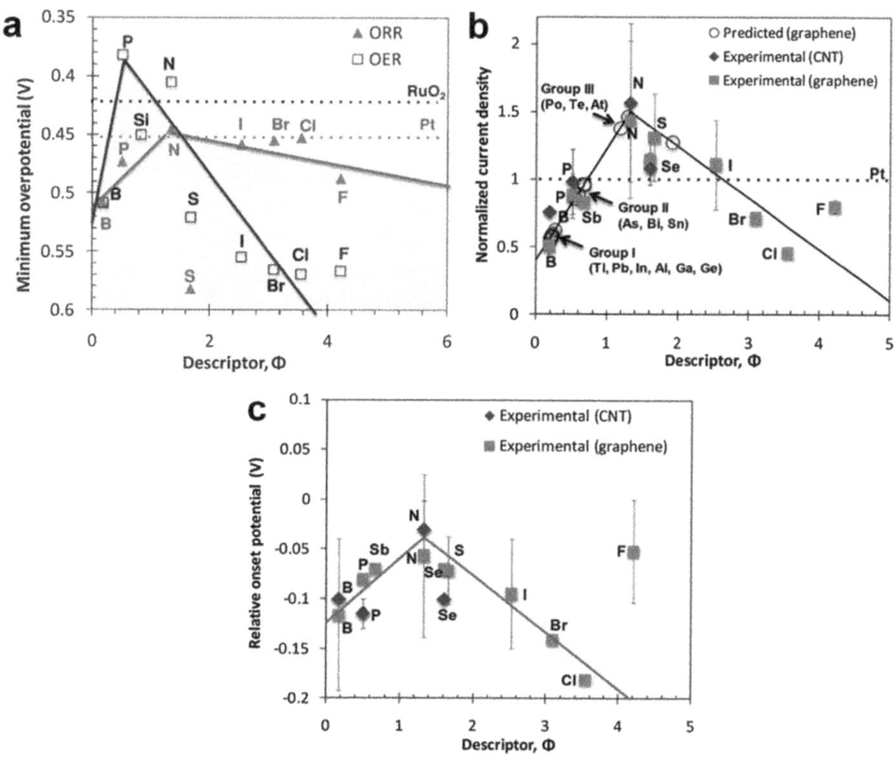

Figure 8.4 Activity descriptor for the ORR and OER. (a) Minimum ORR/OER overpotential *vs.* the descriptor Φ. (b) Normalized limiting current density and (c) relative onset potential based on Pt/C obtained from experimental linear sweep voltammogram (LSV) curves *vs.* the descriptor Φ. (Reprinted with permission from ref. 70. © 2015 WILEY-VCH Verlag GmbH & Co. KGaA, Weinheim.)

while those with $\Phi > 1$ led to a positive charge on adjacent carbon atoms but a negative charge on the dopants. Relatively positively charged atoms were suggested to serve as active sites for the ORR. This descriptor comprehensively reveals the relationship between intrinsic properties of dopants or carbon atoms and resultant reactivity, attributable to the competition between the tendency to transfer electrons and the ability to form covalent bonds.

8.3.1.3 Molecular Doping

Heteroatom doping introduced during *in situ* synthesis or post-treatment of carbon nanomaterials has been well-demonstrated to alter the electron/spin delocalization and resultant ORR reactivity contributed by the intramolecular charge transfer. Contrastively, another approach, named "molecular doping", has also been proposed to modulate the electronic structure of pure carbon materials and enhance the ORR activity *via* intermolecular charge transfer.[93] Wang *et al.* first reported the functionalization of CNTs with a strong electron-withdrawing component, poly(diallyldimethylammonium chloride) (PDDA),[94] to withdraw electrons from the carbon atoms in CNTs and induce a net positive charge (Figure 8.5(a)). As a result, the nitrogen-free CNT functionalized by adsorbed PDDA exhibited an obviously increased current density and positively shifted onset potential (Figure 8.5(b)). In recent work, it has been reported that other alternatives, such as nitrobenzene,[93] and even graphene quantum dots,[48] can also serve as electron-withdrawing and molecular doping components to withdraw electrons from the conjugated nanocarbon plane to these electron-accepting groups (Figure 8.5(c)). This generates a net positive charge on carbon atoms, leading to an enhanced ORR catalytic performance (Figure 8.5(d)), analogous to the case of heteroatom doping.[41,93] These results unambiguously demonstrate the critical role of charge transfer and electron/spin structure modification for metal-free functionalized carbons on ORR activities, regardless of the material structure or doping type.

8.3.2 Edge Effects or Edge Sites

When considering doping configurations and locations, edge effects or edge sites need to be specially acknowledged. Although edge sites are not ORR active sites, doping near the edge plays a more critical role in enhancing reactivity compared with routine basal doping.[62] To produce edge-selectively doped nanocarbon, Baek and co-workers reported an effective and scalable method *via* ball-milling of pristine graphite in the presence of S, Cl_2, Br_2, *etc.*[58,95] The resulting edge-sulfurized graphene nanoplatelets were revealed to exhibit improved ORR activity with long-term stability, even surpassing the Pt/C electrocatalyst after oxidation. The covalently bonded sulfur (S) or oxidized sulfur (O=S=O) at the edges were believed to induce a high positive charge and spin densities on graphene, serving as active sites and strongly promoting electrocatalytic activity.[58]

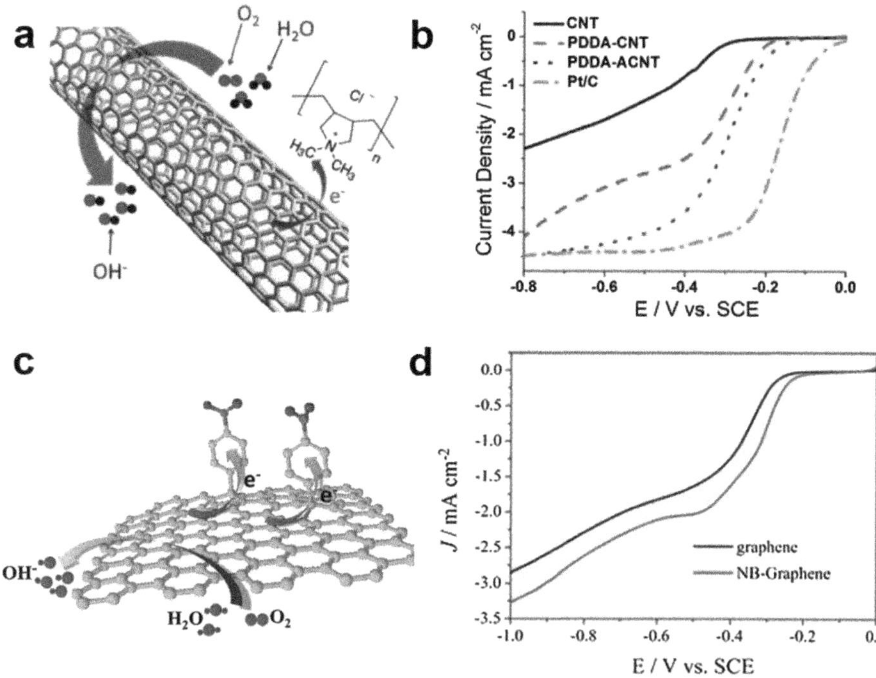

Figure 8.5 The ORR on a molecular doped carbon electrocatalyst. (a) Scheme of charge transfer on PDDA-CNT and (b) resultant LSV curves for the ORR in O_2-saturated 0.10 M KOH. (Reprinted with permission from S. Wang, D. Yu and L. Dai, *J. Am. Chem. Soc.*, 2011, **133**, 5182. Copyright 2011 American Chemical Society.[94]) (c) Illustration of charge transfer between nitrobenzene and graphene, and (d) resultant ORR activities in O_2-saturated 0.10 M KOH. (Reproduced from ref. 93 with permission from The Royal Society of Chemistry.)

Nanocarbon structures with a greater edge exposure can afford preferable doping locations for higher catalytic activities experimentally and theoretically.[57] Xia and co-workers systematically investigated the doping and edge effects of N-doped graphene on ORR and OER activities by density functional theory (DFT) calculations based on the four-electron pathway in an acid electrolyte (Figure 8.6(a)).[57] Nitrogen doping within distance of the edge effect (~2.5 Å) significantly decreases the ORR overpotential attributed to the accelerated adsorption and transformation of O_2 to OOH* (rate determining step). As shown in Figure 8.6(b), doping at armchair graphene edges results in a lower ORR overpotential than the zigzag edge in most cases.[57] Additionally, similar conclusions can be achieved for OER catalysis using metal-free N-doped graphene (Figure 8.6(c)). These observations clearly reveal the importance of edge effects in the engineering of heteroatom-doped nanocarbon catalysts for ORR/OER catalysis.

Notably, the intrinsic chemistry of nanocarbon edges has attracted great attention for its critical impact on electronic properties and surface reactivity.

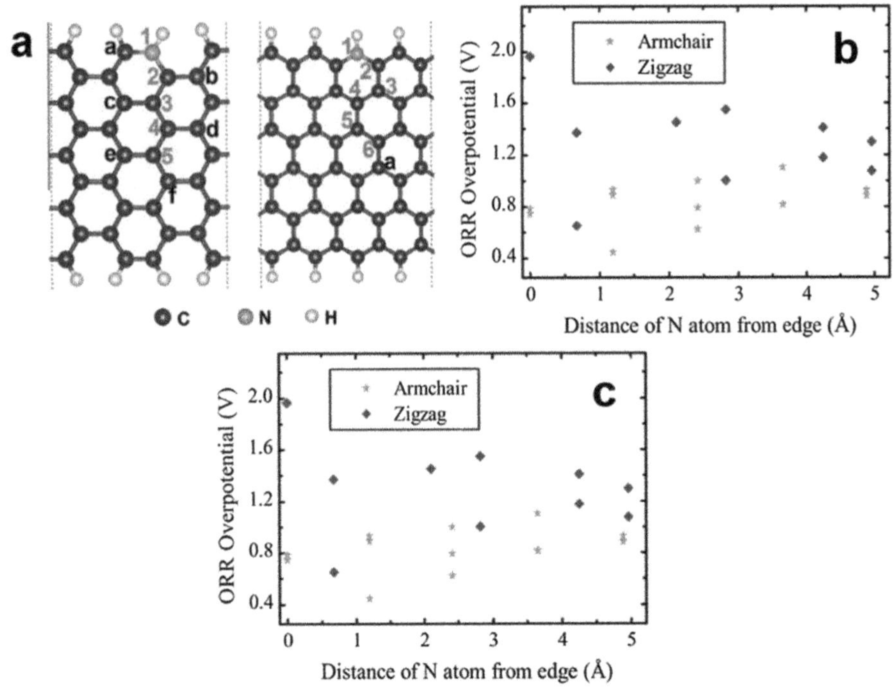

Figure 8.6 Edge effect on the ORR activity of heteroatom-doped graphene. (a) Scheme of model graphene structures with different doping and edges for calculation. (b) ORR and (c) OER overpotential *vs.* the distance of the N atom from the edge obtained by DFT calculations. (Reprinted from *Journal of Catalysis*, **314**, M. Li, L. Zhang, Q. Xu, J. Niu, Z. Xia, N-doped graphene as catalysts for oxygen reduction and oxygen evolution reactions: Theoretical considerations, 66–72, Copyright 2014, with permission from Elsevier.[57])

Compared with bulk atoms, graphene edges deliver twice the reactivity, a specific capacitance four orders of magnitude larger, and a much faster electron transfer rate.[96,97] Dangling bonds at the edges are regarded as high-energy sites and serve as catalytic centers for many acid–base or redox reactions after saturation.[61] It is more convincing and intuitive to characterize the ORR activity of carbon atoms at the edges or basal planes accurately. For this purpose, Wang and co-workers developed a micro apparatus to measure the ORR activity in an electrolyte solution droplet on highly oriented pyrolytic graphite (HOPG), as illustrated in the inset of Figure 8.7(a).[98] A much higher current density is obtained at the edge of graphite compared with the basal plane on the LSV curves (Figure 8.7(a)), suggesting the edges are more active. Furthermore, ball-milled graphite with more exposed edges was demonstrated to render a much better ORR performance than pristine graphite (Figure 8.7(b)), validating the importance of edge sites. This was rationalized by DFT calculations, which revealed a delocalized charge distribution and higher charge densities on the graphene edge atoms.[98]

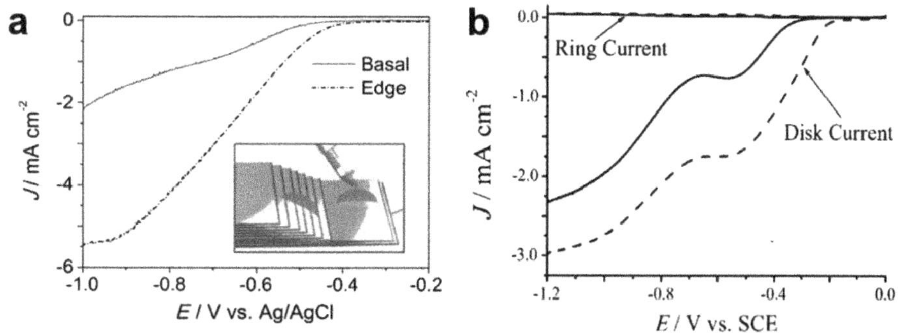

Figure 8.7 ORR activity on the edge sites of a carbon electrocatalyst. (a) LSV curves obtained on the edge or basal plane of HOPG. The inset shows the micro apparatus for ORR tests in a droplet. (b) RRDE results for pristine graphite and ball-milled graphite in O_2-saturated 0.10 M KOH solution. (Reprinted with permission from ref. 98. © 2014 WILEY-VCH Verlag GmbH & Co. KGaA, Weinheim.)

In summary, the doping or edge effects can significantly enhance the ORR activity with a lower overpotential, increased kinetics, and preferential four-electron pathway for metal-free functionalized carbons. The underlying mechanism is elucidated by the modified electron properties of the sp^2 conjugated carbon matrix, regardless of the specific configurations.[99] Atoms (dopants or neighboring carbon atoms) with a higher positive charge or/and spin densities are more likely to serve as active sites, leading to a modulated adsorption mode of O_2, optimized binding of intermediates, weakened O–O bond, and facilitated electron transfer.

8.3.3 Intrinsic Topological Defects

The contribution of intrinsic topological defects of functionalized carbon materials to ORR activity has been strongly considered recently.[63–65,68,69,100] In fact, disorder or defects are unavoidable in crystalline materials, ascribed to the second law of thermodynamics. Even when energetically unfavourable, a certain amount of extrinsic or intrinsic defects can be introduced into various nanocarbon materials during synthesis or post-treatment.[101,102] Besides, heteroatom-doping and edge regions are not sole configurations, and are always concomitant with other crystalline disorders and defects, such as dangling groups, single/multiple vacancies, non-hexagonal topologies, and lattice reconstructions. These intrinsic defects themselves are considered to be active sites for adsorption and chemical applications.[77,101,102] Consequently, it is essential to probe the role of such defects in ORR/OER electrocatalytic performances. On the one hand, a comprehensive understanding of the active sites will help to elucidate different experimental results. For instance, the controversial conclusion of the favourable N-doping configurations is ascribed to an ignorance of the vital contribution of intrinsic defects.[63]

On the other hand, an insight into intrinsic topological defects is supposed to accelerate further material innovation and performance optimization.

8.3.3.1 Theoretical Investigation of the Role of Defects

In-plane topological defects, such as pentagons, heptagons, octagons, and other non-hexagonal rings, can be presented in the form of point defects (*e.g.*, Stone–Wales defect, single vacancy, and multiple vacancy) or line defects (*e.g.*, grain boundary). They can lead to various modifications of electron trajectories[103] and thereby alter the ORR activities as discussed above. Systematically, Xia and co-workers studied the ORR catalytic properties of graphene with different types of point/line defects by DFT calculations.[77] The charge or spin densities of graphene clusters with defects didn't exhibit obvious differences compared with perfect graphene in most cases. However, with a pentagon carbon ring at the zigzag edge, high spin densities can be obtained at the opposite zigzag edge (C #87: 0.41, Figure 8.8(a)), while high charge densities

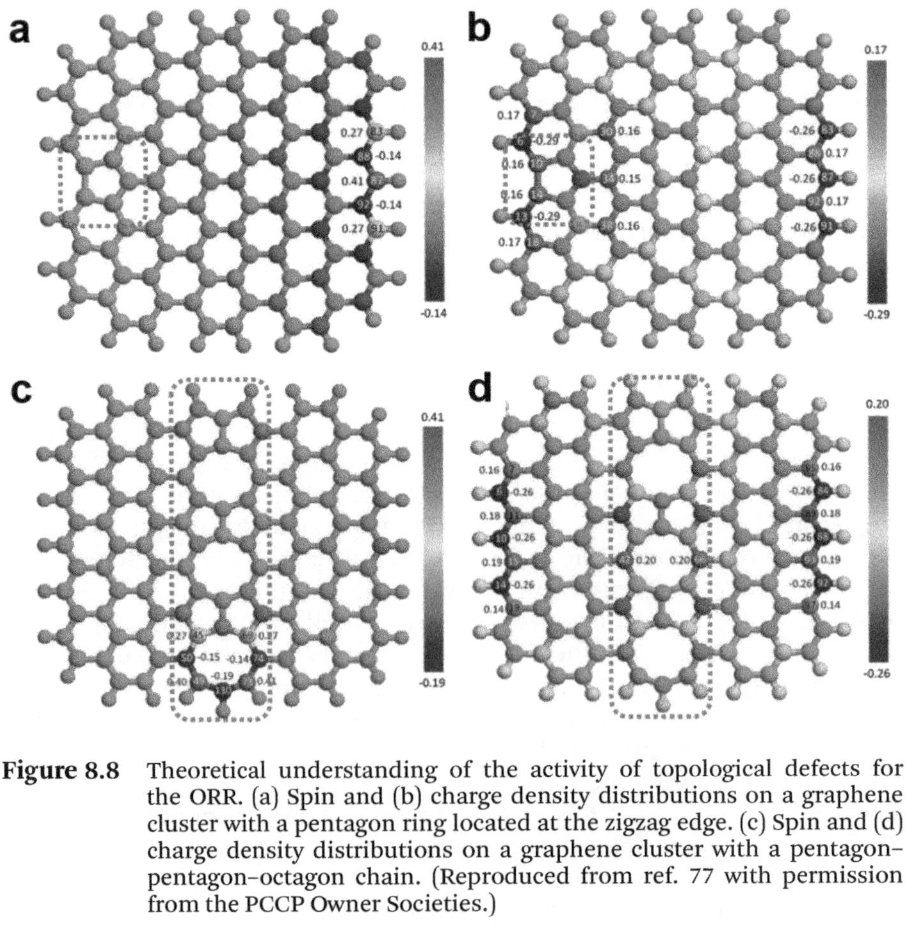

Figure 8.8 Theoretical understanding of the activity of topological defects for the ORR. (a) Spin and (b) charge density distributions on a graphene cluster with a pentagon ring located at the zigzag edge. (c) Spin and (d) charge density distributions on a graphene cluster with a pentagon–pentagon–octagon chain. (Reproduced from ref. 77 with permission from the PCCP Owner Societies.)

appear at two zigzag edges (Figure 8.8(b)). In the case of line defects, the pentagon–pentagon–octagon chain leads to the highest spin density (0.41) in the octagon ring (Figure 8.8(c)) and high charge densities are mainly distributed along zigzag edges or octagon rings (Figure 8.8(d)). The local charge and spin polarization are mainly located at the zigzag edges or some defects themselves, which are suggested to be catalytically active. When the pentagon defect is located at armchair edges or graphene planes, there is no spin density as introduced in the case of the zigzag edge. Therefore, with the synergistic effect between intrinsic topological defects and favourable edge structures, some types of defects can serve as ORR active sites, similar to heteroatom dopants.[77] Analogous to the conventional "foreign doping" as highlighted above (*e.g.*, heteroatom doping, molecular doping), the intrinsic defects can modify the local electronic properties of nanocarbon and improve the catalytic activity, which is identified as "self-doping".[101,103] Remarkably, although the defects are demonstrated to facilitate ORR catalysis, the activity is considerably limited due to the small population of effective defects at the edge.

In fact, the first experimental evidence and calculation results pointing out the role of defects in the ORR can be traced back to 1999. Ajayan and co-workers reported improved catalytic behavior at CNT electrodes for oxygen reduction.[104] With the assistance of DFT calculations and molecular dynamics simulations, they revealed that pentagons at the CNT tips, pentagon–heptagon defect pairs in the lattice (Stone–Wales defect), and curvature, to a lesser extent, could generate a higher density of states and was responsible for enhanced ORR activity.[104,105]

Owing to the non-hexagonal feature, topological defects can alter bond lengths and locally induce Gaussian curvatures.[101] It has been reported that the curvature around active sites can also significantly tune the adsorption and reduction of oxygen, during theoretical studies comparing CNTs and graphene,[106] and CNTs with different diameters.[107] With the curvature of the carbon surface varying from positive (inside CNTs) to negative values (outside CNTs), the adsorption energy of O_2 involved in the ORR becomes smaller. When compared with Si-doped graphene, the free energy change of the rate determining step for the ORR on the concave inner surface of Si-doped CNTs is smaller, while that on the convex outer surface is larger.[106] These observations provide fresh insight into the ORR mechanism based on metal-free functionalized carbons and also new directions for the further development of effective electrocatalysts.

8.3.3.2 *Defective Nanocarbons as Superior ORR Catalysts*

To verify the ORR activity contributing to intrinsic defects in nanocarbon, it is a prerequisite to obtain certain functionalized carbon materials with specific defects in the absence of heteroatom doping and metal species. However, various types of intrinsic topological defects, in combination with doping, edge or metal residues, are generated easily but uncontrollably during the formation or post-treatment process of the carbons. Recently, some impressive research has shown the successful fabrication of metal-free and dopant-free nanocarbon materials as excellent ORR catalysts *via* metal-/heteroatom-free growth or post-treatment.[63–66]

In 2014, Song and co-workers fabricated topological-defect CNTs for ORR catalysis *via* CVD using ethanol as a precursor.[108] Ethanol was decomposed into various carbon-containing gases and led to the different growth rates of CNTs at the same cross-section level. This novel growth behavior can deform the graphite structure of CNTs and introduce curvatures with pentagon, heptagon, and octagon defects (Figure 8.9(a)). LSV curves reveal that the steady-state diffusion current density for topological-defect CNTs, routine CNTs fabricated using benzene as a precursor, and Pt/C catalysts are 4.68, 0.98, and 4.18 mA cm^{-2}, respectively (Figure 8.9(b)), suggesting the great potential of topological defects for activity improvement. Furthermore, DFT calculations show that the introduction of more pentagons or heptagons breaks the delocalization of orbitals and promotes the formation of a localized orbital on CNTs, which is favorable for O_2 adsorption with small positive binding energy values (0.16/0.21 eV, Figure 8.9(c)) and the subsequent O_2 dissociation.

Metal oxide templated growth of 3D porous graphene is an effective strategy to controllably fabricate defective and metal-/dopant-free nanocarbon materials.[109] Recently, Hu's group reported the fabrication of dopant-free carbon nanocages containing various defects (*e.g.*, pentagons at the corners for a positive curvature, edge defects at broken fringes and hole defects in the microporous shells) to address the roles of intrinsic carbon defects in ORR activity.[63] The carbon nanocages were synthesized by the *in-situ* MgO template method using benzene as the precursor, presenting a cuboidal hollow structure 10–20 nm in size with abundant defects (Figure 8.9(d), inset of Figure 8.9(e)). Compared with N-doped CNTs, this material delivered superior ORR activity with a high onset potential of about 0.11 V *vs.* a normal hydrogen electrode, a much larger current density, and an electron transfer number of ~2.90 ± 0.10 in 0.10 M KOH (Figure 8.9(e)). DFT calculations indicate that the pentagon and zigzag edge defects serve as the main active sites owing to the favorable electronic structures. For example, the zigzag edge defects provide a portion of active unpaired π electrons locally, which can substantially decrease the reaction free energy and facilitate the electron transfer for OOH* formation (Figure 8.9(f)). Similarly, Zhang and co-workers also employed MgO templates to fabricate a defect-rich graphene material as bifunctional ORR/OER catalysts *via* direct carbonization of carbon/nitrogen precursors.[64] The as-obtained graphene mesh was predominantly a few layers thick with abundant nanopores in plane (Figure 8.9(g)), and self-assembled into a hierarchical porous scaffold. Unexpectedly, the graphene mesh (GM) with trace N (0.41 at%) delivered a substantially higher current density than the N-doped GO with a much higher nitrogen content (~7.48 at%) but few in-plane holes (Figure 8.9(h)). All possible activity origins, such as nitrogen dopants, edge effects, and topological defects were systematically studied by DFT investigations (inset of Figure 8.9(i)). Doping at the edge was

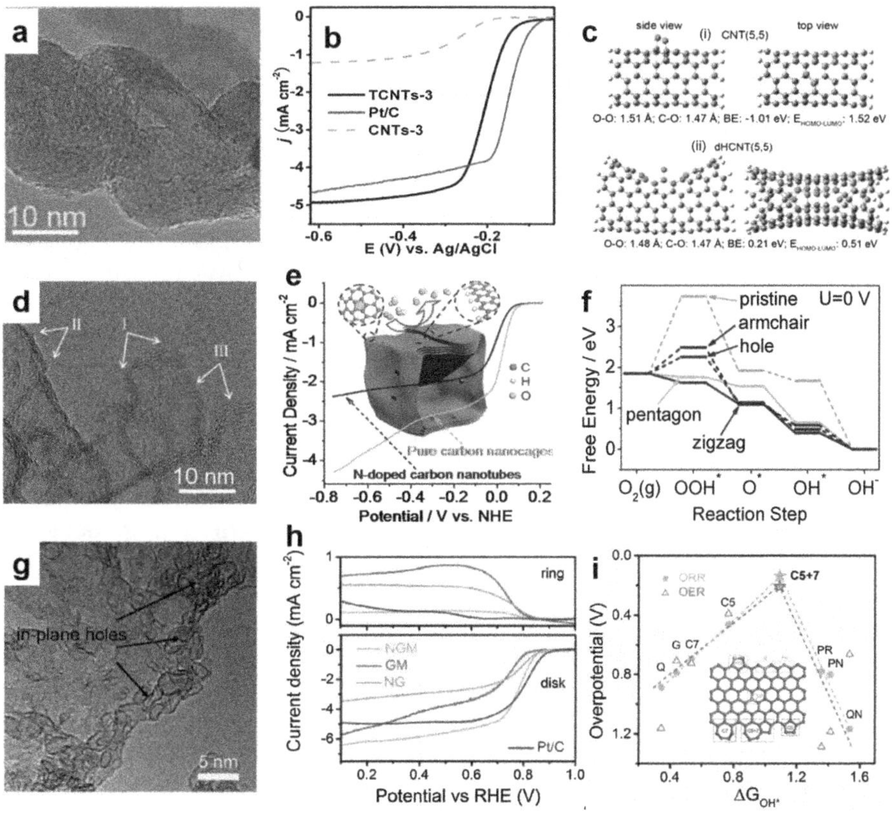

Figure 8.9 Metal-free carbon electrocatalysts with intrinsic defects for the ORR. (a) TEM image of the topological-defect CNTs fabricated by CVD using ethanol as a precursor. (b) LSV curves obtained in O_2-saturated 0.10 M NaOH with a rotation speed of 2500 rpm. (c) The calculated parameters for O_2 adsorption on a single-walled CNT(5,5) and a CNT(5,5) with double heptagon defects (dHCNT). ((a)–(c) are reproduced from ref. 108 with permission from The Royal Society of Chemistry). (d) TEM image of the carbon nanocage grown at 700 °C on MgO templates. (e) The LSV curves of a pure carbon nanocage compared with N-doped CNTs obtained in O_2-saturated 0.10 M KOH with a rotating speed of 2500 rpm. The inset shows the scheme of the carbon nanocage with pentagon and zigzag edge defects, and (f) shows the calculated free energy diagrams for the ORR of graphene with different defects. ((d)–(f) are reprinted with permission from Y. Jiang, L. Yang, T. Sun, J. Zhao, Z. Lyu, O. Zhuo, X. Wang, Q. Wu, J. Ma, Z. Hu, *ACS Catal.*, 2015, **5**, 6707. Copyright 2015 American Chemical Society.[63]) (g) High-resolution TEM image of a nitrogen-doped graphene mesh (NGM). (h) LSV curves of NGM obtained in O_2-saturated 0.10 M KOH. (i) The obtained volcano plots of overpotential *vs.* adsorption energy of OH*. The inset represents the graphene nanoribbon with different kinds of active sites used for DFT calculations. ((g)–(i) are reprinted with permission from ref. 64. © 2016 WILEY-VCH Verlag GmbH & Co. KGaA, Weinheim.)

demonstrated to be more effective than those in bulk, while the topological defects (such as pentagons and heptagons) at the edge were revealed to be even more important than doping and the edge effect. Specifically, a nitrogen-free topological defect with an adjacent pentagon and heptagon was proposed and identified as the optimal configuration at the peak of volcano plots for both the ORR and OER (Figure 8.9(i)).

In addition to the *in situ* fabrication of dopant-free but defect-rich functionalized carbons, another effective strategy is post-treatment to generate topological defects[65] or remove heteroatom dopants.[69] Wang and co-workers used Ar plasma technology to etch and modify graphene to produce dopant-free and edge-rich nanocarbon (P–G) materials, as shown in Figure 8.10(a).[65] Many holes with diameters around 15 nm were generated in the graphene plane without macroscopic structure damage (Figure 8.10(b)). The holes and defects can be well-controlled by tuning the treatment time and plasma power. LSV curves showed that the ORR activity of plasma-treated graphene was obviously enhanced with a positively shifted half-wave potential and increased current density (Figure 8.10(c)). Higher charge densities were observed on the edge carbon atoms than the basal plane carbon by DFT calculations, thereby rendering the edge-rich graphene more active for ORR catalysis.[65] Zn removal methods were also very versatile to create pores and defects on carbon without dopants or metal residues for ORR catalysts.[66,110] Recently, Yao and co-workers reported a facile nitrogen removal procedure to render defective graphene from an N-doped precursor.[69,111] During the high-temperature annealing at 1150 °C for 2 h, the doped N-atoms are subtracted from N-doped graphene and various defects are produced from atom reconstruction near vacancies (Figure 8.10(d)). XPS analysis confirmed the removal of surface nitrogen from N-doped precursors after heat treatment (inset of Figure 8.10(d), with 0.7 at% nitrogen content determined by element analysis). Both Raman spectra and TEM images suggested the formation of more defective and porous domains. More clearly, aberration-corrected high-resolution TEM was used in this work to visualize the topological defects in graphene. As shown in Figure 8.10(e), various intrinsic defects consisting of pentagons, heptagons, and octagons, were observed to be proximal to the edges or vacancies. Furthermore, both experimentally and theoretically, the defects in the materials obtained were revealed to be more effective in enhancing the ORR activity (Figure 8.10(f)). Trifunctional catalytic activities for the ORR, OER, and HER based on such defective graphene were also presented.[69]

8.3.3.3 Correlation Among Dopants, Edges, and Defects

The theoretical insights and experimental advances highlighted above mutually prove the intrinsic catalytic activity for the ORR on metal-free functionalized carbons, which is attributed to doping (heteroatom or molecular doping), edge (edge effects or edge sites), and topological defects. Various research has been conducted to probe the specific reactivity induced by certain kinds of

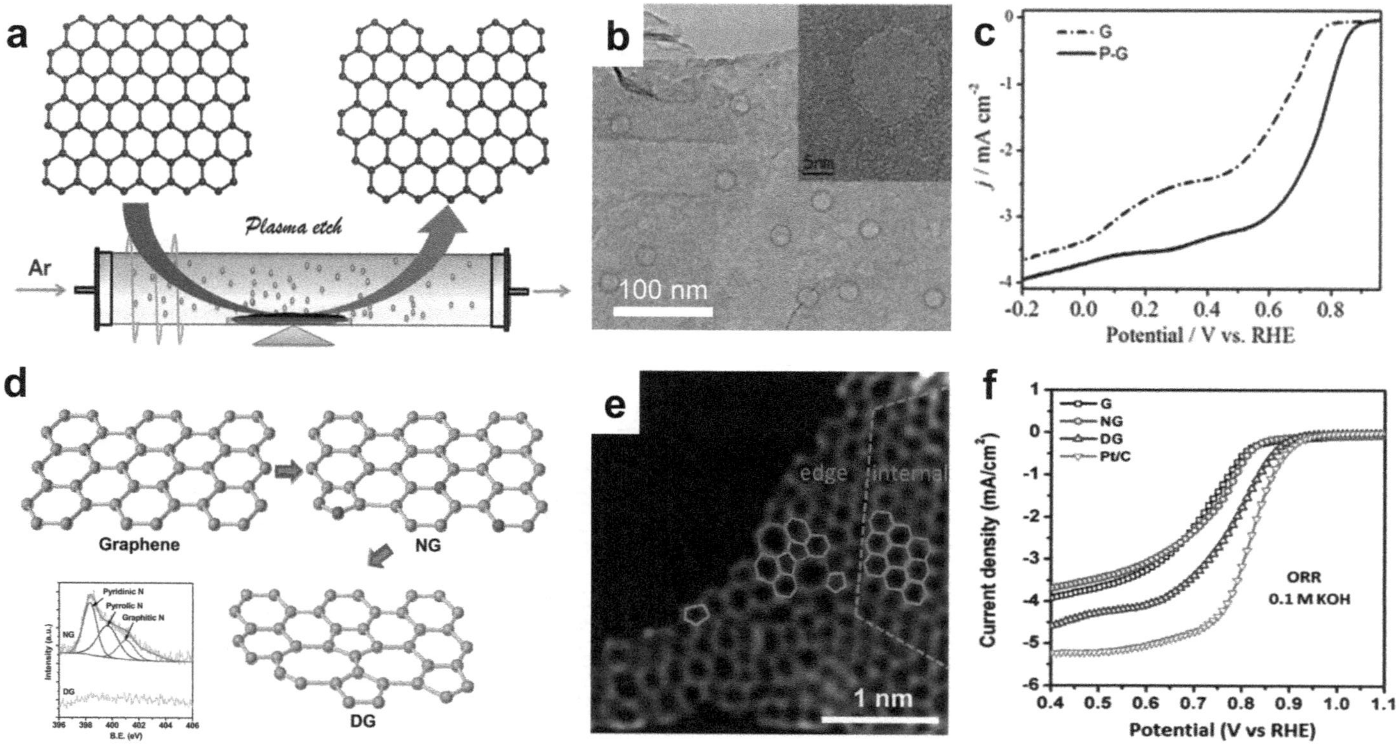

Figure 8.10 Dopant- and metal-free defect-rich carbon electrocatalysts. (a) Schematic of the preparation of defective graphene using Ar plasma etching. (b) TEM image of the Ar plasma-treated graphene and (c) LSV curves obtained in O_2-saturated 0.10 M KOH. ((a)–(c) are reproduced from ref. 65 with permission from The Royal Society of Chemistry). (d) Illustration of the nitrogen removal procedure to make defective graphene, with high-resolution N 1s XPS spectra. (e) HAADF image of the graphene material with carbon defects (DG). (f) ORR LSV curves for the pristine graphene, nitrogen-doped graphene and NG. ((d)–(f) are reprinted with permission from ref. 69. © 2016 WILEY-VCH Verlag GmbH & Co. KGaA, Weinheim.)

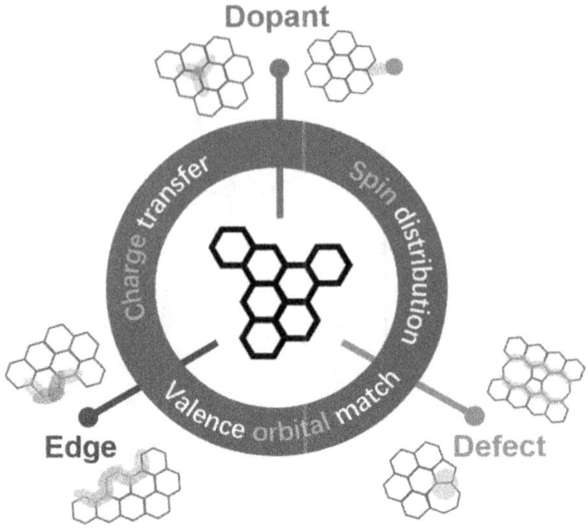

Figure 8.11 Comparison and correlation among different ORR activity origins for metal-free functionalized carbons, including dopants, edges, and defects. (Reprinted with permission from ref. 67. © 2017 WILEY-VCH Verlag GmbH & Co. KGaA, Weinheim.)

active sites. This is helpful for understanding the mechanism and an optimal material design (Figure 8.11). On the one hand, all these activity origins have been well-demonstrated to effectively modify the charge/spin distribution on an sp^2 conjugated carbon matrix. The resultant positive charge or/and higher spin densities at the surrounding carbon atoms or heteroatoms can optimize the chemisorption of intermediates and the subsequent electron transfer for bond breaking, leading to enhanced catalytic performances. Consequently, it is rational to hypothesize that specific heteroatom doping at defective edges is the optimal strategy to alter electronic structures and achieve the most effective active sites for the ORR on metal-free functionalized carbons.

On the other hand, various types of active sites always appear together, which is unfavorable for a fundamental study, but beneficial for practical applications. The synergistic effects between them always result in more significantly modified electron structures, a higher density of active sites, and thus superb electrocatalytic activities. For instance, doping within distance of edge effects,[31,58,62] topological defects combined with favourable edge sites,[63] and intrinsic defects with a very low doping content[64,69] are demonstrated to render the corresponding metal-free carbon materials as superb electrocatalysts for the ORR. Most recently, Chen and co-workers systematically investigated the correlation and interplay between chemical doping and topological defects in porous graphene for HER catalysis.[68] They found that the curvature and topological defects of nanoporous graphene can effectively accommodate and stabilize more chemical dopants with more active sites, favourable local electronic structures, and thus improve catalytic performance.

8.4 Metal-free Functionalized Carbons for OER Catalysis

Compared with the intensive research focusing on metal-free carbon catalysts for the ORR, the direct applications of functionalized carbon materials for OER catalysis are fewer and appear much later.[71] As the reverse process of the ORR, the O_2 molecule is generally inclined to generate from metal oxide surfaces, especially transition-metal-based materials (Ni, Fe, Co, Mn, *etc.*) and precious metals (Ir, Ru, *etc.*).[112,113] Firstly, nanocarbon materials were employed as multifunctional substrates to strongly couple with other active phases and enhance the OER performance.[113–115] During the development of carbon-based ORR catalysts, some researchers found that metal-free carbons can also serve as OER catalysts with bifunctional catalytic activities.[71,116] With further investigations, both theoretically and experimentally, the type of active sites and activity origins were demonstrated to be similar in the ORR and OER.[57,64,117] Therefore, in this section, we will only focus on several types of effective OER catalysts, such as N-doped, oxidized, and dual-doped carbons. Other advances in ORR/OER bifunctional catalytic activities or multifunctional activity including the HER will be discussed in detail in the next section.

8.4.1 N-doped Carbons for OER Catalysis

Analogously to ORR catalysis, N-doped carbon materials were also revealed as the most promising OER catalysts. In 2013, Wong and co-workers first reported the use of nitrogen-doped graphene (NG) as an ORR/OER bifunctional catalyst.[116] The NG sample was fabricated by simple pyrolysis of graphene oxide and polyaniline, resulting in a nitrogen content of 2.4 at% (inset of Figure 8.12(a)). Electrochemical characterizations revealed that NG exhibited a good catalytic activity toward the OER in addition to the ORR, and was better than undoped graphene or the Pt/C catalyst (Figure 8.12(a)), indicating the critical role of nitrogen-doping on OER catalytic activity.[116] Zhang's group also reported several metal-free nitrogen-doped nanocarbon materials as ORR/OER bifunctional catalysts, and the OER performance based on metal-free carbons is becoming more and more promising (Figure 8.12(b)).[27,28,64,118]

Remarkably, Hashimoto and co-workers reported an efficient non-metal nitrogen-doped carbon material (N/C) as a superior OER electrocatalyst in an alkaline electrolyte, which was comparable to iridium and cobalt oxide catalysts.[55] The sample was synthesized by pyrolyzing a hybrid of melamine formaldehyde polymer and nickel nitrate at 700 °C, followed by acid leaching to remove the metal species. TEM images exhibit that the N/C materials are nanoparticles with a diameter of 30–40 nm in the absence of metal particles (Figure 8.12(c) and (d)). The nitrogen content was about 4.1 at% determined by XPS with negligible nickel species (0.002 at%). Impressively, the N/C material afforded a current density of 10 mA cm^{-2} at a low overpotential

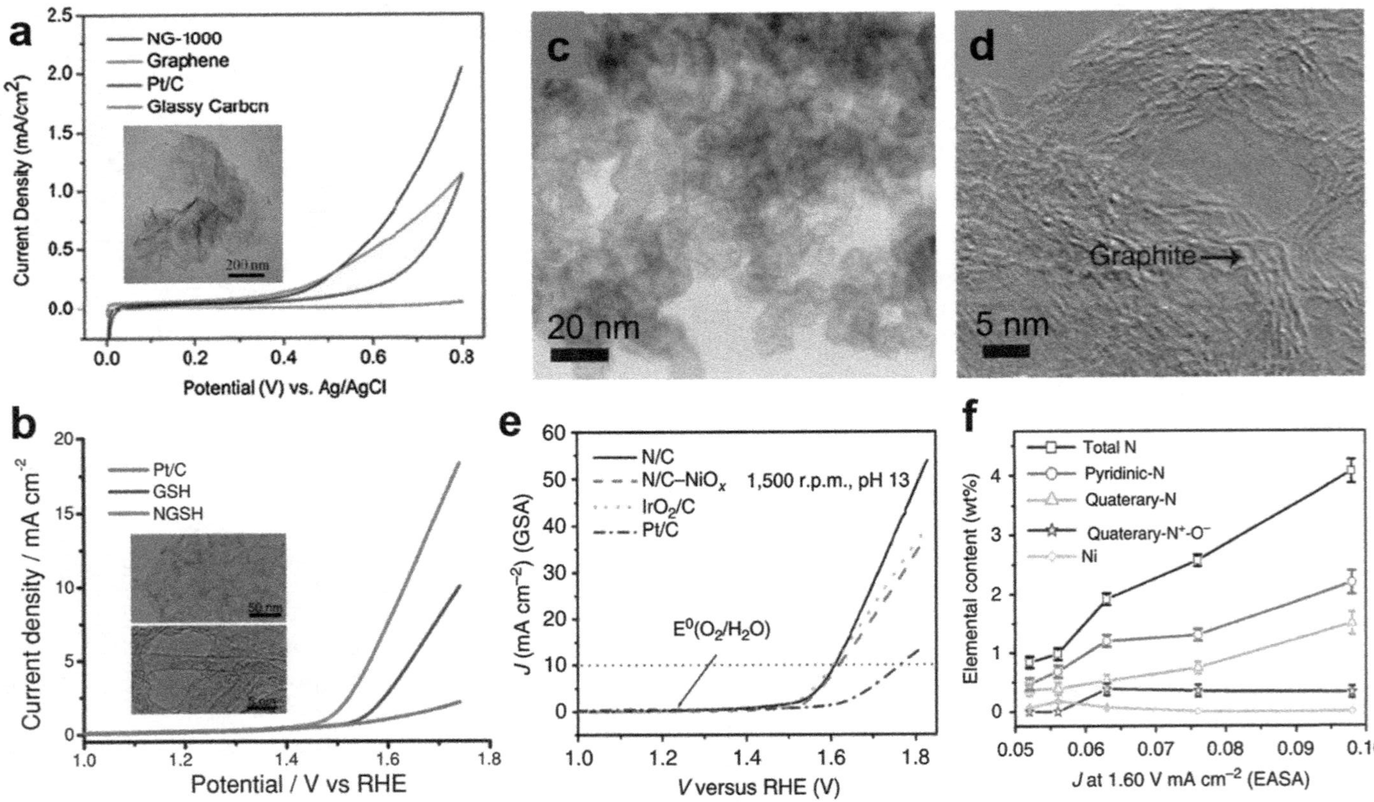

Figure 8.12 Nitrogen-doped carbon for the OER. (a) LSV curve of nitrogen-doped graphene showing OER activity in 0.10 M KOH solution at a scan rate of 10 mV s^{-1}. (Reprinted from *Carbon*, **53**, Z. Lin, G. H. Waller, Y. Liu, M. Liu and C.-P. Wong, Simple preparation of nanoporous few-layer nitrogen-doped graphene for use as an efficient electrocatalyst for oxygen reduction

of 380 mV, which was comparable with the state-of-the-art commercial IrO_2/C catalyst (Figure 8.12(e)). With a systematical examination of the relationship between different species and OER activities, the high OER activity of N/C was ascribed to the pyridinic-N and quaternary-N-related active sites, similar to those in N-doped carbon catalysts for the ORR (Figure 8.12(f)). Positively charged carbon atoms adjacent to the nitrogen to adsorb OH^- are considered as the active sites for the ORR, while the nitrogen species is hypothesized to participate in the OER catalysis due to the density of state near the Fermi level.[55]

With a graphene-like framework, graphitic carbon nitride (g-C_3N_4) is an especially attractive precursor for the fabrication of N-doped carbon catalysts, due to the high level of nitrogen content and largely exposed active centers. However, the poor electrical conductivity significantly limits the electrocatalytic performance of g-C_3N_4. Qiao's group reported several effective strategies to assemble active g-C_3N_4 with conductive CNTs[53] or graphene[119] into strongly coupled catalysts. The OER activity and durability obtained were demonstrated to be superior to those of the IrO_2 catalysts, which could be attributed to the large amount of N-related active sites and 3D interconnected porous carbon frameworks with improved mass and electron transport.

8.4.2 Oxidized Carbons for OER Catalysis

In addition to the intentional heteroatom doping, the formation of oxygen-containing functional groups is almost inevitable on carbon materials, during oxidation, functionalization, or even preservation. As a highly electronegative atom, oxygen induces positive charges on adjacent carbon atoms, facilitates intermediate adsorption and thus improves electrocatalytic activity.[36,120] Zhao and co-workers revealed for the first time the critical role of oxygen-containing functional groups on OER activities.[121] They synthesized surface-oxidized CNTs with a relatively higher amount of oxygen groups and largely maintained morphologies, by sequential acid oxidation, and hydrothermal and *in-situ* electrochemical activation treatments, which exhibited dramatically enhanced OER activity with an onset overpotential of 0.3 V in 0.10 M KOH. It was demonstrated that the oxygen-containing groups,

---

and oxygen evolution reactions, 130–136, Copyright 2012, with permission from Elsevier.[116]) (b) OER currents of Pt/C, graphene/CNT hybrid and N-doped graphene/CNT hybrid electrodes in 0.10 M KOH solution at a scan rate of 5 mV s^{-1}. (Reprinted with permission from ref. 28. © 2014 WILEY-VCH Verlag GmbH & Co. KGaA, Weinheim.) (c), (d) TEM images of the N/C sample showing a small size and no metal particles. (e) OER LSV curves of N/C materials compared with other samples. (f) Relationship between different elemental contents and OER activities for the N/C electrocatalysts. ((c)–(f) are reprinted by permission from Macmillan Publishers Ltd: *Nature Communications*, Y. Zhao, R. Nakamura, K. Kamiya, S. Nakanishi and K. Hashimoto, *Nat. Commun.*, 2013, **4**, 2390.[55] Copyright 2013.)

especially ketonic (C=O) groups, altered the electronic structure of the surrounding carbon and facilitated the adsorption of intermediates. Remarkably, it was claimed that the adsorption and stability of OH species were less favorable than the OOH species on adsorption sites, due to the consequent reaction pathway of $H_2O \rightarrow OH \rightarrow O \rightarrow OOH \rightarrow O_2$. The electron-withdrawing ketone group can reduce the electron density at the *meta*-position carbon atoms for the adsorption of OH species, whereas the *ortho*-position and *para*-position atoms are affected less for the adsorption of OOH species.[121]

Most recently, Liu's group investigated the effect of oxygen groups on the adsorption of oxygenated intermediates and OER activities by *ex situ* X-ray photoelectron spectroscopy and *in situ* electrochemical impedance spectroscopy.[122] As shown in Figure 8.13(a), the oxygen plasma oxidized CNTs (O-CNTs) delivered substantially more enhanced activity than pristine CNTs (P-CNTs) and thermal reduced CNTs (R-CNTs), with a low overpotential of 360 mV at a current density of 10 mA cm^{-2}. Detailed characterization revealed a positive correlation with the content of ketone groups (C=O) for all samples at different potentials, suggesting that C=O was the predominant activity origin (Figure 8.13(b)). Similar to previous reports, the positively charged carbon atoms near the C=O group were supposed to adsorb OH^- species easily, and also accelerated the reaction with another OH^- to form O species due to the strong electron-withdrawing capability of ketone groups (Figure 8.13(c)).[122]

8.4.3 Dual-doped Carbons for OER Catalysis

Heteroatom doping or moderate oxidization are demonstrated to be promising methods for improving the OER activities of metal-free functionalized carbons by tuning their electronic structures and chemical states. Particularly, dual-doping with two types of heteroatoms, or the coexistence of several kinds of active sites (such as dopants, edges, and topological defects) can further enhance the OER performance due to the synergistic coupling effects. For example, Qiao's group reported an N, O dual-doped carbon hydrogel film,[123] N, S dual-doped graphene microwires,[124] and N, P dual-doped carbon nanofibers on carbon paper (NPC-CP)[125] as highly active and stable OER catalysts. The NPC-CP material was fabricated *via* a mild oxidation of carbon papers, electrochemically induced polymerization of aniline and phosphonic acid, and subsequent high-temperature carbonization. Randomly oriented N, P dual-doped carbon nanofibers were directly grown on the fibrous carbon paper with a rough surface (Figure 8.14(a)). The addition of P dopants could effectively enhance the textural property and micro/mesoporosity with a higher specific surface area of 473 $m^2 g^{-1}$. When used as the working electrode for the OER directly, the NPC-CP exhibited robust stability and high activity (a low overpotential of 310 mV required for 10 mA cm^{-2}), which was comparable to the state-of-the-art IrO_2 catalyst and superior to most transition-metal electrocatalysts (Figure 8.14(b)). The introduction of N and P atoms can redistribute the conjugate electrons of pristine carbons and facilitate the charge transfer between heteroatoms and adjacent carbon

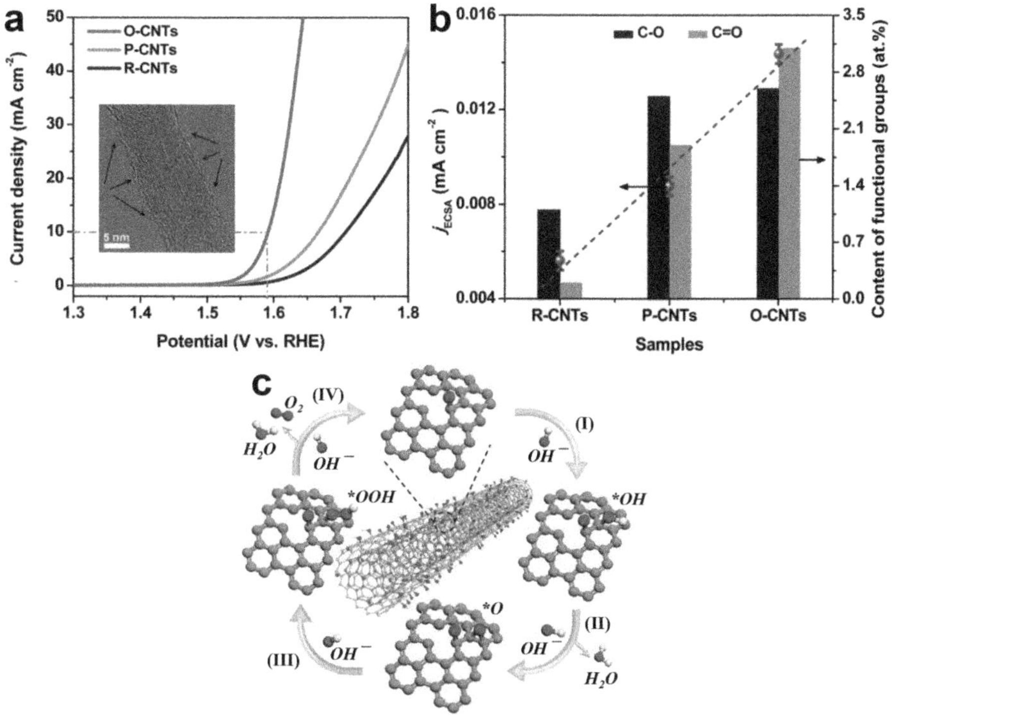

Figure 8.13 Oxygen-rich CNTs for OER. (a) OER LSV curves obtained in 1.0 M KOH, with the TEM image of oxidized CNTs (inset). (b) Correlation between the OER current densities (at an overpotential of 300 mV) and oxygen functional groups. (c) Proposed OER pathway on oxidized CNTs. (Reprinted with permission from L. Li, H. Yang, J. Miao, L. Zhang, H.-Y. Wang, Z. Zeng, W. Huang, X. Dong and B. Liu, *ACS Energy Lett.*, 2017, **2**, 294. Copyright 2017 American Chemical Society.[122])

atoms, which become electron-rich and catalytic active. Furthermore, theoretical calculations revealed that the dual-doping of N and P can significantly reduce the overall free energy with a minimum overpotential of 0.505 V, lower than single-doped and pristine carbons and even superior to IrO_2 (Figure 8.14(c)),[125] suggesting the crucial role of dual-doping on OER activity.

An N, S dual-doped graphene/CNT material was also reported as a metal-free electrocatalyst for the OER due to an enhancement by sulfur doping.[126] DFT calculations suggested that the C–S–C structures could serve as active sites for the OER by facilitating the adsorption of an OH^- intermediate due to the improved spin density and positively charged S atoms. Additionally, the coexistence of heteroatom doping and defective edges was demonstrated to make metal-free functionalized carbons more active for the OER.[127] Wang and co-workers fabricated edge-selectively P-doped graphene by simply ball-milling graphite and red phosphorus.[127] The as-obtained material was determined to be composed only of P, C and O with 0.83, 89.28 and 9.89 at%,

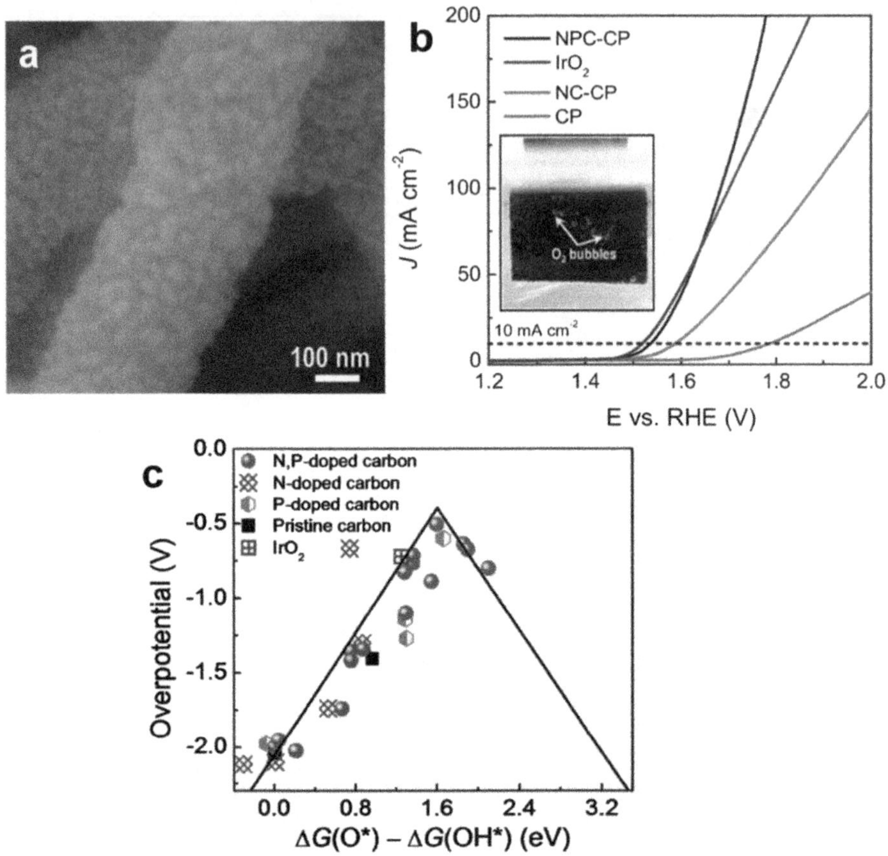

Figure 8.14 Dual-doped carbon for OER. (a) SEM image of NPC-CP. (b) OER curves of NPC-CP, compared with N-doped or pristine carbons, in O_2-saturated 1.0 M KOH at a scan rate of 2 mV s^{-1}. (c) The volcano plot of OER overpotential *vs.* the difference between the adsorption energies of O* and OH*. (Reprinted with permission from ref. 125. © 2017 WILEY-VCH Verlag GmbH & Co. KGaA, Weinheim.)

respectively. Remarkably, the overpotential required to reach a current density of 10 mA cm^{-2} was as low as 330 mV with a small Tafel slope of 62 mV dec^{-1}, which is among the best records for carbon-based electrocatalysts. The intrinsic defects, P doping, and oxygen functional groups introduced during the ball-milling process were supposed to be responsible for the high activity.

In contrast to ORR catalysis, metal-free functionalized carbons always exhibit unsatisfactory durability for the OER due to the oxidization of carbon with unstable active sites at a highly positive potential or the instability of nafion-polymer binder.[55] Therefore, insight into the instability mechanism and strategies for material engineering with improved durability are urgently required. Recently, Suib and co-workers reported a sequential two-step strategy to introduce stable C–S–C active sites into CNT-graphene nanolobes for efficient OER and ORR catalysis.[128] They revealed that the sulfur introduced by hydrothermal treatment with thiourea was not strongly bonded to carbon, while the pyrolytic reaction with benzyl disulfide at 1000 °C was effective at introducing high content, stable and active sites. The as-obtained S/C catalyst showed a low overpotential of 350 mV for 10 mA cm^{-2}, and high TOF of 5.87×10^{-5} s^{-1} at an overpotential of 400 mV, comparable with the Ir/C electrocatalyst. More importantly, the metal-free S-doped carbon catalyst maintained 100% of its initial current density above 10 mA cm^{-2} over 75 h, and even exhibited self-healing with a current increase of 20% from its initial value after 5.5 h. This novel sequential two-step doping strategy with favorable precursors provides an emerging avenue to tune the active sites with more stable dopants and defects and fewer unstable oxygen functional groups, which is significantly important for improving the durability of metal-free functional carbons.

In summary, metal-free functional carbons can be used as efficient and effective ORR or OER catalysts with remarkable activities through material engineering, including heteroatom/molecular-doping, generation of more edge regions and intrinsic defects, coexistence of several kinds of active sites, and design of hierarchical porous structures. Additionally, with a further understanding of the active mechanism and favorable active sites, the durability of metal-free functional carbon catalysts is believed to be greatly improved by precisely tuning the doping, defects or functional groups towards more active and stable electrocatalysts.

8.5 Multifunctional Metal-free Carbon Catalysts

Along with intensive research efforts to develop metal-free functionalized carbons for the ORR or OER, aimed at completely replacing precious-metal catalysts, there are more and more reports on the applications of carbon materials as multifunctional catalysts including the ORR, OER and HER.[71,129] ORR/OER bifunctional catalysts are important for rechargeable metal–air batteries or regenerative fuel cells, and OER/HER bifunctional catalysts are critical for water splitting.

8.5.1 ORR/OER Bifunctional Catalysts

8.5.1.1 N-doped Carbons

In 2013, nitrogen-doped graphene was first reported to have the bifunctional activity to catalyze the ORR and OER, but the performance was not satisfactory.[116] Through structural optimization for surface-enriched active sites, the ORR/OER bifunctional activities can be significantly enhanced.[27,28] In addition to the types of nitrogen doping, the structural features of carbon materials, such as the diameter of CNTs[130] and hierarchical porosity of graphene materials,[131] are also very influential on the resultant ORR/OER performances, due to the tuned mass transfer and active loading.

Recently, Cheng's group reported a 3D porous N-doped carbon microtube (NCMT) sponge exhibiting impressive bifunctional activities for the ORR and OER.[132] The NCMT was fabricated *via* a simple pyrolysis of facial cotton under NH_3 at 1000 °C for 1 h. As shown in Figure 8.15(a), the potential difference between the OER current density at 10 mA cm^{-2} and ORR current density at −3 mA cm^{-2} was as low as 0.63 V, which is the best among carbon-based catalysts to date. This excellent electrocatalytic activity was demonstrated to be ascribed to the unique dictyophora-like structure with a micron-scale hollow core, porous well-graphitized and interconnected tube walls, and high contents of pyridinic and quaternary nitrogen (Figure 8.15(b) and (c)). The hierarchical porous framework (a high surface area of 2358 m^2 g^{-1} and a pore volume of 1.34 cm^3 g^{-1}) endowed the NCMT with highly accessible active sites, sufficient mass transport pathways, and rapid electron transfer capability.

Together with the 3D structural design, a mechanistic understanding of activity sites is also very beneficial to guide the development of bifunctional metal-free carbon catalysts for the ORR/OER. For example, Liu and

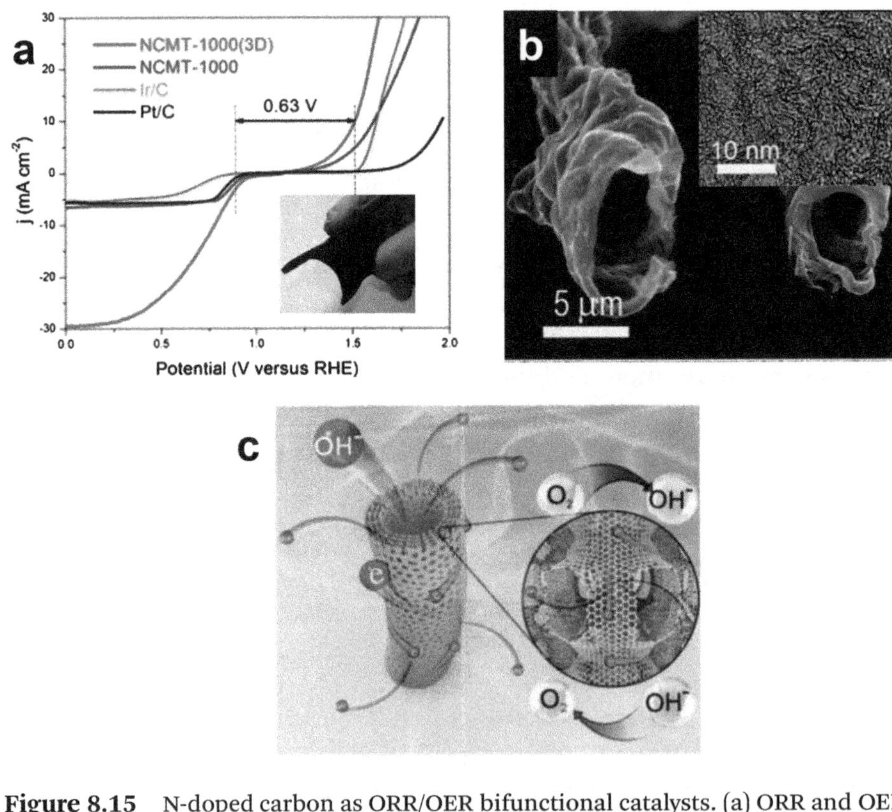

Figure 8.15 N-doped carbon as ORR/OER bifunctional catalysts. (a) ORR and OER performances of NCMT. Inset is the optical image. (b) SEM and TEM (inset) images, and (c) schematic illustration of the structure of NCMT. (Reproduced from ref. 132 with permission from The Royal Society of Chemistry.)

co-workers experimentally demonstrated the specific active sites responsible for the ORR and OER, respectively, in N-doped graphene materials.[84] The N-doped graphene nanoribbons were fabricated *via* a two-step carbonization (polymerization at 600 °C and pyrolysis at 1000 °C under Ar) of a mixture of melamine and L-cysteine, exhibiting a 3D interconnected porous structure with a nitrogen content of 6.0 at%. Electrochemical tests exhibited a half-wave potential of 0.84 V for the ORR, an overpotential of 360 mV for 10 mA cm^{-2} OER current density, and a potential difference of 0.75 V. The superb bifunctional activities rendered this material highly efficient as an air electrode for rechargeable zinc–air batteries, with a specific capacity of 873 mA h g^{-1} and a peak power density of 65 mW cm^{-2}. To identify the ORR and OER catalytic sites, they performed X-ray absorption near-edge structure (XANES) spectroscopic measurements on the working electrode before and after reaction. The nitrogen K-edge XANES spectra of N-doped graphene nanoribbons after the ORR exhibits a new peak at the lower energy side of graphitic N attributed to the adsorption of O* and OOH* intermediates on the carbon atoms next to graphitic N, while after the OER, the pyridinic N peak is changed with an increased width and a new peak at the higher energy side. This experimental evidence directly reveals the different N species for the ORR and OER. The graphitic N atoms afford electrons to the adjacent carbon (n-type doping) to enhance the O_2 adsorption and accelerate the ORR process, while the electron-withdrawing pyridinic N accepts electrons from the adjacent carbon (p-type doping) and facilitates the adsorption of OH^- and OOH^- for the OER in alkaline media.[84]

8.5.1.2 Dual-doped Carbons

To date, many reports have strongly demonstrated that dual-doped metal-free carbons are more efficient ORR/OER bifunctional catalysts than single-doped carbons, and are even superior to precious-metal catalysts. In 2015, Dai and co-workers reported a 3D N, P co-doped mesoporous nanocarbon (NPMC) foam fabricated by pyrolysis of polyaniline aerogels synthesized from aniline and phytic acid, which exhibited excellent electrocatalytic properties for both the ORR and OER, and also promising application in rechargeable zinc–air batteries.[26] The NPMC was highly interconnected into a hierarchical porous network with a large surface area of ~1663 m^2 g^{-1} (Figure 8.16(a)). The contents of C, N, P, and O were determined to be 90.8, 3.2, 1.1, and 4.9 at%, respectively, for the NPMC pyrolyzed at 1000 °C. The LSV curves obtained in O_2-saturted 0.10 M KOH exhibit an ORR half-wave potential of 0.85 V comparable to Pt/C, and a lower OER onset potential than RuO_2 nanoparticles, suggesting attractive bifunctional activities (Figure 8.16(b)). To gain insights into the catalytic mechanism, DFT calculations were performed to investigate the role of different doping structures. As a result, the N-dopant is identified as the most active sites for the ORR, while the N, P co-doped graphene is revealed to be more active for the OER with a minimum overpotential of 0.39 V (Figure 8.16(c)). Considering that

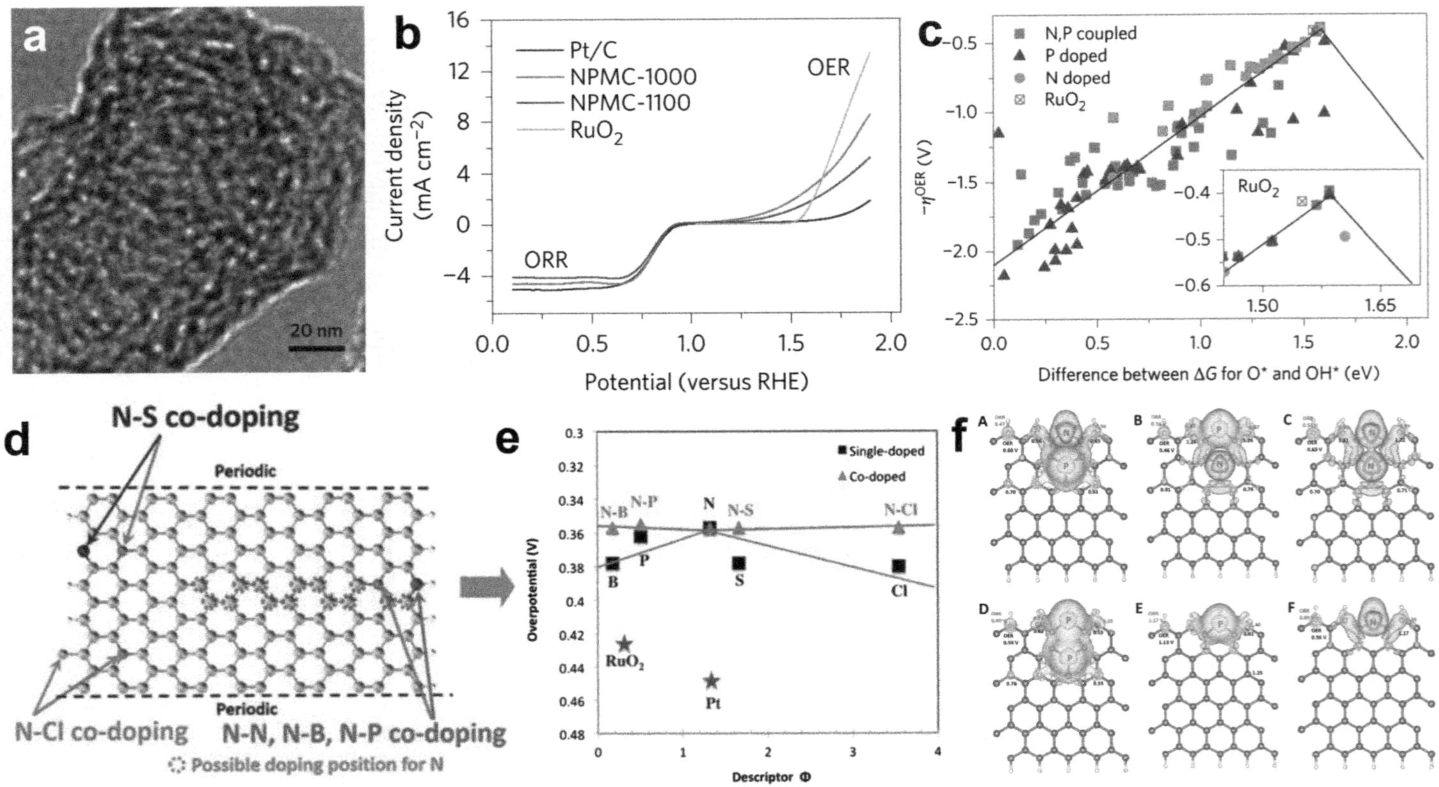

Figure 8.16 Dual-doped carbon as an ORR/OER bifunctional catalyst. (a) TEM image of NPMC. (b) ORR/OER LSV curves obtained in O_2-saturted 0.1 M KOH. (c) OER volcano plot of the calculated overpotential *vs.* the difference between the adsorption energy of O* and OH*. ((a)–(c) are reprinted by permission from Macmillan Publishers Ltd: *Nature Nanotechnology*, J. Zhang, Z. Zhao, Z. Xia and L. Dai, *Nat. Nanotechnol.*, 2015, **10**, 444.[26] Copyright 2015.) (d) Scheme of the dual-doped graphene nanoribbon used for DFT calculations. (e) The volcano plots obtained by correlating between overpotential and descriptor Φ. (f) Differential charge density and overpotential distribution on dual- or single-doped graphene. ((d)–(f) are reprinted with permission from Z. Zhao and Z. Xia, *ACS Catal.*, 2016, **6**, 1553. Copyright 2016 American Chemical Society.[117])

various reported metal-free functionalized carbons can deliver outstanding ORR performances, the significant improvement on OER activity ascribed to dual-doping is very attractive for developing effective ORR/OER bifunctional catalysts. Recently, a series of metal-free dual-doped carbon materials have been reported to exhibit remarkable bifunctional electrocatalytic activities, even approaching the benchmark Pt/C for the ORR and IrO_2/C for the OER, such as N, S co-doped carbon nanosheets,[133] P-doped g-C_3N_4,[29] N, P co-doped graphene/carbon nanosheets,[134] N, P co-modified carbon nanodots,[135] and so on.

Similar to single-element-doped carbon catalysts, Xia and co-workers extended the proposed descriptor: $\Phi = (E_X/E_C) \times (A_X/A_C)$,[70] to dual-element-doped carbons for ORR/OER bifunctional catalysis.[117] They probed all possible active sites on dual-doped graphene with N and X (X = N, B, P, S, Cl, *etc.*) and also considered the edge and distance effect (Figure 8.16(d)). When correlating the calculated lower overpotential *versus* descriptor Φ, a very shallow inverted volcano can be obtained (Figure 8.16(e)). Notably, the overpotentials for both the ORR and OER for dual-element-doped carbon materials are significantly decreased by 10–40% compared with single-element-doped counterparts, and also much lower than precious-metal catalysts. This enhancement was rationalized by the interaction between dual dopants, which generated overlapped p-electron clouds and more active sites (Figure 8.16(f)). The distance between two heteroatoms was also influential on the resultant synergistic effects and activity improvement. Therefore, several design principles were provided: (i) choose dual dopants with relatively different properties such as electronegativity and electron affinity; (ii) keep the distance between two heteroatoms smaller if Φ is much smaller than N, and *vice versa*; (iii) dope near the edge to make use of edge effects.

8.5.2 HER-containing Multifunctional Catalysts

In addition to the ORR and OER, the HER is also an essential reaction for the future of sustainable energy to produce clean hydrogen *via* water splitting. In contrast to the great achievement of metal-free carbon catalysts for the ORR/OER exhibiting comparable performance to precious-metal catalysts, the performance of carbon materials for the HER is still limited and is far from satisfactory. To probe the underlying mechanism and guide the development of highly active HER catalysts, Qiao's group have reported theoretical and experimental investigations on a series of heteroatom-doped graphene materials.[136,137] DFT calculations revealed that heteroatom doping could improve the HER activity of graphene owing to the tuned H* adsorption abilities. However, even climbing to the volcano's peak, the HER activity of single-heteroatom-doped graphene is still less than the Pt/C and MoS_2 benchmark (Figure 8.17(a)) due to the weak hydrogen adsorption on graphene.[137] By introducing secondary elements with synergistically coupling effects to further modify the electron acceptor–donor properties, the activity can be greatly enhanced, attributed to the lowered adsorption energy of H* (Figure 8.17(b) and (c)).

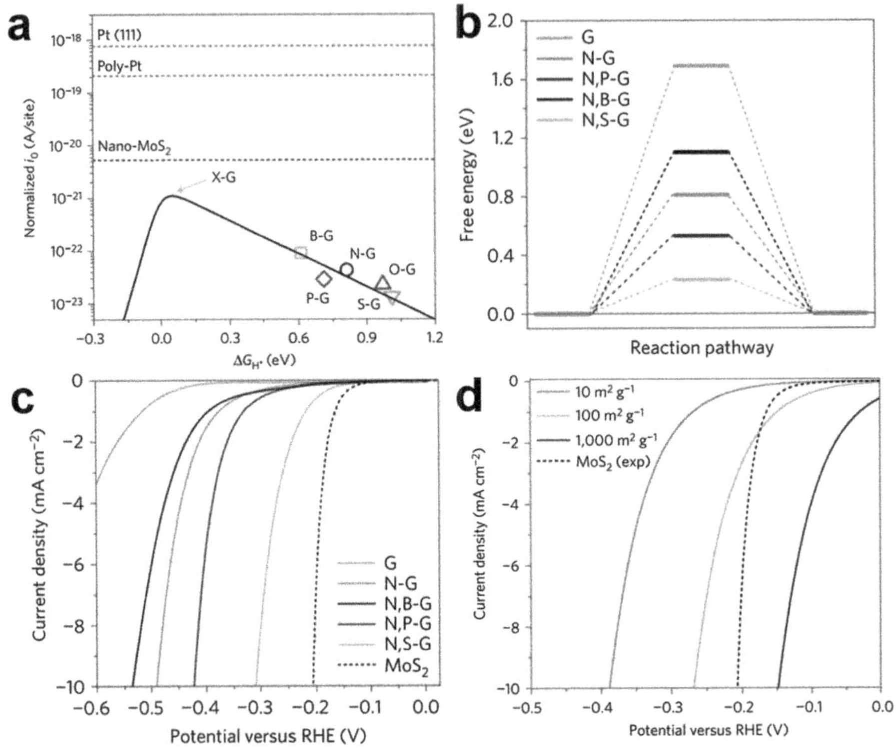

Figure 8.17 Heteroatom-doped carbon catalysts for the HER. (a) Volcano plot between the normalized theoretical exchange current per active site and free energy change of hydrogen adsorption. (b) A free energy diagram for different graphene models. (c) LSV curves obtained in 0.5 M H_2SO_4. (d) Predicted HER performance of doped graphene materials with different surface areas. (Reprinted by permission from Macmillan Publishers Ltd: *Nature Energy*, Y. Jiao, Y. Zheng, K. Davey and S. Z. Qiao, *Nat. Energy*, 2016, **1**, 16130.[137] Copyright 2016.)

In addition to the intrinsic electronic structure, the extrinsic physicochemical properties, such as surface area/doping concentration, can also significantly govern the HER overpotential. With a higher active concentration of dopants and a higher surface area improved to 1000 $m^2 g^{-1}$, the ultimate HER performance of dual-doped metal-free carbon was predicted to be superior to that of MoS_2 materials (Figure 8.17(d)).[137] Consequently, N, S co-doped[137] or N, P co-doped graphene[136] have been investigated as highly active metal-free catalysts for the HER. These metal-free dual-doped carbons are also reported as excellent catalysts for the OER,[26,29,126,133,134] thereby suggesting promising OER/HER bifunctional catalytic activities for overall water splitting. However, an excellent HER performance is mostly achieved in acid and effective OER catalysts are only available under alkaline or neutral conditions. Thus, there is still a long way to go to develop effective OER/HER bifunctional catalysts based on metal-free functionalized carbons working in the same pH range.[129]

In addition to ORR/OER bifunctional catalysts for overall water splitting, the development of ORR/HER bifunctional metal-free carbon catalysts to replace Pt-based catalysts for regenerative fuel cells is also attractive. Recently, several effective carbon catalysts have been reported to exhibit extraordinary electrocatalytic activity for the ORR and good activity for the HER in an acidic or alkaline medium, such as an N, P dual-doped carbon network,[138] N, S dual-doped carbon tubes,[139] and an N-doped tubular carbon.[140] In alkaline media, the ORR activity of metal-free heteroatom-doped carbon materials is comparable to, or even superior to, the state-of-the-art Pt/C catalyst, while the HER performance is limited with a high overpotential.[138] Contrastively, the HER activity of metal-free carbon catalysts in acidic conditions can be considerably improved, while the ORR performance is always poor with a negatively shifted onset potential and half-wave potential.[139] Great efforts are still required to improve the ORR and HER performances of metal-free functionalized carbons under the same operation conditions. Recently, Dai and co-workers further employed multi-doped carbon materials (*e.g.*, N, P, F tri-doped graphene,[141] N, S dual-doped porous graphitic sheets[142]) as trifunctional catalysts for the ORR, OER, and HER, which can be used for self-powered electrochemical water splitting with the same catalyst. The defect graphene without heteroatom doping was also demonstrated to be effective for the ORR, OER, and HER in an alkaline medium, both experimentally and theoretically.[69]

8.6 Conclusions

Metal-free functionalized carbons have been well-demonstrated as a family of sustainable electrocatalysts to replace their precious-metal counterparts for vast electrocatalysis reactions, including the ORR, OER, and HER. During the last decade, significant progress has been achieved in both materials science and fundamental understanding, resulting in various metal-free carbon materials that can compete with precious-metal-based catalysts. Generally, the superb catalytic activity is correlated with the altered charge/spin distribution on the conjugated sp^2 carbon matrix, which is attributed to the heteroatom-/molecular-doping, edge sites or the edge effect, and intrinsic topological defects. N-doped carbon materials were first reported and widely demonstrated as the most effective catalysts for the ORR and OER. Furthermore, dual-doping, such as N/P or N/S, is regarded as a critical strategy to improve the OER and HER activities towards the benchmark precious-metal-based catalysts. In addition to the modification of intrinsic electron structures, the optimization of extrinsic physicochemical properties (*e.g.*, surface area, porosity hierarchy, doping content, conductivity, *etc.*) is also beneficial for the improvement of catalytic performances. Meanwhile, more effects are highly expected to promote further research and push metal-free carbon catalysts towards practical applications.

A fundamental understanding of the activity origin and degradation mechanism is very important for further optimization towards highly active and stable catalysts. On the one hand, the effect of electrolytes is expected to be taken into consideration for a more precise description, calculation, and prediction of working electrocatalysts. On the other hand, more powerful and effective characterizations, such as advanced electron microscopies, *operando* and *in situ* techniques are highly recommended to clearly identify different types of active configurations and probe the catalytic behaviours. Additionally, the catalytic activity and durability should be further improved to meet the requirements of practical applications for clean energy conversion and storage. Guided by theoretical predictions, new methodologies are required to fabricate metal-free functionalized carbons with more active and stable active sites. Besides, the structure of carbon-based catalysts is suggested to be engineered towards 3D hierarchical frameworks with high conductivity, large surface areas, tuned surface hydrophilicity, interconnected pores, and flexible or even freestanding scaffolds. Furthermore, for device-targeted applications, more attention should be paid to the exploration of multifunctional carbon catalysts under specific working conditions. ORR/OER bifunctional catalysts active in alkaline media are required for the assembly of metal–air batteries, while OER/HER bifunctional catalysts active in alkaline media are expected for water splitting. Especially, the development of carbon-based HER catalysts with a high performance in alkaline conditions and ORR catalysts in acidic media is significantly desirable.

Along with the incorporation of present achievements and further research in this field, metal-free functionalized carbons will undoubtedly have a promising future with practical applications in energy-related electrocatalysis.

Acknowledgements

This work was supported by the National Key Research and Development Program (Nos. 2016YFA0202500, 2016YFA0200102, and 2015CB932500).

References

1. I. Katsounaros, S. Cherevko, A. R. Zeradjanin and K. J. J. Mayrhofer, *Angew. Chem., Int. Ed.*, 2014, **53**, 102.
2. M. G. Walter, E. L. Warren, J. R. McKone, S. W. Boettcher, Q. X. Mi, E. A. Santori and N. S. Lewis, *Chem. Rev.*, 2010, **110**, 6446.
3. Y. Jiao, Y. Zheng, M. T. Jaroniec and S. Z. Qiao, *Chem. Soc. Rev.*, 2015, **44**, 2060.
4. C. C. L. McCrory, S. Jung, J. C. Peters and T. F. Jaramillo, *J. Am. Chem. Soc.*, 2013, **135**, 16977.
5. M. Gong and H. J. Dai, *Nano Res.*, 2015, **8**, 23.
6. J. O. M. Bockris, *Science*, 1972, **176**, 1323.
7. J. O. M. Bockris, *Int. J. Hydrogen Energy*, 2002, **27**, 731.
8. Y. C. Ling and Y. Li, *Part. Part. Syst. Charact.*, 2014, **31**, 1113.
9. X. Zhang, X. Cheng and Q. Zhang, *J. Energy Chem.*, 2016, **25**, 967.

10. C. Tang and Q. Zhang, *J. Mater. Chem. A*, 2016, **4**, 4998.
11. M. K. Debe, *Nature*, 2012, **486**, 43.
12. J. Suntivich, H. A. Gasteiger, N. Yabuuchi, H. Nakanishi, J. B. Goodenough and Y. Shao-Horn, *Nat. Chem.*, 2011, **3**, 546.
13. Y. Gorlin and T. F. Jaramillo, *J. Am. Chem. Soc.*, 2010, **132**, 13612.
14. Y. Y. Liang, Y. G. Li, H. L. Wang, J. G. Zhou, J. Wang, T. Regier and H. J. Dai, *Nat. Mater.*, 2011, **10**, 780.
15. F. Y. Cheng and J. Chen, *Chem. Soc. Rev.*, 2012, **41**, 2172.
16. L. Grande, E. Paillard, J. Hassoun, J. B. Park, Y. J. Lee, Y. K. Sun, S. Passerini and B. Scrosati, *Adv. Mater.*, 2015, **27**, 784.
17. Y. G. Li and H. J. Dai, *Chem. Soc. Rev.*, 2014, **43**, 5257.
18. P. G. Bruce, S. A. Freunberger, L. J. Hardwick and J. M. Tarascon, *Nat. Mater.*, 2012, **11**, 19.
19. W. Xia, A. Mahmood, Z. Liang, R. Zou and S. Guo, *Angew. Chem., Int. Ed.*, 2016, **55**, 2650.
20. Y. Lee, J. Suntivich, K. J. May, E. E. Perry and Y. Shao-Horn, *J. Phys. Chem. Lett.*, 2012, **3**, 399.
21. H. Over, *Chem. Rev.*, 2012, **112**, 3356.
22. J. Rossmeisl, Z. W. Qu, H. Zhu, G. J. Kroes and J. K. Norskov, *J. Electroanal. Chem.*, 2007, **607**, 83.
23. Y. Nie, L. Li and Z. Wei, *Chem. Soc. Rev.*, 2015, **44**, 2168.
24. J. I. Jung, H. Y. Jeong, J. S. Lee, M. G. Kim and J. Cho, *Angew. Chem., Int. Ed.*, 2014, **53**, 4582.
25. K. L. Pickrahn, S. W. Park, Y. Gorlin, H.-B.-R. Lee, T. F. Jaramillo and S. F. Bent, *Adv. Energy Mater.*, 2012, **2**, 1269.
26. J. Zhang, Z. Zhao, Z. Xia and L. Dai, *Nat. Nanotechnol.*, 2015, **10**, 444.
27. G. L. Tian, Q. Zhang, B. S. Zhang, Y. G. Jin, J. Q. Huang, D. S. Su and F. Wei, *Adv. Funct. Mater.*, 2014, **24**, 5956.
28. G. L. Tian, M. Q. Zhao, D. S. Yu, X. Y. Kong, J. Q. Huang, Q. Zhang and F. Wei, *Small*, 2014, **10**, 2251.
29. T. Y. Ma, J. Ran, S. Dai, M. Jaroniec and S. Z. Qiao, *Angew. Chem., Int. Ed.*, 2015, **54**, 4646.
30. Y. Zhao, K. Kamiya, K. Hashimoto and S. Nakanishi, *J. Phys. Chem. C*, 2015, **119**, 2583.
31. L. B. Ma, X. P. Shen, G. X. Zhu, Z. Y. Ji and H. Zhou, *Carbon*, 2014, **77**, 255.
32. S. Dou, L. Tao, J. Huo, S. Wang and L. Dai, *Energy Environ. Sci.*, 2016, **9**, 1320.
33. K. Gong, F. Du, Z. Xia, M. Durstock and L. Dai, *Science*, 2009, **323**, 760.
34. L. J. Yang, S. J. Jiang, Y. Zhao, L. Zhu, S. Chen, X. Z. Wang, Q. Wu, J. Ma, Y. W. Ma and Z. Hu, *Angew. Chem., Int. Ed.*, 2011, **50**, 7132.
35. L. F. Lai, J. R. Potts, D. Zhan, L. Wang, C. K. Poh, C. H. Tang, H. Gong, Z. X. Shen, L. Y. Jianyi and R. S. Ruoff, *Energy Environ. Sci.*, 2012, **5**, 7936.
36. D. W. Wang and D. S. Su, *Energy Environ. Sci.*, 2014, **7**, 576.
37. D. S. Geng, Y. Chen, Y. G. Chen, Y. L. Li, R. Y. Li, X. L. Sun, S. Y. Ye and S. Knights, *Energy Environ. Sci.*, 2011, **4**, 760.

38. Z. W. Liu, X. Fu, M. Li, F. Wang, Q. D. Wang, G. J. Kang and F. Peng, *J. Mater. Chem. A*, 2015, **3**, 3289.
39. Z. Yang, Z. Yao, G. F. Li, G. Y. Fang, H. G. Nie, Z. Liu, X. M. Zhou, X. Chen and S. M. Huang, *ACS Nano*, 2012, **6**, 205.
40. C. Z. Zhang, N. Mahmood, H. Yin, F. Liu and Y. L. Hou, *Adv. Mater.*, 2013, **25**, 4932.
41. L. M. Dai, Y. H. Xue, L. T. Qu, H. J. Choi and J. B. Baek, *Chem. Rev.*, 2015, **115**, 4823.
42. J. Duan, S. Chen, M. T. Jaroniec and S. Z. Qiao, *ACS Catal.*, 2015, **5**, 5207.
43. D. S. Yu, L. Wei, W. C. Jiang, H. Wang, B. Sun, Q. Zhang, K. L. Goh, R. M. Si and Y. Chen, *Nanoscale*, 2013, **5**, 3457.
44. W. Wei, H. W. Liang, K. Parvez, X. D. Zhuang, X. L. Feng and K. Mullen, *Angew. Chem., Int. Ed.*, 2014, **53**, 1570.
45. L. T. Qu, Y. Liu, J. B. Baek and L. M. Dai, *ACS Nano*, 2010, **4**, 1321.
46. S. B. Yang, L. J. Zhi, K. Tang, X. L. Feng, J. Maier and K. Mullen, *Adv. Funct. Mater.*, 2012, **22**, 3634.
47. S. Y. Wang, L. P. Zhang, Z. H. Xia, A. Roy, D. W. Chang, J. B. Baek and L. M. Dai, *Angew. Chem., Int. Ed.*, 2012, **51**, 4209.
48. H. Jin, H. Huang, Y. He, X. Feng, S. Wang, L. Dai and J. Wang, *J. Am. Chem. Soc.*, 2015, **137**, 7588.
49. J. Liang, Y. Zheng, J. Chen, J. Liu, D. Hulicova-Jurcakova, M. Jaroniec and S. Z. Qiao, *Angew. Chem., Int. Ed.*, 2012, **51**, 3892.
50. Y. G. Li, W. Zhou, H. L. Wang, L. M. Xie, Y. Y. Liang, F. Wei, J. C. Idrobo, S. J. Pennycook and H. J. Dai, *Nat. Nanotechnol.*, 2012, **7**, 394.
51. D. Yu, Q. Zhang and L. Dai, *J. Am. Chem. Soc.*, 2010, **132**, 15127.
52. C.-Y. Chen, C. Tang, H.-F. Wang, C.-M. Chen, X. Zhang, X. Huang and Q. Zhang, *ChemSusChem*, 2016, **9**, 1194.
53. T. Y. Ma, S. Dai, M. Jaroniec and S. Z. Qiao, *Angew. Chem., Int. Ed.*, 2014, **53**, 7281.
54. L. Wang, Y. Huang, C. Li, J. J. Chen and X. Sun, *Synth. Met.*, 2014, **198**, 300.
55. Y. Zhao, R. Nakamura, K. Kamiya, S. Nakanishi and K. Hashimoto, *Nat. Commun.*, 2013, **4**, 2390.
56. M. Zhou, H.-L. Wang and S. Guo, *Chem. Soc. Rev.*, 2016, **45**, 1273.
57. M. Li, L. Zhang, Q. Xu, J. Niu and Z. Xia, *J. Catal.*, 2014, **314**, 66.
58. I.-Y. Jeon, S. Zhang, L. Zhang, H.-J. Choi, J.-M. Seo, Z. Xia, L. Dai and J.-B. Baek, *Adv. Mater.*, 2013, **25**, 6138.
59. L. Zhang and Z. Xia, *J. Phys. Chem. C*, 2011, **115**, 11170.
60. Y. Jiao, Y. Zheng, M. Jaroniec and S. Z. Qiao, *J. Am. Chem. Soc.*, 2014, **136**, 4394.
61. D. S. Su, S. Perathoner and G. Centi, *Chem. Rev.*, 2013, **113**, 5782.
62. E. J. Biddinger and U. S. Ozkan, *J. Phys. Chem. C*, 2010, **114**, 15306.
63. Y. Jiang, L. Yang, T. Sun, J. Zhao, Z. Lyu, O. Zhuo, X. Wang, Q. Wu, J. Ma and Z. Hu, *ACS Catal.*, 2015, **5**, 6707.
64. C. Tang, H.-F. Wang, X. Chen, B.-Q. Li, T.-Z. Hou, B. Zhang, Q. Zhang, M.-M. Titirici and F. Wei, *Adv. Mater.*, 2016, **28**, 6845.

65. L. Tao, Q. Wang, S. Dou, Z. Ma, J. Huo, S. Wang and L. Dai, *Chem. Commun.*, 2016, **52**, 2764.
66. X. Wang, X. Li, C. Ouyang, Z. Li, S. Dou, Z. Ma, L. Tao, J. Huo and S. Wang, *J. Mater. Chem. A*, 2016, **4**, 9370.
67. C. Tang and Q. Zhang, *Adv. Mater.*, 2017, **29**, 1604103.
68. Y. Ito, Y. H. Shen, D. Hojo, Y. Itagaki, T. Fujita, L. H. Chen, T. Aida, Z. Tang, T. Adschiri and M. W. Chen, *Adv. Mater.*, 2016, **28**, 10644.
69. Y. Jia, L. Z. Zhang, A. J. Du, G. P. Gao, J. Chen, X. C. Yan, C. L. Brown and X. D. Yao, *Adv. Mater.*, 2016, **28**, 9532.
70. Z. Zhao, M. Li, L. Zhang, L. Dai and Z. Xia, *Adv. Mater.*, 2015, **27**, 6834.
71. X. Liu and L. Dai, *Nat. Rev. Mater.*, 2016, **1**, 16064.
72. J. Shui, M. Wang, F. Du and L. Dai, *Sci. Adv.*, 2015, **1**, e1400129.
73. U. Martinez, J. H. Dumont, E. F. Holby, K. Artyushkova, G. M. Purdy, A. Singh, N. H. Mack, P. Atanassov, D. A. Cullen, K. L. More, M. Chhowalla, P. Zelenay, A. M. Dattelbaum, A. D. Mohite and G. Gupta, *Sci. Adv.*, 2016, **2**, e1501178.
74. W. Ding, Z. D. Wei, S. G. Chen, X. Q. Qi, T. Yang, J. S. Hu, D. Wang, L. J. Wan, S. F. Alvi and L. Li, *Angew. Chem., Int. Ed.*, 2013, **52**, 11755.
75. W. Ding, L. Li, K. Xiong, Y. Wang, W. Li, Y. Nie, S. Chen, X. Qi and Z. Wei, *J. Am. Chem. Soc.*, 2015, **137**, 5414.
76. J. Masa, W. Xia, M. Muhler and W. Schuhmann, *Angew. Chem., Int. Ed.*, 2015, **54**, 10102.
77. L. Zhang, Q. Xu, J. Niu and Z. Xia, *Phys. Chem. Chem. Phys.*, 2015, **17**, 16733.
78. S. Buller and J. Strunk, *J. Energy Chem.*, 2016, **25**, 171.
79. M. Qiao, C. Tang, G. He, K. Qiu, R. Binions, I. P. Parkin, Q. Zhang, Z. Guo and M. M. Titirici, *J. Mater. Chem. A*, 2016, **4**, 12658.
80. J.-L. Shi, G.-L. Tian, Q. Zhang, M.-Q. Zhao and F. Wei, *Carbon*, 2015, **93**, 702.
81. X.-B. Cheng, Q. Zhang, H.-F. Wang, G.-L. Tian, J.-Q. Huang, H.-J. Peng, M.-Q. Zhao and F. Wei, *Catal. Today*, 2015, **249**, 244.
82. D. Guo, R. Shibuya, C. Akiba, S. Saji, T. Kondo and J. Nakamura, *Science*, 2016, **351**, 361.
83. D. Geng, Y. Chen, Y. Chen, Y. Li, R. Li, X. Sun, S. Ye and S. Knights, *Energy Environ. Sci.*, 2011, **4**, 760.
84. H. B. Yang, J. Miao, S.-F. Hung, J. Chen, H. B. Tao, X. Wang, L. Zhang, R. Chen, J. Gao, H. M. Chen, L. Dai and B. Liu, *Sci. Adv.*, 2016, **2**, e1501122.
85. K.-H. Wu, D.-W. Wang, X. Zong, B. Zhang, Y. Liu, I. R. Gentle and D.-S. Su, *J. Mater. Chem. A*, 2017, **5**, 3239.
86. T. Ishizaki, S. Chiba, Y. Kaneko and G. Panomsuwan, *J. Mater. Chem. A*, 2014, **2**, 10589.
87. R. S. Zhong, Y. H. Qin, D. F. Niu, J. W. Tian, X. S. Zhang, X. G. Zhou, S. G. Sun and W. K. Yuan, *J. Power Sources*, 2013, **225**, 192.
88. L. Zhang, J. Niu, M. Li and Z. Xia, *J. Phys. Chem. C*, 2014, **118**, 3545.
89. B. Hammer and J. K. Norskov, *Nature*, 1995, **376**, 238.

90. J. Y. Cheon, J. H. Kim, J. H. Kim, K. C. Goddeti, J. Y. Park and S. H. Joo, *J. Am. Chem. Soc.*, 2014, **136**, 8875.
91. Y. Zheng, Y. Jiao and S. Z. Qiao, *Adv. Mater.*, 2015, **27**, 5372.
92. D. Geng, N. Ding, T. S. A. Hor, Z. Liu, X. Sun and Y. Zong, *J. Mater. Chem. A*, 2015, **3**, 1795.
93. S. Dou, A. Shen, L. Tao and S. Wang, *Chem. Commun.*, 2014, **50**, 10672.
94. S. Wang, D. Yu and L. Dai, *J. Am. Chem. Soc.*, 2011, **133**, 5182.
95. I.-Y. Jeon, H.-J. Choi, M. Choi, J.-M. Seo, S.-M. Jung, M.-J. Kim, S. Zhang, L. Zhang, Z. Xia, L. Dai, N. Park and J.-B. Baek, *Sci. Rep.*, 2013, **3**, 1810.
96. R. Sharma, J. H. Baik, C. J. Perera and M. S. Strano, *Nano Lett.*, 2010, **10**, 398.
97. W. Yuan, Y. Zhou, Y. Li, C. Li, H. Peng, J. Zhang, Z. Liu, L. Dai and G. Shi, *Sci. Rep.*, 2013, **3**, 2248.
98. A. Shen, Y. Zou, Q. Wang, R. A. W. Dryfe, X. Huang, S. Dou, L. Dai and S. Wang, *Angew. Chem., Int. Ed.*, 2014, **53**, 10804.
99. L. Zhang, J. Niu, L. Dai and Z. Xia, *Langmuir*, 2012, **28**, 7542.
100. D. Yan, Y. Li, J. Huo, R. Chen, L. Dai and S. Wang, *Adv. Mater.*, 2017, **29**, 1606459.
101. F. Banhart, J. Kotakoski and A. V. Krasheninnikov, *ACS Nano*, 2011, **5**, 26.
102. L. Vicarelli, S. J. Heerema, C. Dekker and H. W. Zandbergen, *ACS Nano*, 2015, **9**, 3428.
103. A. H. Castro Neto, F. Guinea, N. M. R. Peres, K. S. Novoselov and A. K. Geim, *Rev. Mod. Phys.*, 2009, **81**, 109.
104. P. J. Britto, K. S. V. Santhanam, A. Rubio, J. A. Alonso and P. M. Ajayan, *Adv. Mater.*, 1999, **11**, 154.
105. D. L. Carroll, P. Redlich, P. M. Ajayan, J. C. Charlier, X. Blase, A. DeVita and R. Car, *Phys. Rev. Lett.*, 1997, **78**, 2811.
106. P. Zhang, X. L. Hou, J. L. Mi, Y. Q. He, L. Lin, Q. Jiang and M. D. Dong, *Phys. Chem. Chem. Phys.*, 2014, **16**, 17479.
107. G.-L. Chai, Z. Hou, D.-J. Shu, T. Ikeda and K. Terakura, *J. Am. Chem. Soc.*, 2014, **136**, 13629.
108. S. J. Jiang, Z. Li, H. Y. Wang, Y. Wang, L. N. Meng and S. Q. Song, *Nanoscale*, 2014, **6**, 14262.
109. Q. Wu, L. J. Yang, X. Z. Wang and Z. Hu, *Acc. Chem. Res.*, 2017, **50**, 435.
110. X. J. Zhao, X. Q. Zou, X. C. Yan, C. L. Brown, Z. G. Chen, G. S. Zhu and X. D. Yao, *Inorg. Chem. Front.*, 2016, **3**, 417.
111. H. Y. Zhao, C. H. Sun, Z. Jin, D. W. Wang, X. C. Yan, Z. G. Chen, G. S. Zhu and X. D. Yao, *J. Mater. Chem. A*, 2015, **3**, 11736.
112. B.-Q. Li, C. Tang, H.-F. Wang, X.-L. Zhu and Q. Zhang, *Sci. Adv.*, 2016, **2**, e1600495.
113. C. Tang, H. F. Wang, X. L. Zhu, B. Q. Li and Q. Zhang, *Part. Part. Syst. Charact.*, 2016, **33**, 473.
114. M. Gong, Y. Li, H. Wang, Y. Liang, J. Z. Wu, J. Zhou, J. Wang, T. Regier, F. Wei and H. Dai, *J. Am. Chem. Soc.*, 2013, **135**, 8452.
115. H. F. Wang, C. Tang, X. L. Zhu and Q. Zhang, *J. Mater. Chem. A*, 2016, **4**, 3379.

116. Z. Y. Lin, G. H. Waller, Y. Liu, M. L. Liu and C. P. Wong, *Carbon*, 2013, **53**, 130.
117. Z. Zhao and Z. Xia, *ACS Catal.*, 2016, **6**, 1553.
118. H.-F. Wang, C. Tang and Q. Zhang, *Catal. Today*, 2018, **301**, 25.
119. S. Chen, J. J. Duan, J. R. Ran and S. Z. Qiao, *Adv. Sci.*, 2015, **2**, 1400015.
120. C. Su, M. Acik, K. Takai, J. Lu, S.-J. Hao, Y. Zheng, P. Wu, Q. Bao, T. Enoki, Y. J. Chabal and K. P. Loh, *Nat. Commun.*, 2012, **3**, 1298.
121. X. Y. Lu, W. L. Yim, B. H. R. Suryanto and C. Zhao, *J. Am. Chem. Soc.*, 2015, **137**, 2901.
122. L. Li, H. Yang, J. Miao, L. Zhang, H.-Y. Wang, Z. Zeng, W. Huang, X. Dong and B. Liu, *ACS Energy Lett.*, 2017, **2**, 294.
123. S. Chen, J. J. Duan, M. Jaroniec and S. Z. Qiao, *Adv. Mater.*, 2014, **26**, 2925.
124. S. Chen, J. Duan, Y. Zheng, X. Chen, X. W. Du, M. Jaroniec and S.-Z. Qiao, *Energy Storage Mater.*, 2015, **1**, 17.
125. Y. P. Zhu, Y. Jing, A. Vasileff, T. Heine and S. Z. Qiao, *Adv. Energy Mater.*, 2017, **7**, 1602928.
126. J. Zhao, Y. Liu, X. Quan, S. Chen, H. Zhao and H. Yu, *Electrochim. Acta*, 2016, **204**, 169.
127. Z. Xiao, X. Huang, L. Xu, D. Yan, J. Huo and S. Wang, *Chem. Commun.*, 2016, **52**, 13008.
128. A. M. El-Sawy, I. M. Mosa, D. Su, C. J. Guild, S. Khalid, R. Joesten, J. F. Rusling and S. L. Suib, *Adv. Energy Mater.*, 2016, **6**, 1501966.
129. C. G. Hu and L. M. Dai, *Angew. Chem., Int. Ed.*, 2016, **55**, 11736.
130. R. M. Yadav, J. Wu, R. Kochandra, L. Ma, C. S. Tiwary, L. Ge, G. Ye, R. Vajtai, J. Lou and P. M. Ajayan, *ACS Appl. Mater. Interfaces*, 2015, **7**, 11991.
131. S. Liu, H. Zhang, Q. Zhao, X. Zhang, R. Liu, X. Ge, G. Wang, H. Zhao and W. Cai, *Carbon*, 2016, **106**, 74.
132. J.-C. Li, P.-X. Hou, S.-Y. Zhao, C. Liu, D.-M. Tang, M. Cheng, F. Zhang and H.-M. Cheng, *Energy Environ. Sci.*, 2016, **9**, 3079.
133. K. Qu, Y. Zheng, S. Dai and S. Z. Qiao, *Nano Energy*, 2016, **19**, 373.
134. R. Li, Z. Wei and X. Gou, *ACS Catal.*, 2015, **5**, 4133.
135. J. Liu, S. Zhao, C. Li, M. Yang, Y. Yang, Y. Liu, Y. Lifshitz, S.-T. Lee and Z. Kang, *Adv. Energy Mater.*, 2016, **6**, 1502039.
136. Y. Zheng, Y. Jiao, L. H. Li, T. Xing, Y. Chen, M. Jaroniec and S. Z. Qiao, *ACS Nano*, 2014, **8**, 5290.
137. Y. Jiao, Y. Zheng, K. Davey and S. Z. Qiao, *Nat. Energy*, 2016, **1**, 16130.
138. J. Zhang, L. Qu, G. Shi, J. Liu, J. Chen and L. Dai, *Angew. Chem., Int. Ed.*, 2016, **55**, 2230.
139. T. Sun, Q. Wu, Y. Jiang, Z. Zhang, L. Du, L. Yang, X. Wang and Z. Hu, *Chem.–Eur. J.*, 2016, **22**, 10326.
140. D. K. Singh, R. N. Jenjeti, S. Sampath and M. Eswaramoorthy, *J. Mater. Chem. A*, 2017, **5**, 6025.
141. J. T. Zhang and L. M. Dai, *Angew. Chem., Int. Ed.*, 2016, **55**, 13296.
142. C. Hu and L. Dai, *Adv. Mater.*, 2017, **29**, 1604942.

CHAPTER 9

Metal-free Functionalised Carbons in Photocatalysis

REN SU

Syncat@Beijing, Synfuels China Technology Co. Ltd., Leyuan South Street II, No.1, Yanqi Economic Development Zone C# Huairou District, Beijing, 101407, P. R. China
*E-mail: suren@synfuelschina.com.cn

9.1 Overview, Basic Concepts, and Key Terminologies

9.1.1 Overview

Photocatalysis is a catalytic reaction driven by photons. Photocatalysts need two functions: harvesting photons (light absorption) and performing catalytic reactions. Since one hour of the solar irradiation received by the Earth is more than the current annual energy consumption of the world, using solar energy by means of photocatalysis to drive some very important reactions, for example, hydrogen production, CO_2 reduction, water/air purification and organic synthesis, is of great interest. Therefore, as a branch of heterogeneous catalysis, photocatalysis has received great attention over the last decades, both in application oriented and fundamental investigations. The growth of both scientific and engineering interest shows an exponential trend, and the total amount of publications to date exceeds 120 000.

Catalysis Series No. 31
Metal-free Functionalized Carbons in Catalysis: Synthesis, Characterization and Applications
Edited by Alberto Villa and Nikolaos Dimitratos

Published by the Royal Society of Chemistry, www.rsc.org

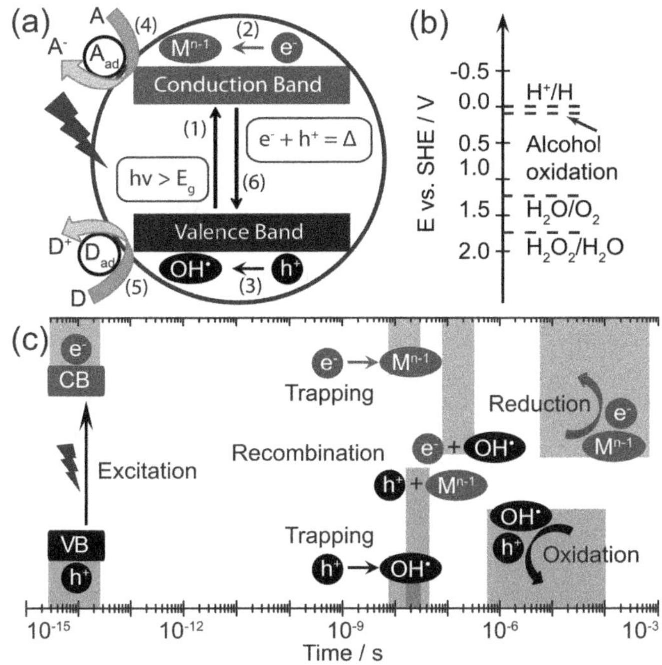

Figure 9.1 (a) Basic steps in a semiconductor-based photochemical mechanism: (1) formation of e^-–h^+ pairs; (2) and (3) trapping of an e^- in a dangling surficial bond to yield M^{2n-1} and a h^+ in the form of a OH$^•$ radical; (4) and (5) initiation of a redox reaction by the e^-–h^+ pairs or their corresponding radicals; (6) charge carrier recombination. (b) The CB and VB positions required for different photocatalytic reactions, and (c) a typical timescale for elementary steps in photocatalysis.

The photocatalysts discussed in this chapter are all semiconductors, which have their electronic structure characterised by an empty conduction band (CB) and a valence band (VB) filled with electrons (e^-) in the dark. The energy difference between the conduction band minimum (CBM) and the valence band maximum (VBM) is defined as the bandgap (E_g) of the semiconductor. This definition is roughly the same as that of LUMO (lowest unoccupied molecular orbital) and HOMO (highest occupied molecular orbital) in orbital theory. A photocatalyst plays dual roles in the process: (I) light absorber and (II) reactor. Figure 9.1(a) presents the basic steps involved in the photochemical reaction with a semiconductor photocatalyst. The first step involves a photon with energy ($h\nu$) matching the E_g of the semiconductor (equal to or greater) being absorbed, resulting in an e^- been excited from the VB to the CB and leaving a hole (h^+) in the VB, as shown in process 1. Then the photon-generated e^-–h^+ pairs can either be trapped into metastable surface states (process 2 and 3) or react with electron acceptors (A) and electron donors (D) that are adsorbed on the semiconductor surface (process 4 and 5). The e^-–h^+ pairs can also recombine and release the photon energy as heat, as shown in process 6.

The CBM and VBM of the semiconductor determine the reduction and oxidation potential (powers) of the photon-generated e^-–h^+ pairs to drive certain redox reactions, as depicted in Figure 9.1(b). A CBM that is no larger than 0 *vs.* a standard hydrogen electrode (SHE) is required for the reduction of proton (H_2 evolution), whereas a VBM greater than that of a H_2O or H_2O_2 oxidation potential is required for O_2 evolution. A more negative CBM and positive VBM would be needed for CO_2 reduction and decomposition of organic molecules, respectively. However, it is not necessary to have a very negative CBM and positive VBM for a photocatalyst, which will increase the E_g that hinders visible light absorption. Considering that UV light only accounts for 4% of the solar spectrum, it is vital to optimise the bandgap and the band positions of the photocatalyst for specific reactions.

Figure 9.1(c) reveals the elementary steps involved in photocatalytic reactions from a kinetic perspective. Whilst the excitation of e^-–h^+ pairs *via* irradiation normally takes place within a femtosecond scale, the remaining elementary steps are much slower. In the absence of A and D both the surface charge trapping (step 2 and 3) and recombination of e^-–h^+ pairs (step 6) take place within a timescale of 10–100 ns, whereas the interfacial charge transfers to the surface-adsorbed electron acceptor (A_{ad}) and donor (D_{ad}) generally occur within microseconds to milliseconds (step 4 and 5). The competition between charge trapping and recombination, along with the slow redox reaction rates, explain the intrinsic low efficiency of photocatalysis. Therefore, another crucial issue in photocatalysis is engineering the surface properties of the material to slow down the recombination kinetics or speeding up the interfacial charge transfers to improve the QE.

The most extensively investigated photocatalyst materials are inorganic semiconductors. These materials range from simple metal oxides, nitrides, and sulfides that consist of d0 or d10 cations together with non-metal anions (*i.e.*, TiO_2, ZnO, Ta_3N_5, CdS), to more complicated compounds such as perovskites, zeolites, and metal–organic frameworks (MOFs). In order to prolong the lifetime of photogenerated charge carriers and accelerate preferred redox reactions, noble metal nanoparticles as promoters are frequently employed to decorate the surface of photocatalysts. Although the advances in using inorganic photocatalyst is obvious (high stability, easy synthesis, high performance), the high cost of metal-containing precursors also needs to be considered for practical reasons.

Recently, the development of the graphitic carbon nitride (g-C_3N_4) materials has attracted great attention due to their tuneable electronic properties and chemical and thermal stability. Depending on the synthesis method and crystallinity, the HOMO–LUMO gaps of the g-C_3N_4 vary from 2.4–2.9 eV, making it feasible for harvesting visible light. Besides, the redox potentials of the e^-–h^+ pairs are capable of driving water splitting reactions. Most importantly, the g-C_3N_4 can easily be synthesised by the polymerisation of cyanamide, dicyandiamide, melamine, or even urea, making the cost of the g-C_3N_4 photocatalyst only ~1–10% of most conventional metal oxide photocatalysts. Therefore, considering the variety of organic material and tremendous possibility of fine tuning the electronic properties, there should be a great future for polymer-based photocatalyst materials.

9.1.2 Key Terminologies

The following terminologies will appear frequently in this chapter and journal papers discussing photocatalysis, therefore, it is necessary to introduce these concepts for beginners:

Quantum efficiency (QE)/quantum yield: the catalytic performance of a photocatalyst material in certain photocatalytic reactions can be evaluated by the QE, as shown in eqn (9.1):

$$\mathrm{QE} = \sum\nolimits_i n_{\mathrm{e},i} * n_{\mathrm{p},i} * N_\mathrm{A} \big/ Q * t * 100\% \tag{9.1}$$

where $n_{\mathrm{e},i}$ is the number of electrons to generate the product i; $n_{\mathrm{p},i}$ is molar number of product i evolved within the irradiation time; N_A is the Avogadro constant; Q is the photon flux of the light source; t is the irradiation time. It should be noted that the QE discussed here is more close to the so-called apparent quantum efficiency (AQE), and more detailed estimations of quantum efficiency can be found in a recent review by Qureshi and Takanabe.[1]

Determination of bandgap and band positions: the bandgap can be measured by either diffuse reflectance spectrometry (DRS) or electrochemistry. While the optical bandgap (light absorption edge) can be extracted by applying the Kubelka–Munk theory (see an example of TiO_2 by Su *et al.*[2]), the electrochemical bandgap can be estimated by performing a voltammetry measurement. The energy gap is the difference between the redox potentials (CBM and VBM) multiplied by the charge on the electron. Note that the optical bandgap of a material is not necessarily identical to the electrochemical bandgap (in most cases, optical bandgaps are smaller than the electrochemical bandgaps). This is because once the semiconductor is in contact with the electrolyte, the charge transfer between the semiconductor and the electrolyte will take place to maintain electrostatic equilibrium, which will align the Fermi level of the photocatalyst and the redox potential of the redox species in the electrolyte at the interface of the photocatalyst and the redox species. The alignment introduces a downward band bending of both the CBM and VBM in an n-type semiconductor photocatalyst and an upward band bending in a p-type semiconductor photocatalyst. Besides, the valence band structure of the photocatalyst can also be characterised by ultraviolet photoelectron spectrometry (UPS). By coupling the optical bandgap determined by the DRS, the CBM can, therefore, be derived (see an example of TiO_2 by Scanlon *et al.*[3]).

Sacrificial agents: Sacrificial agents (or scavengers) are frequently used in heterogeneous photocatalysis when completing the catalytic cycle and studying reduction and oxidation half reactions. For the photocatalytic hydrogen evolution half reaction, alcohols (*i.e.*, methanol, ethanol, glycerol) and triethylamine (TEA) are often used as scavengers to donate electrons, where corresponding aldehydes are formed accompanied with H_2 evolution. Similarly,

$AgNO_3$ and Cr^{6+} are used as scavengers to accept electrons when investigating photocatalytic oxygen evolution half reactions. Nevertheless, reduction and oxidation half reactions (*i.e.*, H_2 evolution, O_2 evolution) can also be investigated by electrochemical methods by preparing the photoelectrode using photocatalyst materials.

Light sources: The holy grail of photocatalysis is using solar light to drive reactions. Therefore, light sources that can simulate the solar spectrum are ideal. Gas discharge lamps (*i.e.*, mercury-vapour lamp), arc lamps (*i.e.*, the Xe arc lamp), and light-emitting diode (LED) lamps are the most often used light sources in the lab. Low-pressure mercury-vapour lamps are rarely used as they only have emission lines at 184 nm and 253 nm. Medium-pressure mercury-vapour lamps can be constructed to emit primarily in the UV-A (~400 nm) or UV-C (~250 nm) range. High-pressure mercury-vapour lamps can emit primarily in the blue and green region (440–550 nm). Xe arc lamps are the most frequently used light sources as they can provide a high light intensity and the raw spectrum matches reasonably well to sunlight. LED lamps are getting popular as they can be considered as "monochromatic light" as the emission bandwidth is ~10–20 nm (0.1–0.2 eV). This is especially interesting for research purposes since estimating the QE of certain reactions at a specific wavelength range became more precise.

9.2 Organic Polymer Photocatalysts

9.2.1 Linear Structured Polymers

The first application of an organic polymer as a photocatalyst material can be dated back to 1985, when Yanagida *et al.* used poly(*p*-phenylene) (PPP) for photocatalytic H_2 evolution in the presence of an electron donor under UV irradiation ($\lambda > 290$ nm).[4] The linear, conjugated PPP (Figure 9.2(a)) that consists of several phenyl monomers and a 4-bromophenyl group as the terminator can be synthesised *via* polymerisation of the Grignard reagent from 1,4-dibromobenzene or 4,4′-di-bromobiphenyl, as demonstrated in eqn (9.2) and (9.3), which was described by Yamamoto *et al.* in 1978.[5]

$$nC_6H_4Br_2 + 2nMg \xrightarrow{\text{Grignard reaction}} nC_6H_4Mg_2Br_2 \quad (9.2)$$

$$nC_6H_4Mg_2Br_2 \xrightarrow{\text{Transition metal catalyst}} (C_6H_4Br)(C_6H_4)_{n-2}(C_6H_4Br) \quad (9.3)$$

The conjugated organic semiconductors are characterised by an optical bandgap of 2–5 eV depending on the structures, making it ideal and tuneable to harvest visible to UV light. Besides, the conjugated configuration makes the photogenerated charge carriers (*i.e.*, electrons) delocalised (mobile), which is essential for photocatalysis. Although the quantum yield (QE) for H_2 production is low (0.0335 under 313 nm irradiation), and deactivation was observed during the reaction, it demonstrates the concept of using organic polymers as photocatalyst materials.

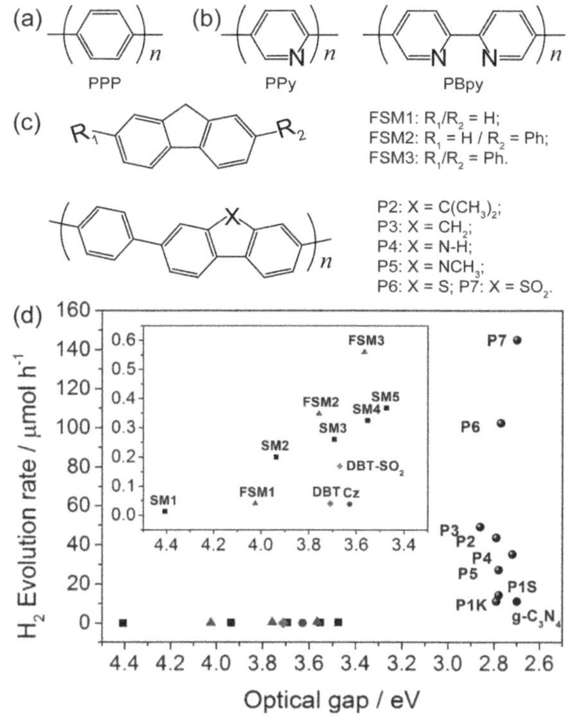

Figure 9.2 Molecular structure of (a) poly(*p*-phenylene); (b) poly(pyridine-2,5-diyl) and poly(2,2′-bipyridine-5,5′-diyl), reproduced with permission from T. Yamamoto, T. Maruyama, Z.-H. Zhou, T. Ito, T. Fukuda, Y. Yoneda, F. Begum, T. Ikeda and S. Sasaki, *J. Am. Chem. Soc.*, 1994, **116**, 4832, Copyright 1994 American Chemical Society.[7] (c) Planarized polymers with carbon or heteroatom substitution. (d) Comparison of photocatalytic H_2 evolution rates of various linear polymer photocatalysts correlated with their bandgap (SM1–SM5: PPP with 2–6 *p*-phenylene, Cz: 9*H*-carbazole, DBT: dibenzo-[*b*,*d*]thiophene) under broad-spectrum irradiation (>295 nm). Adapted from ref. 8, http://dx.doi.org/10.1002/anie.201510542. Published under the terms of the CC BY 4.0 licence, https://creativecommons.org/licenses/by/4.0/.

Later, it was discovered that the efficiency of photocatalytic H_2 production under visible light irradiation can be further enhanced by controlling the number of phenyl monomers in the PPP and the addition of noble metal nanoparticles (NPs) on the surface. Interestingly, the PPP that contains 11 phenyl monomers with Ru NPs showed a QE of 0.015 at 405 nm irradiation.[6] Meanwhile, the use of PPP for the photocatalytic reduction of carbonyl compounds and olefins has also been explored. The carbonyls can be photoreduced to their corresponding alcohols and/or pinacols in methanol solution with TEA as the electron donor. The selectivity to alcohols or diols depends on the reduction potential of the diols, which are the intermediates of this two-electron process. The photoreduction of olefins results in the formation of corresponding dihydro compounds, however, rapid *cis–trans* isomerisation of the olefins is observed during the process. Furthermore, the length of the linear polymer also influences the photocatalytic activity. Matsuoka *et al.* found that at least three *p*-phenylene units in the PPP polymer are necessary for the separation of photogenerated charge carriers to initiate H_2 evolution, and the hydrogen evolution efficiency increases with increasing chain length of the PPP polymer. The efficiency can also be enhanced by improving the solubility of the polymer, suggesting the homogeneous catalysis may overcome the heterogeneous system. However, the stability will become an issue for applications.

Meanwhile, the use of heterocyclic polymers as photocatalyst materials to extend light absorption into the visible light region has also been developed.[7] Poly(pyridine-2,5-diyl) (PPy) and poly(2,2′-bipyridine-5,5′-diyl) (PBpy) (Figure 9.2(b)) have been synthesised by Yamamoto *et al.* using the zero valent Ni complexes:

$$n\text{X–Ar–X} + n\text{Ni(0)L}_m \rightarrow (\text{–Ar–})_n + n\text{NiX}_2\text{L}_m \qquad (9.4)$$

where X–Ar–X is the corresponding halogenated aromatic precursor (*i.e.*, Br–Py–Br, Cl–Py–Cl, and Br–Bpy–Br). The molecular weights of the PPy and PBpy polymers varied depending on the monomer identity and the ligand identity of the Ni complex, however, all the synthesised PPy and PBpy polymers showed a similar optical bandgap of 2.8 eV, indicating an absorption edge of 440 nm. Both the PPy and PBpy polymers can be applied for the photocatalytic H_2 evolution reaction (HER) when a proper electron donor (TEA) and the $RuCl_3$ promoter (photoreduced to metallic Ru during the reaction) are present. The following reaction steps (eqn (9.5)–(9.9)) in the PPy systems have been confirmed by combining γ-radiolysis, pulse radiolysis, and laser flash photolysis analysis:[6]

$$\text{Excitation: PPy} \xrightarrow{h\nu} {}^1\text{PPy}^* \rightarrow {}^3\text{PPy}^* \qquad (9.5)$$

$$\text{Radical formation: } {}^1\text{PPy}^* \text{ (or } {}^3\text{PPy}^*) + \text{TEA} \rightarrow \text{PPy}^{\bullet-} \qquad (9.6)$$

$$\begin{gathered}\text{Charge transfer (formation of promoter):}\\ \text{PPy}^{\bullet-} \rightarrow \text{PPy} + e^- \xrightarrow{\text{Ru}^{3+}} \text{Ru}^0 - \text{PPy}\end{gathered} \qquad (9.7)$$

$$\text{Charge transfer (H}_2\text{ evolution): } 2e^- + 2\text{H}^+ \xrightarrow{\text{Ru}^0 - \text{PPy}} \text{H}_2 \qquad (9.8)$$

$$\text{Protonation: } {}^1\text{PPy}^* \text{ (or } {}^3\text{PPy}^*) + \text{H}^+ \rightarrow \text{PPy} - \text{H}^\bullet \qquad (9.9)$$

It should be noted that the protonation of the anion radicals (eqn (9.9)) is competing with the radical formation process (eqn (9.6)), which results in photodegradation of the polymer and a poor performance Although the PBpy polymers showed a performance 3.5 times higher than that of the PPy, the QE is still very low for applications (~0.21% under visible light irradiation).

Steady-state fluorescence analysis showed the emission maximum at 440 nm (~2.8 eV), revealing that both the PPy and PBpy polymers exhibit a direct recombination of the CBM e^- with the VBM h^+. This indicates that the lifetime of the charge carrier is very short, resulting in a poor photocatalytic performance.

The photocatalytic H_2 evolution performance of the PPP polymer has been further scaled-up by introducing bridging groups between the phenyls in the polymer, as shown in Figure 9.2(c).[8] It is considered that a better charge delocalisation can be achieved by lowering the torsional angle. Meanwhile the bandgap can also be manipulated by tuning the identity of the bridging group. Sprick *et al.* also compared the photocatalytic H_2 evolution performance of a series of linear polymer photocatalysts as a function of their bandgap (Figure 9.2(d)). Obviously, conjugated co-polymers with planarized units of dibenzo[*b*,*d*]thiophene (P6) and dibenzo[*b*,*d*]thiophene sulfone (P7) exhibit an extremely high performance compared to that of the pristine PPP polymer under broad-spectrum irradiation (>295 nm). Besides, the bandgaps of these co-polymers have been narrowed down to the visible light region (~2.8 eV), making them capable of performing H_2 production reactions under visible light irradiation (92 μmol h^{-1} under $\lambda > 420$ nm).

9.2.2 Polymeric Networks

The exploration of conjugated systems for the application of photocatalysis has been further expanded to more complicated molecular structures. Schwab *et al.* developed a conjugated poly(azomethine) material with a three-dimensional structure.[9] The azomethine is a special type of Schiff base (named after Hugo Schiff), which can be considered as secondary aldimines with a general structure of $R_1N{=}CR_2R_3$. The synthesis route of azomethine is described in Figure 9.3(a). An aliphatic or aromatic amine (electron-rich reactant) first reacts with a carbonyl compound (electron-deficient reactant) *via* nucleophilic addition (step 1), forming the hemiaminal intermediate (–C(OH)(NR_2)–), which then results in the formation of azomethine *via* a dehydration process (step 2). Poly(azomethine) photocatalyst materials can, therefore, be synthesised using a similar approach by coupling an amine building block (A3-type cross-linker, 1,3,5-tris(4-aminophenyl)-benzene, Figure 9.3(b)) with different bifunctional (B2) aromatic aldehydes (*i.e.*, anthracene-2,6-dicarbaldehyde and anthracene-9,10-dicarbaldehyde, Figure 9.3(b)). In this case, each amino group will react with one carbonyl group in the acetic–dioxane mixture under deaerated conditions, resulting in the formation of the corresponding poly(azomethine) network.

One big advantage of poly(azomethine) materials is the ability to absorb visible light. By increasing the size of the B2 building blocks (from terephthalaldehyde to anthracene-2,6-dicarbaldehyde), the light absorption can be extended to 620 nm, indicating an optical bandgap of ~2 eV. Accordingly, the photocatalytic performance also improved as the bandgap reduced. The optimized poly(azomethine) that is synthesised from A3 and B2a showed

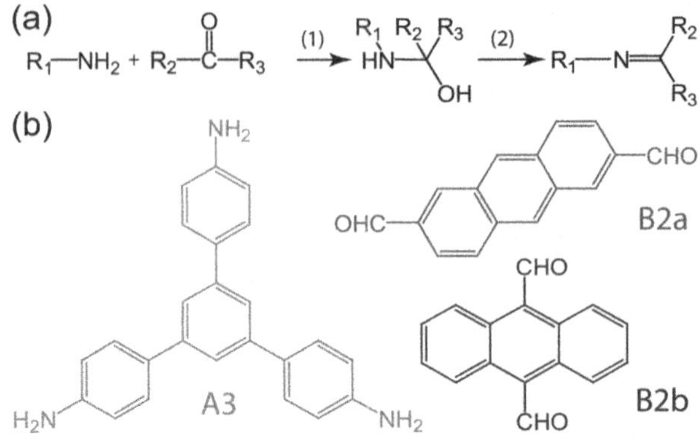

Figure 9.3 (a) Generalised reaction equation of azomethine and poly(azomethine) compounds from an amine and a carbonyl compound; (b) molecular structures of the building units for the poly(azomethine) networks, reproduced from ref. 9 with permission from The Royal Society of Chemistry.

a H_2 evolution rate of 7 μmol h^{-1} ($\lambda > 300$ nm), which is ~3–4 times higher than that of the PPP polymer. In contrast, the poly(azomethine) synthesised from A3 and B2b showed no activity. This can be explained by the structure difference of the network. Compared to the A3 + B2a network, the A3 + B2b polymer exhibits a strong distortion of the network due to an increased torsion angle (<2° for A3 + B2a and ~30° for A3 + B2b) between the imine group and the neighbouring aromatic core. The distorted structure of A3 + B2b reduces the charge mobility through the material thus resulting in a decreased catalytic performance. Another noticeable enhancement is the thermal and photo-stability of the poly(azomethine) materials compared to that of the PPP or PPy polymers. While the protonation of PPP during photocatalytic H_2 evolution leads to the deactivation of the material, such a process has not been observed on the poly(azomethine) materials after being irradiated for 20 hours. However, the complicated synthesis and the need of using a Pt promoter limit the further development of poly(azomethine) photocatalysts.

To overcome the disadvantages of the poly(azomethine) materials, Sprick *et al.* have built conjugated microporous polymers (CMPs) *via* statistical co-polymerisation.[10] The effect of the aromatic core size on both bandgap and photocatalytic performance has also been evaluated by using either a single benzene ring or four benzene ring monomers (pyrene). The synthesis strategy is the same as the Suzuki–Miyaura coupling reaction that takes place under basic conditions:

$$R_1 - BY_2 + R_2 - X \xrightarrow{\text{Pd catalyst}} R_1 - R_2 \tag{9.10}$$

where R_1–BY_2 and R_2–X are the organoboron reactants and halide species, respectively. Here, for the CMP synthesis, 1,4-benzene diboronic acid

(1) and/or 1,3,6,8-tetraboronic pinacol ester of pyrene (3) and/or 1,2,4,5-tetrabromobenzene (2) and/or 1,3,6,8-tetrabromopyrene (4) are used for the Suzuki–Miyaura coupling reaction. By tuning the molar ratios of the monomers, a series of CMPs with different physical properties can be synthesised. The reactants (1), (2), (4), and two representative polymers (CP-CMP1 and CP-CMP10) are shown in Figure 9.4.

All these polymers are characterised by relatively large surface areas (~600–1700 m^2 g^{-1}) with Type-I sorption isotherms, indicating the microporous structure (pores <2 nm). The optical bandgaps of the CMPs reduce from ~3 eV to ~1.9 eV upon increasing the pyrene content in the polymer, indicating the possibility of fine-tuning the optical properties of the CMP networks by adjusting the molar ratio of the monomers during synthesis. The stabilities of the CMP materials are as good as the poly(azomethine) compounds, probably due to the reduced protonation sites in the network structures. Compared to other polymer photocatalyst materials, the most striking advantages of the CMP photocatalysts are that noble metal promoters are not needed and the high activity under visible light irradiation (λ > 420 nm). It should be noted that photocatalytic H_2 evolution performances did not improve monotonically following an increase of visible light absorption, but showed a maximum at a bandgap of 2.3–2.4 eV for the CP-CMP10 photocatalyst (~18 μmol h^{-1}). The authors consider that the drop in performance of other polymers besides CP-CMP10 can be rationalised by a non-radiative electron–hole recombination in the pyrene-rich materials, however, no evidence can prove this assumption. Like other polymer photocatalyst materials, the CMP polymer cannot perform full water splitting without a hole scavenger (diethylamine here). According to density functional theory (DFT) calculations, the ionisation

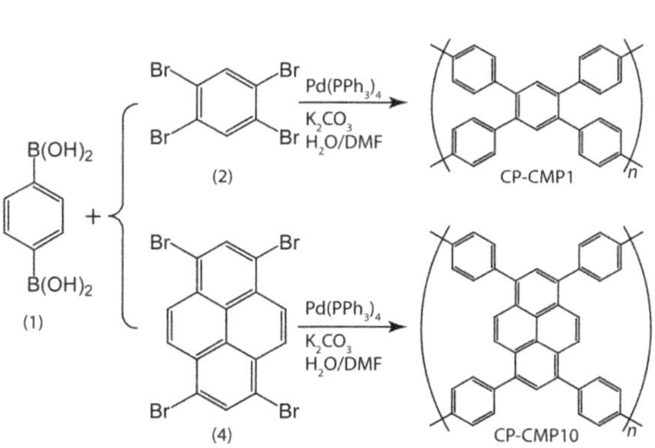

Figure 9.4 Representative synthesis of CP-CMP1 and CP-CMP10 from 1,4-benzene diboronic acid (1) and/or 1,2,4,5-tetrabromobenzene (2) and/or 1,3,6,8-tetrabromopyrene (4), reproduced from ref. 10, http://dx.doi.org/10.1021/ja511552k. Published under the terms of the CC BY 4.0 licence, https://creativecommons.org/licenses/by/4.0/.

potentials (IP, similar to the VBM or HOMO here) of these polymers are inefficient to drive the water oxidation (O_2 evolution, remove electrons from O^{2-}) reaction.

It is also possible to further couple microporous polymer materials with inorganic semiconductor photocatalysts to further improve the photocatalytic performance. In this case, the organic polymer serves as either light absorber or photocatalyst. Park *et al.* first synthesised a visible-light-harvesting microporous organic network (VH-MON) *via* the Knoevenagel condensation process, then prepared a VH-MON/TiO_2–Pt composite by mixing and fixation.[11] As shown in Figure 9.5(a), the Knoevenagel condensation process is a nucleophilic addition reaction, where an active hydrogen compound (CZ_2H_2) connects to a compound with a carbonyl group *via* a dehydration reaction. The Z functional group in the compound is an electron-withdrawing functional group (*i.e.*, halogens, nitriles, and nitro groups) that should be strong enough to facilitate deprotonation to the enolate ion under mild basic conditions, as self-condensation of the aldehyde or ketone will take place under strong basic conditions. For the synthesis here, tri(4-formylphenyl)-amine (TA, the aldehyde) and [1,10-biphenyl]-4,40-diacetonitrile (BN, the active hydrogen compound) are connected to form the VH-MON polymer under NaOH conditions, as depicted in Figure 9.5. The VH-MON polymer is characterised by a relatively large surface area (474 $m^2\ g^{-1}$) with a porosity of 1 $cm^3\ g^{-1}$.

The optical bandgap is ~2.4 eV according to the DRS analysis. The DFT calculation reveals that the HOMO level of the VH-MON (~−5 eV *vs.* vacuum, ~0.6 V *vs.* SHE) is close to that of the TA, whereas the LUMO level of the VH-MON (~−2.2 eV *vs.* vacuum, ~−2.1 V *vs.* SHE) is located near that of the BN. While the LUMO level is sufficient enough to drive the proton reduction half reaction, the HOMO level is insufficient to drive the water

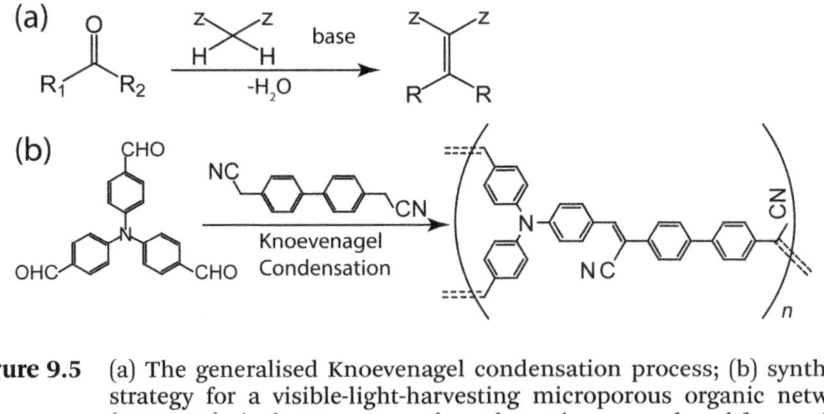

Figure 9.5 (a) The generalised Knoevenagel condensation process; (b) synthetic strategy for a visible-light-harvesting microporous organic network (VH-MON) *via* the Knoevenagel condensation, reproduced from ref. 11 with permission from The Royal Society of Chemistry.

oxidation half reduction. Therefore, a hole scavenger (triethanolamine, TEOA) is used to complete the reaction cycle (redox potential 0.59 V *vs.* SHE). The VH-MON/TiO_2–Pt photocatalyst material polymer showed a better photocatalytic H_2 evolution performance compared to that of other dye-modified TiO_2–Pt photocatalysts, both in efficiency and stability, indicating that the crosslinking reduced the decomposition of the light-harvesting components.

9.2.3 Covalent Organic Frameworks

Covalent organic frameworks (COFs) are 2D or 3D organic frameworks made of building blocks connected *via* strong covalent bonds. Similar to metal–organic frameworks (MOFs),[12] COFs also feature a highly porous structure (large surface area) and long range orders (well-defined crystalline, like diamond and graphite) but without the need of a metal component, making them good candidates for gas storage and separation. Semiconductor-type COFs that emit blue light have also been designed and synthesised.[13] These advantages may provide more active sites and reduce the density of defects in the material to improve the photocatalytic performance. However, the most important advantage of the COF materials is the relatively easy synthesis protocols compared to those of conjugated polymer photocatalysts. While the conjugated polymer photocatalysts mentioned previously normally need several steps involving expensive catalysts and delicate control of the reaction conditions, COFs can be easily synthesised by dehydration reactions between boronic acids,[14] by dynamic trimerization of aromatic nitriles in ionothermal conditions (molten $ZnCl_2$ at ~400 °C),[15] or by imine condensation of aniline with benzaldehyde,[16] which make the COFs ideal for scaling-up purposes. A hydrazone ($R_1R_2C{=}NNH_2$)-linked COF constructed from 1,3,5-tris-(4-formylphenyl)triazine (TFPT) and 2,5-diethoxy-terephthalohydrazide (DETH) building blocks has been developed for photocatalytic H_2 production under visible light irradiation.[17] Besides, the COF can also be coupled with an inorganic semiconductor (CdS) photocatalyst to produce H_2.[18]

Vyas *et al.* reported that azine-linked COFs can be used for photocatalytic H_2 production under visible light irradiation.[19] Azines are organic compounds with a typical structure of $R_1R_2C{=}N{-}N{=}CR_1R_2$, which can be produced *via* the condensation of hydrazine with ketones and/or aldehydes. Here, the N_x-aldehydes are reacted with the hydrazine to yield N_x-COFs (x = 0–3), as demonstrated in Figure 9.6. The core of the N_x-aldehydes, triphenylaryl, is the platform for tuning the nitrogen content, wherein the C–H groups can be gradually replaced by N atoms on the green dot sites (X, Y, and Z). Such variation in the aryl central ring (x = 0: phenyl; x = 1: pyridyl; x = 2: pyrimidyl; and x = 3: triazine) leads to a change in both the steric and electronic properties of the triphenylaryl thus the N_x-COFs. A decrease in the dihedral angle between the aryl ring and the phenyl rings following an increase in the N content is

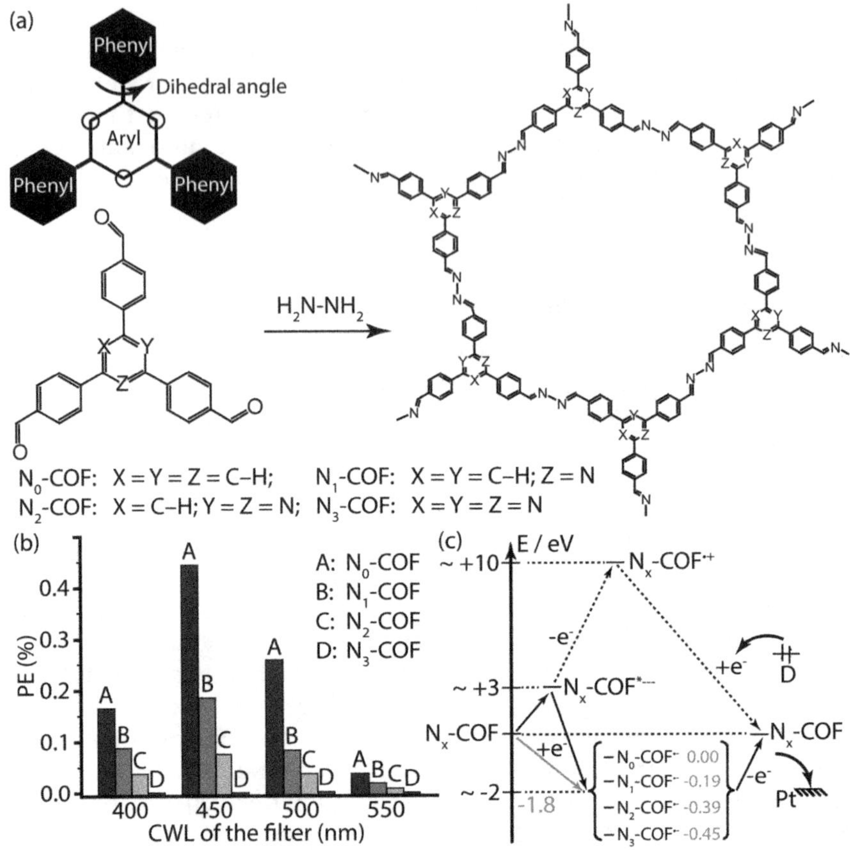

Figure 9.6 (a) The triphenylarene platform for synthesizing N_x-COFs (x = 0–3) with tuneable nitrogen content. The green dot sites are C–H groups or N atoms, which can change the angle between the aryl and phenyl rings. (b) Photonic efficiency (PE) of the N_x-COFs under visible-light irradiation measured using different filters with central wavelengths (CWLs) at 400, 450, 500 and 550 nm. (c) Scheme of the two possible pathways after photoexcitation of N_x-COFs. The energies in red are the differences of the electron affinities between the N_x-COF$^{•-}$ and their neutral states. The * denotes the excited state. Adapted from ref. 19, http://dx.doi.org/10.1038/ncomms9508. Published under the terms of the CC BY 4.0 licence, https://creativecommons.org/licenses/by/4.0/.

expected and confirmed by calculations. An increase in the N content in the aryl ring also results in a gradual decrease in electron density in the resulting aryl ring in the N_x-COFs. The LUMO of all N_x-COFs are within 3.1–2.8 eV below vacuum level, indicating that H^+ can be reduced to H_2 thermodynamically (4.44 eV below vacuum level).

H_2 evolution tests show that the photocatalytic performance of N_x-COFs increases following an increase in N content in the aryl ring of the building block, and the best is observed with N_3-COFs (Figure 9.6(b)). Calculations on a 1248-atom cluster of the N_3-COFs (a system with seven hexagons) reveal that the HOMO and LUMO are located at the terminal hydrazone moieties and the innermost hexagon, respectively. An efficient separation of the HOMO–LUMO in the N_3-COFs and the electron-rich terminal suggest an efficient charge separation and active sites for proton reduction. The performance difference in the N_x-COFs has been further rationalised by calculating the electron affinity of the photogenerated N_x-COF radical anions. As shown in Figure 9.6(c), the calculated energy of possible radicals of the excited N_x-COFs reveals two possible pathways after photoexcitation of the N_x-COFs. Obviously, the realistic pathway is *via* the quenching of the h^+ on the N_x-COF by the electron donor, resulting the formation of N_x-COF$^{\bullet -}$ (red arrow). The formation of the N_x-COF$^{\bullet +}$ cation requires an energy of ~10 eV, making it an unfavourable pathway (dotted black arrows). The differences in the electron affinities between the N_x-COF$^{\bullet -}$ and their corresponding neutral state energies (red values) suggest that the excited N_3-COF photocatalyst (built from triazine units) is more inclined to take electrons from electron donors than other systems, to form stable anion radicals. Such stable anion radicals will facilitate the charge transfer to the platinum co-catalyst, thus improving the photocatalytic performance.

As one type of COF, covalent triazine frameworks (CTFs) have also been used as photocatalysts for H_2 production. The classical CTF-1 synthesised from the trimerisation process of terephthalonitrile catalysed by trifluoromethanesulfonic acid exhibits an optical bandgap of ~2.9 eV and has shown a capability of performing photocatalytic H_2 evolution and O_2 evolution separately in the presence of a proper co-catalyst (Pt and RuO_2) and scavengers (TEA and $AgNO_3$).[20] Shortly, it was realised that by reducing the ionothermal temperature (300 °C) and extending the reaction time (168 h), highly active phenyl-triazine oligomers (PTOs) with higher crystallinity compared to that of the CTF-1 could be synthesised.[21] The carbon-to-nitrogen weight ratios of the PTOs are close to the theoretical value of 3.43 for CTF-1, which is due to the low temperature synthesis that prevents carbonisation. Such defect-less PTO oligomers are beneficial for prolonging the lifetime of the photo-excited charge carriers, thus enhancing the photocatalytic performance.

9.3 Graphitic Carbon Nitrides (g-C_3N_4)

g-C_3N_4-based materials have attracted huge attention since Wang *et al.* reported their photocatalytic properties in 2008,[22] although such a material was reported as early as 1834 by Berzelius and Liebig (in German).[23] g-C_3N_4-based materials can still be considered as π-conjugated polymers composed of triazine units (nitrogen-containing heterocycles with the molecular formula $C_3H_3N_3$), however, the easy synthesis from various cheap precursors (*i.e.*, melamine, cyanamide or even urea), the tenability in geometric and electronic structure, the versatility in many photocatalytic reactions (*i.e.*, H_2 evolution, O_2 evolution, CO_2 reduction, organic synthesis) and the high stability make g-C_3N_4-based materials more promising than the traditional conjugated polymers or COFs.

9.3.1 Geometric and Electronic Structures

It should be made aware that the extensively used "g-C_3N_4" in photocatalysis is a very ambiguous term that consists of a branch of different materials with slight variation in composition but a huge difference in structure. The most commonly mentioned g-C_3N_4 has two possible ideal structures, as shown in Figure 9.7(a) and (b). One possible structure of the g-C_3N_4 photocatalysts is built by tri-*s*-triazine (heptazine or melem) rings, wherein the tri-*s*-triazine rings are connected *via* trigonal nitrogen atoms (Figure 9.7(a)). This is a highly crystalline structure and only the edge sites are terminated by H atoms. The tri-*s*-triazine building block can also form the g-C_3N_4 structure shown in Figure 9.7(b), where the incomplete deamination (removal of NH_3) of the precursors results in the formation of in-plane, infinite one-dimensional chains of NH-bridged melem oligomers, which are connected *via* hydrogen bonds. The g-C_3N_4 with such a structure is also called Liebig's melon. Obviously, the H content of the melon-structured g-C_3N_4 is much higher than that of the perfectly crystalline g-C_3N_4. Another possible structure of the g-C_3N_4 is similar to that of graphite, where melamine (or *s*-triazine, $C_3N_3H_6$) serves as a building block (Figure 9.7(c)). It is generally considered that the melon-structured g-C_3N_4 is energetically more favourable, and most of the g-C_3N_4 polymeric photocatalysts that are synthesised from pyrolysis processes show such structures. However, a very recent study by Algara-Siller *et al.* has shown that a g-C_3N_4 single crystal with a melamine building block is also possible.[24] Moreover, it should also be noted that the g-C_3N_4 photocatalyst materials synthesised by conventional methods are not perfectly crystalline and, therefore, have a lot of defects, which result in poor crystallinity and protonation of the material (terminated by H). The synthesis of g-C_3N_4 will be discussed later in this chapter.

The corresponding electronic structures of g-C_3N_4 with different geometric structures have been calculated by Lin *et al.*, as shown in Figure 9.7(d)–(f).[25] While the VBMs of all g-C_3N_4 materials are mainly originated from the N p orbitals, the CBMs of all g-C_3N_4 materials are from the contribution of hybridised C p and N p orbitals. The calculated bandgaps of the tri-*s*-triazine-based g-C_3N_4, melon, and triazine-based g-C_3N_4 are ~1.2, 2.4, and 1.5 eV, which are slightly small compared to those of their optical bandgaps (*i.e.*, 2.6–2.8 eV for melon). Although the smaller bandgap of the tri-*s*-triazine-based g-C_3N_4 and triazine-based g-C_3N_4 will improve the visible light absorption to the near infrared region (>800 nm), it is unlikely to enhance the photocatalytic performance of the materials as the redox potentials are too low to drive many reactions. Furthermore, the VBM and CBM potentials of the g-C_3N_4 with a melon structure are ~1.5 V and ~−1.1 V *vs.* SHE,

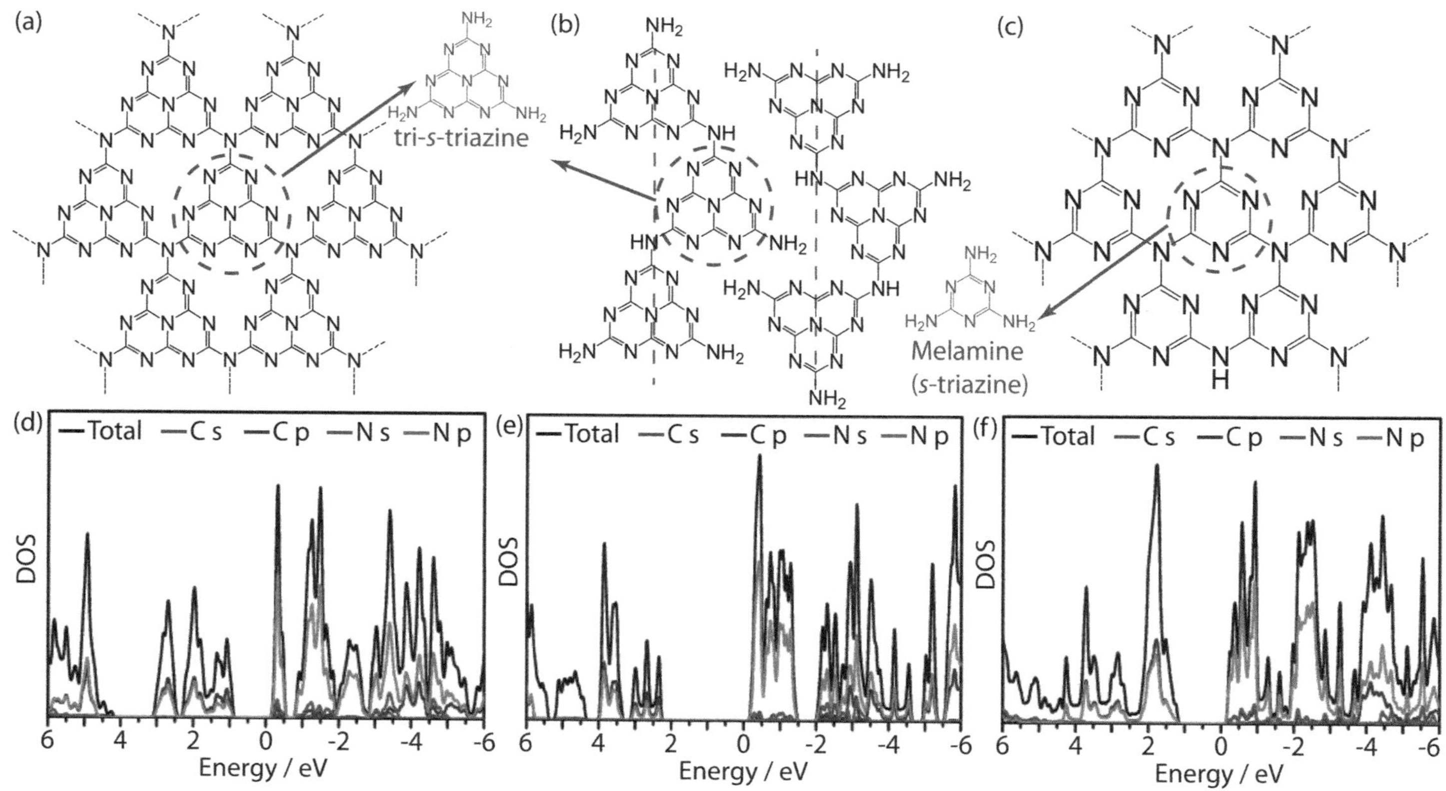

Figure 9.7 Ideal structures of g-C_3N_4 built from a tri-*s*-triazine unit (a) and (b), or a melamine unit (c). (d)–(f) Calculated density of states (DOS) of the corresponding g-C_3N_4. Reproduced with permission from L. Lin, H. Ou, Y. Zhang and X. Wang, *ACS Catal.*, 2016, **6**, 3921, Copyright 2016 American Chemical Society.[25]

indicating a sufficient reduction potential for the proton reduction reaction but a relative weak oxidation potential for the oxygen evolution reaction (OER). More importantly, the electronic structure of g-C_3N_4 can be adjusted by tuning the composition of the tri-*s*-triazine and the compression of aromatic planes. For example, a DFT calculation reveals that the CBM of g-C_3N_4 can be more negative (referenced to SHE) by deprotonation of the material (removing H atoms in the structure), thus resulting in a larger driving force for reduction reactions. Engineering of the electronic structure will be discussed later in this chapter.

9.3.2 Pristine Bulk g-C_3N_4

g-C_3N_4 can be synthesised *via* a condensation reaction from various N-containing organics. The most simple, direct condensation process can be realised by simple pyrolysis of the N-containing precursors, including cyanamide (CN_2H_2), dicyandiamide ($C_2N_4H_4$, a dimer of cyanamide) or melamine ($C_3N_6H_6$, a trimer of cyanamide). Figure 9.8 demonstrates the formation process of g-C_3N_4 from cyanamide. According to thermogravimetric analysis, the cyanamide powders will first melt at 47 °C, and then convert into dicyandiamide dimers at 137 °C and subsequently convert into melamine at 234 °C. Both the dicyandiamide and melamine will then condense into melam *via* release of an NH_3, wherein the generated melam intermediate will continue losing one more NH_3 to form the melem at an elevated temperature. A further increase of the reaction temperature to 389 °C will promote release of ammonia gases, thus resulting in the formation of polymetric melem, which eventually condenses into bulk, non-porous g-C_3N_4 at a temperature higher than 525 °C.[22] Later, the synthesis of g-C_3N_4 was further extended to cheap precursors, *i.e.*, urea [$CO(NH_2)_2$] and thiourea [$CS(NH_2)_2$],[26,27] which first form cyanamide *via* release of H_2O or H_2S. It should be noted that the resulting g-C_3N_4 polymers normally exhibit a C–N ratio that is slightly different to that of the ideal material (normally lower than 0.75). This is because the

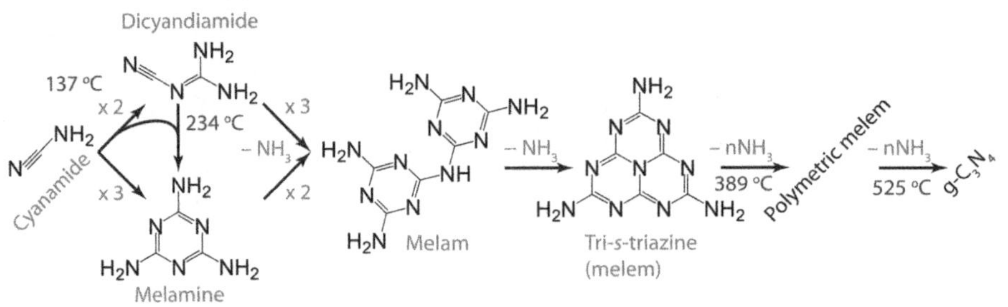

Figure 9.8 Formation mechanisms of g-C_3N_4 *via* condensation reactions from cyanamide. Reprinted by permission from Macmillan Publishers Ltd: *Nature Materials* (X. Wang, K. Maeda, A. Thomas, K. Takanabe, G. Xin, J. M. Carlsson, K. Domen and M. Antonietti, *Nat. Mater.*, 2009, **8**, 76. Copyright (2008)).[22]

ammonia gases are not completely released from the polymer, thus resulting in a defect-rich g-C_3N_4 terminated with amino groups. Such defect-rich g-C_3N_4 may have fast recombination kinetics of the charge carriers, thus limiting the photocatalytic performance.

The challenge of synthesising g-C_3N_4 with fewer defects has recently been undertaken by careful control of the pyrolysis conditions using the dicyandiamide precursor.[28] The g-C_3N_4 with the highest possible crystallinity (according to the X-ray diffraction pattern, XRD) can be obtained at a temperature of 520 °C (labelled as D52), where a further increase in the calcination temperature results in the formation of $-NH/NH_2$ terminal groups that reduce the crystallinity. The pale yellow g-C_3N_4 gets darker following an increase in annealing temperature, indicating that the excessive H atoms protonate the g-C_3N_4. Photoluminescence (PL) analysis has further confirmed an increase in defect concentration in the material upon raising the annealing temperature (Figure 9.9(a)). The g-C_3N_4 with the highest crystallinity (D52) exhibits a relatively narrow emission peak centred at 450 nm, whereas the g-C_3N_4 samples synthesised at elevated temperatures (D55: 550 °C, D60: 600 °C, D64: 640 °C) show a red shift and broadening effect of the emission peaks. While the emission peak at 450 nm (~2.75 eV) corresponds to the radiative

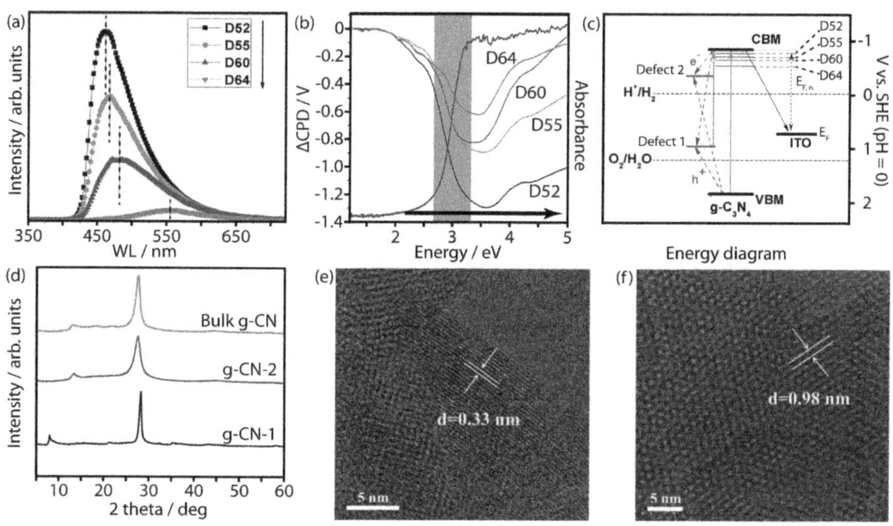

Figure 9.9 (a) PL spectra of g-C_3N_4 suspensions (excitation wavelength: 350 nm). (b) SPV spectra of g-C_3N_4 films on ITO glass. The absorption spectrum of D52 (dark yellow line) is also plotted for reference. (c) Energy diagram of g-C_3N_4 and ITO glass. D52, D55, D60, and D64 denote the samples calcined at 520 °C, 550 °C, 600 °C, and 640 °C. Reproduced from ref. 28 with permission from the Royal Society of Chemistry. (d) XRD patterns of g-C_3N_4 with different crystallinities. (e) and (f) Transmission electron microscopic (TEM) images of the g-CN-1 sample. Reproduced with permission from L. Lin, H. Ou, Y. Zhang and X. Wang, *ACS Catal.*, 2016, **6**, 3921. Copyright 2016 American Chemical Society.[25]

recombination of the CBM and VBM charge carriers, the broad, redshifted emission peaks (~470 nm and ~560 nm) indicate the presence of sub-gap defects in the materials synthesised at higher temperatures. The authors consider that the gradual decrease of the emission intensity upon increasing the annealing temperature can be attributed to the increased non-radiative recombination rates. Surface photo-voltage (SPV) analysis was used to study these g-C_3N_4 samples further. SPV measures the potential change caused by the charge separation in the semiconductor material *via* a vibrating Kelvin probe as a function of the irradiation wavelength, which is very powerful in characterising the bandgap, type of charge carrier, defects, and Fermi levels of semiconductor materials. As shown in Figure 9.9(b), all g-C_3N_4 samples are n-type semiconductors, as they all generate a negative voltage upon irradiation. The dominant signals observed at 2.80–3.50 eV can be assigned to the bandgap excitation of g-C_3N_4, and the maximum voltages of the g-C_3N_4 samples decrease upon an increase in defect concentrations in the materials (D52: −1.29 V, D55: −0.90 V, D60: −0.82 V, and D64: −0.62 V). It should also be noted that the photo-voltage can be generated even at sub-gap photon energies (1.80–2.80 eV), which suggests the excitation of the negative charge from the mid-gap defect states to the CBM. By coupling the PL and SPV analyses, the authors proposed the energy diagrams of the g-C_3N_4 samples with different crystallinities and the positions of two possible defect states (Figure 9.9(c)). An increase in the crystallinity of g-C_3N_4 will negatively shift the CBM *vs.* SHE, thus providing a stronger driving force for photocatalytic H_2 evolution. Besides, the onset photo-energy excited at 1.8 eV suggests a defect state (Defect 1) at ~0.95 V *vs.* SHE. Such a state is detrimental to the photocatalytic OER, which explains the poor performance of g-C_3N_4 for pure water splitting. The other defect state (Defect 2) is positioned at ~−0.38 V, which corresponds to the emission peak at 560 nm in the PL spectrum of D64. Such a defect state will reduce the reduction power of the g-C_3N_4 for photocatalytic HER.

Lin *et al.* claimed the successful synthesis of crystalline, tri-*s*-triazine-based g-C_3N_4 *via* further pyrolysis of melon (from melamine) in the presence of LiCl and KCl at 550 °C for 4 h under a N_2 atmosphere.[25] The as-synthesised crystalline g-C_3N_4 (g-CN-1) exhibits much shaper and intense diffraction peaks compared to those of g-C_3N_4 synthesised without the metal salt (g-CN-2) or synthesised in air (bulk-CN), as shown in Figure 9.9(d). While the diffraction peaks of g-CN-2 and bulk-CN can be assigned to that of the melon-structured g-C_3N_4, the shifted peaks of g-CN-1 indicate a different structure of the as-synthesised material. The most intense peak in bulk g-CN (27.4°) has shifted to 28.3° for g-CN-1, suggesting the interlayer distance has reduced from 0.326 nm to 0.316 nm due to an enhanced interaction between layers. On the other hand, the low angle diffraction peak of the g-CN-1 has significantly shifted to 8.0° from 13.0° of the bulk g-CN, indicating the in-plane repeat unit has increased to 1.099 nm from 0.618 nm due to the perfection of the conjugated framework. The TEM images have further confirmed the interlayer distance and the in-plane periodicity (Figure 9.9(e) and (f)). The photocatalytic H_2

evolution performance of the g-CN-1 sample has further confirmed that the crystalline g-C_3N_4 is much better than that of the melon-structured g-C_3N_4.

On the contrary, the amorphous carbon nitride material has also been explored, as it may further extend the visible light absorption into the green or yellow light region.[29] The *ab initio* molecular dynamics simulation suggests that the melon monolayer (see Figure 9.7(b)) tends to form a disordered, twisted structure after annealing at 900 K for 4 ps (Figure 9.10(a)). Such outward twisting of the melon units is caused by the outward movement of the NH_2 groups, which have lost confinement due to breaking of the in-plane H-bonds between the melon strands. However, the coordination number of all atoms within each strand remains unchanged, suggesting that the short-range order of GCN could be retained. Experimentally, amorphous g-C_3N_4 (ACN) powders have been synthesised from melon-structured g-C_3N_4 (GCN) by annealing for 2 h under an Ar atmosphere. The XRD pattern of the ACN reveals that the sharp diffraction peaks of melon-structured g-C_3N_4 (see red curve in Figure 9.9(d) for reference) disappeared and left a very broad weak peak at 27.2°, indicating that the long-range order of the GCN has been destroyed. Meanwhile, the vibrational spectra suggest that the heptazine ring and the trigonal C–N(–C)–C/bridging C–NH–C units of melon remain, suggesting the retention of short-range order in the amorphous material. The optical analyses demonstrate that the ACN can absorb visible light up to ~700 nm (460 nm for GCN), but almost does not emit fluorescence compared

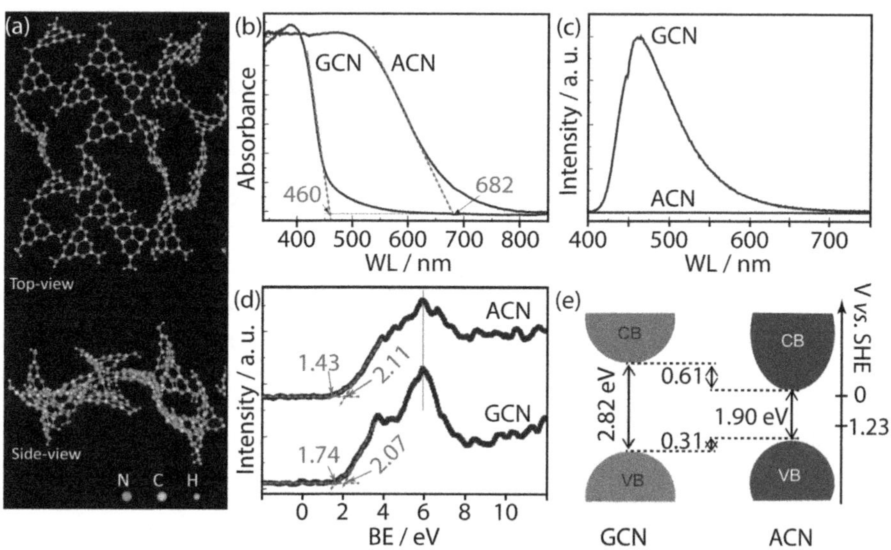

Figure 9.10 (a) Simulated structure of amorphous g-C_3N_4 (ACN). (b) and (c): Light absorption spectra and PL spectra of ACN and melon-structured g-C_3N_4 (GCN) (excitation wavelength: 350 nm). (d) XPS valence band spectra of ACN and GCN. (e) Energy diagram of ACN and GCN. Reproduced with permission from ref. 29, © 2015 WILEY-VCH Verlag GmbH & Co. KGaA, Weinheim.

to the GCN (Figure 9.10(b) and (c)). The authors claim that the low emission of ACN suggests a low recombination kinetics of the photogenerated charge carriers, however, the non-radiative recombination remains unknown. X-Ray photoelectron spectrometry (XPS) valence band spectra reveal that the ACN shows two different features compared to GCN (Figure 9.10(d)). While the broad N 2p peak (~6 eV) can be assigned to the absence of long-range atomic order of ACN, the extended tail to 1.43 eV (1.74 eV for GCN) can be associated to the presence of dangling bonds in ACN. This also indicates that the VBM of the ACN has been negatively shifted by 0.31 eV. By coupling the optical bandgap (2.82 eV for GCN and 1.90 eV for ACN) and the valence band spectra, the CBM and VBM positions of GCN and ACN can be extracted, as shown in Figure 9.10(e). Since the VBM of the ACN has been negatively shifted by 0.31 eV compared to that of the GCN, the remaining 0.61 eV gap narrowing of the ACN is contributed from the positive shift of the CBM. Therefore, the CBM and VBM positions of the ACN reference to the SHE can be drawn using the known band position of the melon-structured g-C_3N_4 (GCN). Although the band positions of the ACN sample are still suitable for both H_2 and O_2 evolution half reactions thermodynamically, the driving forces are also significantly reduced from the kinetic aspect.

Besides the traditional pyrolysis approach for the synthesis of pristine bulk g-C_3N_4 powders, the electrodeposition of a g-C_3N_4 film on a Si(100) surface from a saturated acetone solution of cyanuric trichloride and melamine at room temperature has also been reported.[30] The g-C_3N_4 thin film can also be deposited on an alumina surface at 400–600 °C by *in situ* chemical vapour deposition (CVD) that employs a melamine and uric acid mixture.[31] These approaches may provide well-defined g-C_3N_4 films for surface science studies.

9.3.3 Modified g-C_3N_4 and Carbon Nitride Derivatives

9.3.3.1 Porous Structure

Traditional carbon nitrides synthesised by pyrolysis of N-containing precursors normally result in bulk, non-porous materials with very small surface areas (5–15 $m^2\ g^{-1}$). Therefore, the synthesis of porous structures of g-C_3N_4 is of great interest for catalysis. The preparation of mesoporous carbon nitrides (mpg-C_3N_4) can be realised by employing hard templates (*i.e.*, SiO_2, SBA-15)[32,33] or soft templates (*i.e.*, surfactants, sugars).[34] The classic method to prepare mpg-C_3N_4 involves mixing colloidal silica with a precursor and etching the SiO_2 template by HF. By controlling the architecture of the template, shape-controlled mpg-C_3N_4 can be prepared, as representatively shown in Figure 9.11(a)–(c).[33,35] Hollow-structured g-C_3N_4 spheres (HCNS) were synthesised by employing a dense, monodisperse silica core covered with a thin mesoporous silica shell SiO_2 template (Figure 9.1(a)). Such a core–shell-structured silica template can be prepared by first synthesising the dense SiO_2 core *via* the Stöber method from a tetraethoxysilane (TEOS) precursor.

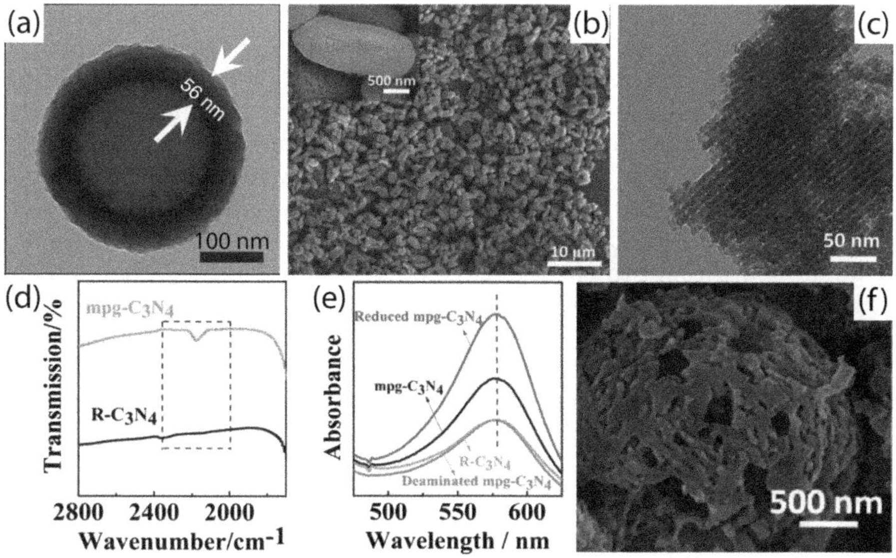

Figure 9.11 (a) TEM image of a hollow g-C_3N_4 sphere (HCNS-3). Adapted with permission from Macmillan Publishers Ltd: *Nature Communications* (J. Sun, J. Zhang, M. Zhang, M. Antonietti, X. Fu and X. Wang, *Nat. Commun.*, 2012, **3**, 1139), copyright (2012).[35] (b) and (c) Scanning electron microscopic (SEM) and TEM images of ordered mesoporous g-C_3N_4 (ompg-CN). Reproduced with permission from ref. 33, © 2013 WILEY-VCH Verlag GmbH & Co. KGaA, Weinheim. (d) FT-IR spectra of a normal mpg-C_3N_4 and a defect-repaired mpg-C_3N_4 (R-C_3N_4). (e) UV-vis spectra of solutions from ninhydrin tests using different carbon nitride materials. Reproduced with permission from W. Wu, J. Zhang, W. Fan, Z. Li, L. Wang, X. Li, Y. Wang, R. Wang, J. Zheng, M. Wu and H. Zeng, *ACS Catal.*, 2016, **6**, 3365. Copyright 2016 American Chemical Society.[36] (f) SEM image of a 3D macroporous g-C_3N_4 synthesised by 550 °C annealing of the co-precipitated melamine with cyanuric acid at 30 °C (MCA-DMSO-30-550). Reproduced with permission from ref. 37, © 2013 Wiley-VCH Verlag GmbH & Co. KGaA, Weinheim.

The mesoporous thin layer is then created using TEOS with *n*-octadecyltrimethoxysilane (C_{18}TMOS) as the pore-forming agent. The thickness and specific surface area of the HCNS, therefore, can be controlled by adjusting the amount of TEOS and C_{18}TMOS added to the silica core.

mpg-C_3N_4 materials have been further used as platforms to study the effect of defects on photocatalytic performance.[36] In general, there are two types of defects in an imperfect C_3N_4: free amino ($-NH_2$) and cyano (C=N) groups. These two possible functional groups are the strongest electron donors and acceptors, and their concentrations in mpg-C_3N_4 will influence the recombination kinetics of the photogenerated charge carriers. While a low-temperature synthesis results in cyano-rich carbon nitrides, increasing the annealing temperature leads to the formation of NH_2-rich materials. Cyano-rich mpg-C_3N_4 shows a typical vibrational peak at a wavenumber of ~2150 cm^{-1}, as can be seen in Figure 9.11(d). By co-polymerising the pre-heated cyanamide with melamine, the C=N defects in the mpg-C_3N_4 can be "repaired" (R-C_3N_4), as indicated by the disappearance of the C=N vibrational peak. On the other hand, the concentration of the free NH_2 defect in the mpg-C_3N_4 can be determined by the ninhydrin (2,2-dihydroxyindane-1,3-dione) test. The ninhydrin will only react with ammonia and primary amines that have alpha hydrogens to generate the purple ninhydrin chromophore (2-(1,3-dioxoindan-2-yl)iminoindane-1,3-dione). Therefore, amino defects in the C_3N_4 can be quantified without the interference of NH or NH_2 in the perfect melem structure. As shown in Figure 9.11(e), mpg-C_3N_4 shows a higher absorption peak compared to that of R-C_3N_4, indicating a higher concentration of the –NH_2 defect. This has been further evidenced by the reduced mpg-C_3N_4 (more –NH_2 defects) and the deaminated mpg-C_3N_4 (fewer –NH_2 defects). The defect-repaired mpg-C_3N_4, therefore, shows a better charge separation, thus an enhanced photocatalytic H_2 performance compared to that of traditional mpg-C_3N_4.

mpg-C_3N_4 materials are normally characterised by a low-dimensional, layer-structured morphology, however, it is also possible to create macroporous C_3N_4 nanoparticles, nanotubes, and nanosheets *via* the co-crystallisation process.[37] Since precursors with a triazine-like structure (*i.e.*, cyanuric acid (CA) and trithiocyanuric acid (TCA)) can form hydrogen-bonded networks during pyrolysis, tailoring the microstructural properties and the interlayer interactions of the carbon nitrides can be controlled by tuning the donor–acceptor pair, the crystallisation solvent, and the polycondensation procedures. Figure 9.11(f) representatively shows macroporous C_3N_4 nanoparticles assembled from thermal polycondensation of the co-precipitated melamine with cyanuric acid at 30 °C in DMSO. Such a macroporous g-C_3N_4 shows a larger surface area than that of bulk g-C_3N_4, thus yields an enhanced photocatalytic performance in H_2 evolution.

9.3.3.2 *Shape Engineering of g-C_3N_4*

The intuitive strategy for engineering bulk g-C_3N_4 is to "cut" the thick layer (few hundred nm) into thinner layers (few nm to sub-nm), as exceptional physicochemical properties (electronic, thermal, mechanical, optical properties) of such nanosheets are expected due to the quantum confinement effect. The large specific surface area with abundant reactive sites and the reduced diffusion length of the charge carriers to the electron acceptor/donor are the most significant advantages of nanosheet g-C_3N_4. Besides, the bandgap of the g-C_3N_4 nanosheet is also expected to be larger than that of the bulk g-C_3N_4 due to the quantum confinement effect, thus resulting in a stronger redox potential of the charge carriers. There are several methods to exfoliate bulk g-C_3N_4 into thin layers, including the thermal oxidation etching process, liquid-phase exfoliation using water or alcohols, and the chemical etching process using acids (H_2SO_4) and/or oxidants. Since the aim of all strategies mentioned here is to break the van der Waals interactions between layers while maintaining the planar bonds (C–N, N–H, and H–H), the exfoliation force should be optimised, especially, to prevent the breaking of the medium-strong hydrogen bond. The classical Hummers method used in the preparation of monolayer graphene failed in the synthesis of g-C_3N_4 nanosheets as it tears up the planar hydrogen bonds.

The thermal oxidation etching process employs O_2 to oxidise the hydrogen bond between the polymeric melon layers under elevated temperatures (*i.e.*, 500 °C).[38] As directly evidenced by the atomic force microscopic (AFM) image and height analysis (Figure 9.12(a)), g-C_3N_4 nanosheets with a thickness of ~2 nm have been successfully prepared. The g-C_3N_4 nanosheets show a similar XRD pattern to that of the bulk counterpart, suggesting the crystal structure has been maintained (Figure 9.12(b)). However, the less pronounced peak

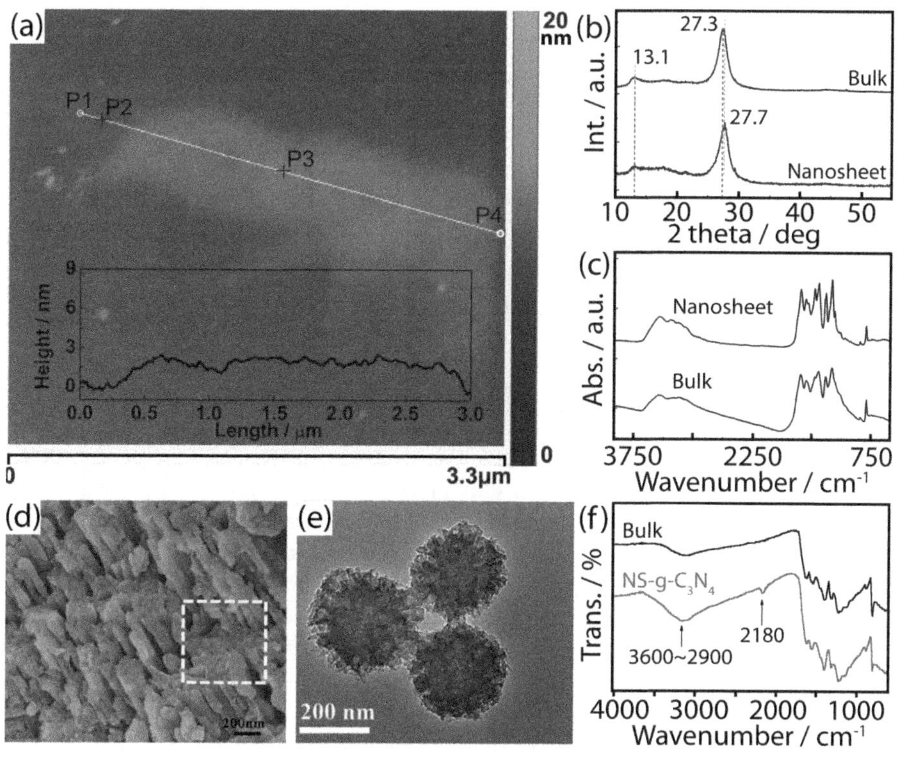

Figure 9.12 (a) AFM image of a single g-C_3N_4 nanosheet deposited on a silicon wafer substrate. The inset is the height curve determined along the line between P1 and P4. (b) and (c) XRD patterns and IR spectra of bulk g-C_3N_4 and g-C_3N_4 nanosheets. Adapted with permission from ref. 38, © 2012 WILEY-VCH Verlag GmbH & Co. KGaA, Weinheim. (d) SEM image of the g-C_3N_4 nanotubes. Reproduced from ref. 41 with permission from The Royal Society of Chemistry. (e) and (f) TEM image and FT-IR spectra of a nanospherical g-C_3N_4 with sharp edges (NS-g-C_3N_4). Reproduced with permission from ref. 42, © 2014 WILEY-VCH Verlag GmbH & Co. KGaA, Weinheim.

at 13.1° and the shifted peak at 27.3° suggest that the g-C_3N_4 nanosheets exhibit a slightly reduced planar size and a denser stacking of the layers compared to those of the bulk material. The characteristic vibrational peaks of the g-C_3N_4 nanosheets are also similar to those of the bulk material (810 cm^{-1}: heptazine ring; 900 to 1800 cm^{-1}: trigonal C–N(–C)–C or bridging C–NH–C units; 3000 to 3600 cm^{-1}: N–H stretching), suggesting the in-plane structure has been maintained (Figure 9.12(c)). The peaks within 900 to 1800 cm^{-1} become sharper and may originate from the more ordered packing of the hydrogen bonds in the layers of the nanosheets. The g-C_3N_4 nanosheets exhibit a significantly enhanced electronic conductivity compared to that of bulk g-C_3N_4 according to the *I–V* measurement, indicating that the charge carriers show a longer mean free path along the in-plane direction. This may improve the spatial charge separation thus slow down the recombination of the photoexcited charge carriers. Noticeably, the surface C–N atomic ratio of the nanosheets is 0.83 (0.65 for bulk g-C_3N_4), indicating N atoms are preferentially etched away and the nanosheet surface is N deficient. Since all C atoms are three-coordinated and N atoms are 50% two-coordinated and 50% three-coordinated, the N atoms are easier to remove. This also indicates that inherent defect sites exist in the exfoliated nanosheets. Although in most cases, enhanced photocatalytic performances are observed for the exfoliated thin-layer g-C_3N_4, the possible negative effect of the defects should be noted. The thermal oxidation etching process is also characterised by a very low yield, as ~94% of the bulk g-C_3N_4 is decomposed into gaseous products. Such a low yield can be overcome by using the liquid-phase exfoliation method using water, alcohols, or diluted acids.[39,40]

Recently, the synthesis of g-C_3N_4 nanotubes and nanoparticles has also been realised. g-C_3N_4 nanotubes can be simply synthesised by pyrolysis of the melamine precursor that is packed in an appropriate compact mode.[41] The resulting g-C_3N_4 nanotubes can grow up to 1 micrometre long with a tube inner diameter of ~18 nm and wall thickness of ~15 nm (Figure 9.12(d)). It is proposed that the g-C_3N_4 nanotubes' formation follows an NH_3-template-assisted rolling-up mechanism. When the melamine is packed with a proper density, the large amount of NH_3 released during the melamine pyrolysis will go through the melamine layers vertically to form slightly rolled g-C_3N_4 sheets. The slightly rolled g-C_3N_4 sheets then tend to further roll into tubular structures to minimise the total surface free energy. Neither too loosely nor too tightly packed melamine layers can form a tubular-structured g-C_3N_4, therefore, careful control of the packing density of melamine is crucial to control the perfection and the tube properties of the product. g-C_3N_4 nanoparticles with sharp edges (NS-g-C_3N_4, Figure 9.12(e)) have been synthesised by employing KCC-1 silica spheres as sacrificial templates.[42] The synthesis strategy is similar to that of the mesoporous g-C_3N_4 discussed previously (Figure 9.11), however, the NS-g-C_3N_4 is considered to have more opened-up surfaces (active sites for reactions), as revealed by the infrared spectra (Figure 9.12(f)). Besides the classical skeletal vibrations of the aromatic CN heterocycles (1200–1600 cm^{-1}) and the breathing vibration of triazine units (810 cm^{-1}), the enhanced vibration of surface-bonded H_2O (2900–3600 cm^{-1}) and the appearance of the vibration of amino groups (2180 cm^{-1}) are observed for the NS-g-C_3N_4, suggesting that the open structure provides more adsorption sites for reactants.

9.3.3.3 *Doping and Surface Engineering*

The optical and electronic properties of pristine g-C_3N_4 have been discussed previously (see Figure 9.7). To further tune the physiochemical properties (*i.e.*, bandgap, band positions, and charge mobility) of the pristine g-C_3N_4, doping non-metal (*i.e.*, P, S, I, B, and Si) and metal (*i.e.*, Na, K, Fe, and Cu) elements into the lattice have been investigated. By employing precursors containing the desired element(s), non-metal element doping can be achieved. For example, P-doped bulk g-C_3N_4 or g-C_3N_4 nanosheets can be synthesised by the co-polymerisation of melamine with 2-aminoethylphosphonic acid [$H_2N(CH_2)_2PO(OH)_2$] in a mass ratio of 60 : 1.[43] XRD and infrared spectra reveal that the backbone of the polymeric g-C_3N_4 remained almost unchanged after P doping. DFT simulation suggests that P atoms are most likely replacing the C atoms in the heptazine at C1 sites, as depicted in Figure 9.13(a). Such a geometry results in an increase of the in-planar heptazine edge from 6.95 Å for pristine g-C_3N_4 to 7.04 Å for P-doped g-C_3N_4. The predicted geometric change caused by P doping has also been proved by the shift of the (100) diffraction peak, which shifted from 13.4° to 13.0° after P doping, corresponding to an increase in the distance of in-planar nitride pores from 6.61 Å to 6.81 Å. The DOS of both pristine g-C_3N_4 and P-doped g-C_3N_4 calculated by DFT have further revealed the effect of P doping on the electronic structure of g-C_3N_4, as shown in Figure 9.13(b). Although it is doubtful that the intrinsic bandgap of g-C_3N_4 is narrowed to 2.66 eV after P doping, the arising of empty mid-gap states below the conduction band (CB) of g-C_3N_4 from the hybridisation of C_{2s2p}, N_{2s2p} and P_{3s3p} is reasonable. It is, therefore, expected that such an empty mid-gap above the Fermi level can trap more photo-excited electrons and also extend the visible light absorption range.

On the other hand, designing or engineering a catalyst to have an optimised surface adsorption energy of the reactants is ideal to accelerate a reaction according to the Sabatier principle. For the case of photocatalytic H_2 evolution, it is essential to optimise the binding energy of the H atoms (ΔG_H) to the surface of the photocatalyst material. The optimised ΔG_H is 0, which means the H atom loosely binds to the catalyst surface (easy to recombine with another H atom and desorbs as gas). Based on this strategy, it is predicted that the ΔG_H of the g-C_3N_4 surface can be tuned by applying either a tensile strain or compressive strain (~5%, depending on the surface coverage of H_{ads}).[44] Such a mechanical strain in g-C_3N_4 can be introduced *via* doping an isoelectronic silicon atom on the bridging carbon site (Figure 9.13(c)), thus realising fine tuning of the ΔG_H and eventually the photocatalytic H_2 evolution performance of g-C_3N_4.

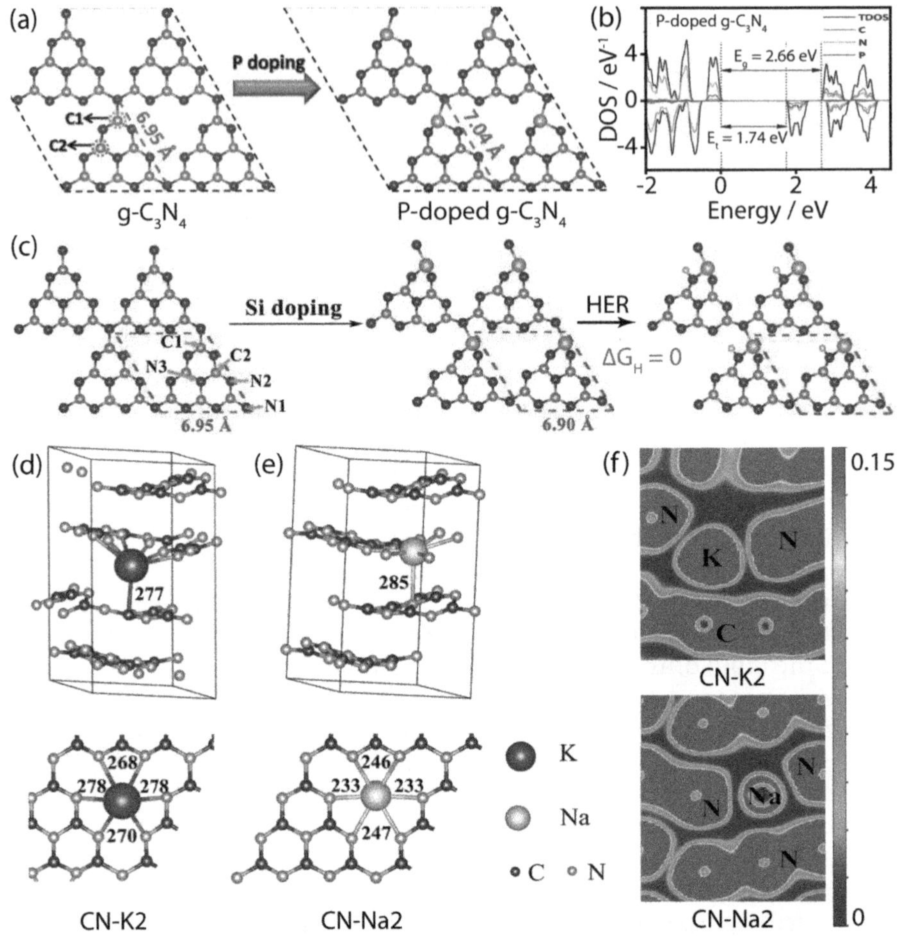

Figure 9.13 (a) Top view of the optimised g-C_3N_4 and P-doped g-C_3N_4 configurations. (b) Total density of states (TDOS) and partial density of states (PDOS) of the P-doped g-C_3N_4. Reproduced from ref. 43 with permission from The Royal Society of Chemistry. (c) Top view of the energetically favoured structures of g-C_3N_4, Si-doped g-C_3N_4, and H adsorbed on Si-doped g-C_3N_4 configurations. Adapted from *Journal of Catalysis*, **332**, G. Gao, Y. Jiao, F. Ma, Y. Jiao, E. Waclawik and A. Du, Metal-free graphitic carbon nitride as mechano-catalyst for hydrogen evolution reaction, 149–155, Copyright 2015, with permission from Elesvier.[44] (d) and (e) Side- and top-view of the calculated crystal structures of K-doped g-C_3N_4 (CN-K2) and Na-doped g-C_3N_4 (CN-Na2). Selected distances are marked in pm. (f) Electronic location function (ELF) analysis of CN-K2 and CN-Na2. Reprinted with permission from T. Xiong, W. Cen, Y. Zhang and F. Dong, *ACS Catal.*, 2016, **6**, 2462, Copyright 2016 American Chemical Society.[45]

The introduction of metal atoms into g-C_3N_4 was also reported to improve the photocatalytic performance of g-C_3N_4. Xiong *et al.* studied the location of K and Na atoms in doped g-C_3N_4 and the effect of the metal dopant on the photocatalytic performance.[45] While DFT calculations suggest that the K atoms tend to locate in between two g-C_3N_4 layers, the Na atoms prefer to sit inside the pore of the three heptazines in the conjugated plane in the optimised configurations (Figure 9.13(d) and (e)). Furthermore, the crystal cell is smaller for the Na-doped g-C_3N_4 compared to that of the K-doped g-C_3N_4, as the ionic radius of K^+ (2.32 Å) is larger than that of Na^+ (1.86 Å). Therefore, K atoms are the preferred ones to bridge the g-C_3N_4 interlayers, in which an improved charge transfer between layers is expected. Although Na and K atoms will donate electrons to the pristine g-C_3N_4 matrix, they show a distinct influence on the charge distribution properties of the system. Since the K atoms are embedded at interlayers, the 4s electron of K atoms will increase the charge density of both upper layer N (C) atoms and lower layer C (N) atoms that are close to the K atoms, thus forming a 3D charge "channel" that is vertical to the layer structure. To the opposite, the 3s electron of Na atoms will only interact with in-planar N atoms to form ionic bonds. This has been further revealed by electronic location function (ELF) mapping, as shown in Figure 9.13(f). The EFL is a dimensionless indicator (0–1), where ELF = 1 corresponds to the perfect localised electron and ELF = 0 is for the uniform electron gas. The ELF values between K–N and Na–N are both quite small (<0.15), suggesting the ionic interaction between K–N and Na–N in doped g-C_3N_4. However, the doped Na cation only interacts with the neighbouring N atoms in a single layer, whereas a weak covalent interaction of K with the C atom at an adjacent layer is observed. This suggests that the K atoms could bond with atoms at the adjacent two layers to reduce the charge localisation, thus promote the charge mobility and subsequently prolong the mean free path of the charge carrier in the conjugated systems to improve the photocatalytic performance.

The other straightforward strategy to improve the photocatalytic performance of g-C_3N_4 is to spatially separate the charge carriers *via* surface engineering. It should be mentioned that pristine g-C_3N_4 shows a negligible photocatalytic H_2 evolution performance, and only with proper promoters (co-catalysts) on the surface can H_2 evolution be initiated. Noble metal (Pt, Pd, and Au) nanoparticles are the most frequently used promoters to decorate pristine g-C_3N_4 in various photocatalytic reactions. Since the free adsorption energy of H on Pt is estimated to be almost optimised ($\Delta G_H \approx 0$), Pt is widely used to modify g-C_3N_4 in photocatalytic H_2 production.[22] The Pt nanoparticles can be loaded onto the g-C_3N_4 *via* a sol-immobilisation approach, direct precipitation, or more easily, photodeposition under deaerated conditions in the presence of electron donors (*i.e.*, alcohols). The advantage of the photodeposition process is that Pt nanoparticles can be deposited freshly under photocatalytic HER conditions with the desired amount of chloroplatinic acid (H_2PtCl_6) salt. Pt^{4+} will be reduced at the reduction reaction sites directly without interference from any ligands or contaminants.

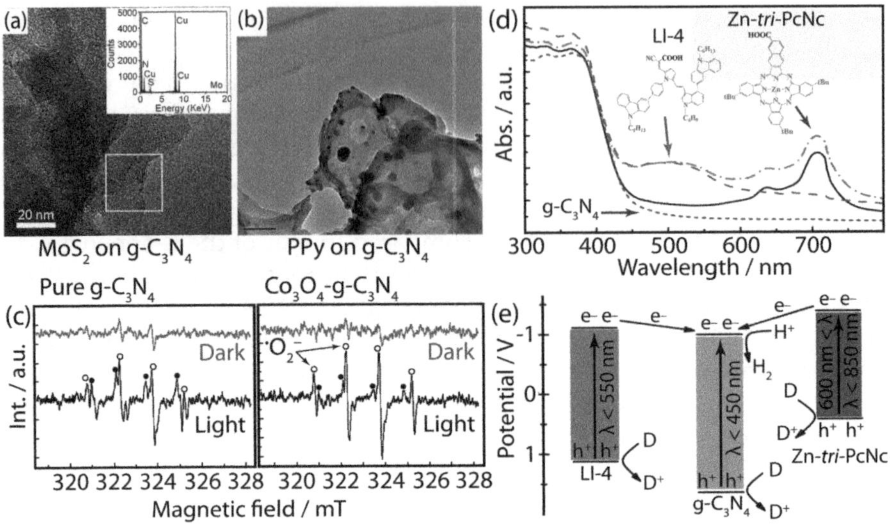

Figure 9.14 (a) TEM image of MoS_2 deposited on g-C_3N_4. Reproduced with permission from ref. 46, © 2013 WILEY-VCH Verlag GmbH & Co. KGaA, Weinheim. (b) TEM image of PPy deposited on g-C_3N_4. Adapted from ref. 47 with permission from The Royal Society of Chemistry. (c) Electron spin resonance (ESR) spectra of g-C_3N_4 and Co_3O_4-decorated g-C_3N_4 under dark and irradiation conditions. Adapted from *Applied Catalysis B: Environmental*, **147**, C. Han, L. Ge, C. Chen, Y. Li, X. Xiao, Y. Zhang and L. Guo, Novel visible light induced Co_3O_4-g-C_3N_4 heterojunction photocatalysts for efficient degradation of methyl orange, 546–553, Copyright 2014, with permission from Elsevier.[50] (d) DRS spectra of g-C_3N_4, LI-4/g-C_3N_4, Zn-*tri*-PcNc/g-C_3N_4, and LI-4/g-C_3N_4/Zn-*tri*-PcNc. (e) Proposed sensitisation mechanisms of LI-4- and Zn-*tri*-PcNc-modified g-C_3N_4 for photocatalytic H_2 production. Reproduced with permission from X. Zhang, T. Peng, L. Yu, R. Li, Q. Li and Z. Li, *ACS Catal.*, 2015, **5**, 504. Copyright 2015 American Chemical Society.[51]

In order to avoid the use of noble metals and to further enhance the photocatalytic performance, earth-abundant metals (*i.e.*, Ni) and semiconductor-type promoters (*i.e.*, MoS_2, CuO, V_2O_5) have also been investigated. Figure 9.14(a) shows a TEM image of a MoS_2-modified g-C_3N_4 photocatalyst for the HER.[46] MoS_2 possesses a similar layered structure to g-C_3N_4, suggesting that a planar growth of MoS_2 slabs on the g-CN surface to form organic–inorganic hybrid heterojunctions can be easily realised with a minimised lattice mismatch. Such a construction of the composite may facilitate interface charge transportation between the g-C_3N_4 and the MoS_2 layers. Besides, the excellent activity of MoS_2 in electrocatalytic H_2 evolution due to the high exposure of the edges implies that the charges injected from the CBM of g-C_3N_4 should also work. Indeed, 0.2 wt% MoS_2 loaded onto mpg-C_3N_4 shows an enhancement in photocatalytic H_2 evolution ~1.3 times that of 2 wt% Pt-loaded mpg-C_3N_4.

Carbon-based materials have also been used as promoters. Polypyrrole (PPy) nanoparticles have been supported on g-C_3N_4 *via* sonochemical synthesis from PPy and g-C_3N_4, as can be visualised by a TEM image (Figure 9.14(b)).[47] It is expected that the conductive polymer will improve the charge mobility on the g-C_3N_4 surface, thus promoting the interfacial charge transfer to the Pt nanoparticles. The authors claim that photocatalytic H_2 evolution can be initiated from pure water with any electron donor, however, oxygen evolution has not been mentioned and which compound has been oxidised remains unclear. In order to further eliminate the use of a Pt promoter, Han *et al.* improved the synthesis strategy and synthesised mesoporous g-C_3N_4 nanofibers with a N-rich carbon shell (g-C_3N_4@C).[48] The construction of such core–shell structured materials was realised by first creating a fiber-like structure of the hydrothermally treated dicyandiamide precursor *via* a freeze-drying process, then loading iodine to polymerise the surface-adsorbed pyrrole monomer into polypyrrole. Eventually, a subsequent calcination of the polypyrrole-coated nanofiber will turn into g-C_3N_4@C. The g-C_3N_4 with such an architecture exhibits a much better photocatalytic performance compared to that of the Pt-loaded g-C_3N_4. Very recently, it has also been found that carbon nanodots exhibit unique catalytic properties when coupled with g-C_3N_4 for pure water splitting.[49] It is considered that the presence of C nanodots facilitates the formation and decomposition of H_2O_2 ($2e^-$ process) while inhibiting direct water splitting ($4e^-$ process), thus resulting in a magnificent QE of 16% under 420 nm irradiation and 2% overall solar energy conversion efficiency. This reaction mechanism was confirmed by electrochemical analysis, which proves that a two-electron transfer process takes place under different irradiation conditions. Besides, the photocurrent measurements demonstrate that while g-C_3N_4 shows a rapid decrease of photocurrent, C nanodots supported on g-C_3N_4 only show current oscillations between 190–240 nA, indicating that the presence of C nanodots accelerates the generation and decomposition of H_2O_2 adsorbed on pristine g-C_3N_4. Unfortunately, we failed to reproduce these results.

It is also possible to tune the identity of active radicals by introducing promoters to improve the photocatalytic performance of g-C_3N_4. Han *et al.* deposited Co_3O_4 nanoparticles on the g-C_3N_4 surface from a $Co(NO_3)_4$ precursor by an evaporation–calcination process.[50] The radical species and concentrations of the pristine and Co_3O_4-modified g-C_3N_4 can be resolved by employing electron spin resonance (ESR) analysis, as shown in Figure 9.14(c). It was found that the superoxide radicals ($^{\bullet}O_2^-$) become the dominant oxidative species with significantly enhanced intensity for the Co_3O_4-g-C_3N_4 photocatalyst. This indicates the photogenerated charge carriers have been well separated in the presence of the Co_3O_4 promoter, thus benefiting the formation of additional $^{\bullet}O_2^-$ radicals from O_2 and photogenerated e^-. The rapid consumption of O_2 along with the very reactive $^{\bullet}O_2^-$ radicals are expected to improve photocatalytic oxidation reactions.

Coupling g-C_3N_4 with organic dye molecules to further extend the visible light absorption is another successful strategy of surface engineering. Figure 9.14(d) representatively shows the sensitisation of g-C_3N_4 with two dyes—LI-4 (indole-based D–π–A organic dye) and Zn-*tri*-PcNc (an asymmetric zinc phthalocyanine derivative).[51] The characteristic absorption peaks of the dyes (~500 nm for LI-4 and ~700 nm for Zn-*tri*-PcNc) can be clearly seen after surface adsorption. Since dyes are normally characterised by huge extinction coefficients (>10 000 M^{-1} cm^{-1} at peak absorption), a very thin layer of dye covering g-C_3N_4 should be able to absorb light efficiently. This may be the essential parameter, as a thick layer of dye could block the diffusion of the reactants and products. It should also be noted that the HOMO–LUMO energy levels of the dye molecules need to match the bandgap of the g-C_3N_4 to ensure electron injection from the dye to g-C_3N_4. As representatively depicted in Figure 9.14(e), the LUMO energy levels of both LI-4 and Zn-*tri*-PcNc dyes are more negative than the CBM of g-C_3N_4, enabling the injection of excited electrons from dyes to g-C_3N_4 to perform photocatalytic reactions. Meanwhile, the HOMO levels of both dyes are sufficient to accept electrons from electron donors (*i.e.*, TEOA and ascorbic acid) to quench the excited dyes, thus creating a cycle of electron flow.

9.3.3.4 Carbon Nitride Derivatives

As mentioned at the beginning of this section, polymeric g-C_3N_4 materials are only a small fraction of the "carbon nitride" family. Wirnhier *et al.* first revealed that the structure of the highly crystalline carbon nitride, namely poly(triazine imide) [PTI], synthesised by thermal condensation using salt melts (*i.e.*, LiCl or NaCl) as solvents is not based on the heptazine-based structure.[52] PTI/Li^+Cl^- (synthesised using LiCl salt) crystallises in the hexagonal space group P63 cm with cell parameters of a = 846.82(10) and c = 675.02(9) pm according to TEM and XRD analysis. As shown in Figure 9.15(a), the layered crystal is composed of a triazine 2D network, in which the triazine units are connected by the imide, resulting in triangular voids. The imide-bridged triazine layers are stacked along the c axis, making the voids in the 2D layer connect into porous channels, where the Li^+ and Cl^- ions fill in these channels.

PTI/Li^+Cl^- shows a slightly improved photocatalytic H_2 evolution performance compared to that of bulk g-C_3N_4. This may be attributed to the 2D architecture of PTI/Li^+Cl^-, which could expand the π-electron system to a lower bandgap compared to that of bulk g-C_3N_4.[53] In order to further improve the photocatalytic performance, defects in the PTI/Li^+Cl^- were created using 4-amino-2,6-dihydroxypyrimidine (4AP) as a dopant, as 4AP possesses a higher carbon and oxygen content and has a similar structure to melamine. Elemental analysis and IR spectra suggest that both C and O atoms have been incorporated into the materials by replacing one of the triazine rings or bridging N atoms, as schematically demonstrated in Figure 9.15(b). The

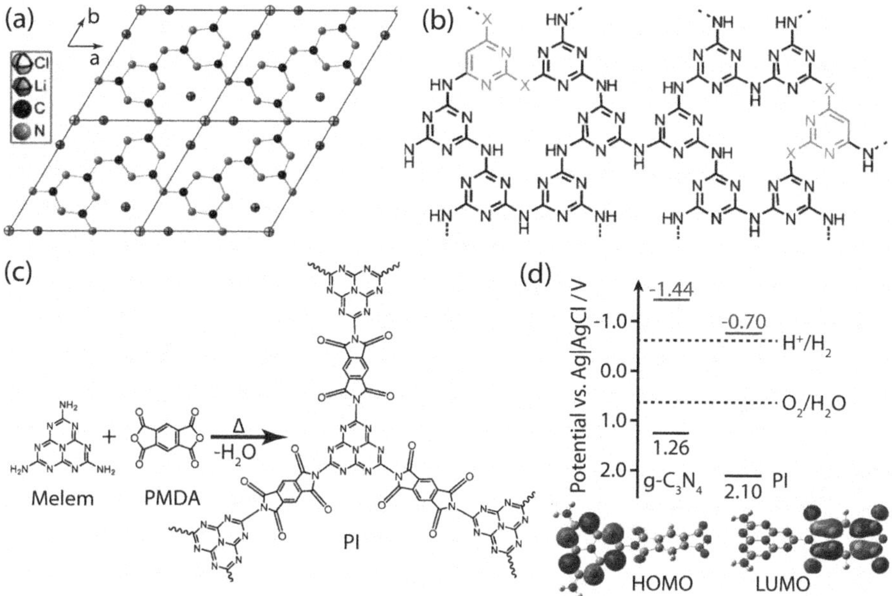

Figure 9.15 (a) Parallel projected one-layer structure of PTI/Li^+Cl^- with Li, C, and N atoms in a plane and Cl atoms shifted by (00 1/4). The depicted Li positions have an occupancy of 1/3. Reproduced with permission from ref. 52, © 2011 Wiley-VCH Verlag GmbH & Co. KGaA, Weinheim. (b) Proposed structure of amorphous PTI with 16% of 4AP dopant ($aPTI\_4AP_{16\%}$), X = NH, O, or OH. Reproduced with permission from ref. 53, © 2013 Wiley-VCH Verlag GmbH & Co. KGaA, Weinheim. (c) Reaction scheme for the synthesis of PI. (d) Band structure of g-C_3N_4 and PI derived from a Mott–Schottky analysis, and the calculated HOMO and LUMO distribution of the optimised PI. Adapted from S. Chu, Y. Wang, Y. Guo, J. Feng, C. Wang, W. Luo, X. Fan and Z. Zou, *ACS Catal.*, 2013, **3**, 912. Copyright 2013 American Chemical Society.[54]

bandgap of the 4AP-doped amorphous PTI can be gradually reduced by increasing the dopant concentration from 2% to 64%. An optimum photocatalytic performance is observed for 16%-4AP-doped amorphous PTI, which can be considered as a balance between the light absorption and the concentration of defects that served as recombination centres in the material.

Pristine g-C_3N_4 is characterised by a relatively negative CBM position (~−1.2 V *vs.* SHE). Such a strong driving force (reduction potential) makes it ideal to drive reduction reactions, especially the HER (0 *vs.* SHE). On the other hand, although the VBM position of pristine g-C_3N_4 makes it thermodynamically possible to drive the OER, the driving force is relatively small (~1.4 V *vs.* SHE). Since the OER (water oxidation) half-reaction is generally considered to be more challenging than the HER, improving the photocatalytic performance in the OER half-reaction is the key to achieving the ultimate goal of overall water splitting. Chu *et al.* have shown a bottom-up strategy to tune the electronic structure of g-C_3N_4 *via* incorporating an electron-deficient pyromellitic dianhydride (PMDA) unit into the network of g-C_3N_4, as depicted in Figure 9.15(c).[54] The melamine (not shown here) will first form the melem and then co-polymerise with the PMDA upon heating (325 °C) *via* a condensation process. Such a polyimide (PI) structure has been confirmed by XRD, an infrared spectrum, and solid-state NMR. The bandgap and band positions of the PI were further revealed by coupling Mott–Schottky analysis with VB XPS, as shown in Figure 9.15(d). Compared to that of the g-C_3N_4, both the CBM and VBM of the PI have been positively shifted without changing the bandgap, suggesting that the introduction of an electron-withdrawing substituent (PMDA here) into the polymer framework has successfully tuned the electronic structure of the material by lowering the HOMO level. Furthermore, DFT calculations reveal that the HOMO and LUMO orbitals are located at the melem moiety and the PMDA moiety, respectively, suggesting that the redox active sites have been spatially separated by incorporation of the electron-withdrawing PMDA unit (Figure 9.15(d)). The dual beneficial effects of the PMDA unit have significantly improved the photocatalytic OER performance.

9.3.4 Other Applications Beyond Photocatalytic HER

The photocatalytic H_2 evolution reaction has been widely used to evaluate the performance of carbon-nitride-based photocatalyst materials. However, it should be noted that other important reactions can also be realised using carbon nitride photocatalysts. The photocatalytic oxidation of toxic NO into NO_2 can be realised by employing porous-structured g-C_3N_4.[55] Porous g-C_3N_4 with an optimised porosity can convert 70% of NO into NO_2 within 20 min under blue light (~450 nm) irradiation. The NO oxidation follows a synergetic mechanism of h^+ and $^{\bullet}O_2^-$, where NO is initially oxidised to NO^+ by the photogenerated h^+, and eventually reacts with the photogenerated $^{\bullet}O_2^-$ (from O_2 reacts with e^-) to produce NO_2. Meanwhile, photocatalytic CO_2 reduction has also been achieved by using a ZnO-nanoparticle-decorated g-C_3N_4 photocatalyst under visible light irradiation.[56] The interaction between the two components is supposed to form a heterojunction structure to slow down the recombination of the charge carriers, thus improving photocatalytic CO_2 reduction. The CO_2 reduction rates can be tuned by adjusting the loading of ZnO, however, CO is found to be the main product.

g-C_3N_4 has also been applied in photocatalytic organic synthesis. For reactants with an aromatic structure, the π-conjugated structure of g-C_3N_4 may facilitate the reactant–catalyst interaction, thus enhancing the surface adsorption and charge transfer of the reactants. Chen *et al.* first reported the photosynthesis of phenol from benzene using Fe-g-C_3N_4 under visible light irradiation with H_2O_2 as the oxidant.[57] The Fe dopant plays a crucial role in activating the benzene molecule, as pristine g-C_3N_4 shows no activity upon benzene oxidation. By incorporating the Fe-g-C_3N_4 with SBA-15, the benzene conversion to phenol can be further improved. Unfortunately, the low conversion (~12%) and the need for H_2O_2 limit such an application.

The bio-inspired photocatalytic regeneration of reduced nicotinamide adenine dinucleotide (NADH) from oxidised NAD (NAD^+) was also realised by using the g-C_3N_4 photocatalyst with a diatom frustule structure.[58] As schematically demonstrated in Figure 9.16, the photogenerated e^- can first reduce the electron mediator ($[Cp^*Rh(bpy)H_2O]^{2+}$ here), which will then react with a proton and eventually donate the H atom to the NAD^+ to form the NADH. The electron donor (TEOA here) will be oxidised and will also donate protons to the electron mediator. Such a process even works in the absence of an electron mediator, indicating that the direct electron transfer from the photocatalyst to NAD^+ can be achieved *via* a π–π interaction between the heptazine unit of the g-C_3N_4 and the adenine sub-unit of NAD^+. However, the low conversion efficiency (50%) in the absence of an electron mediator and the relatively poor selectively to the enzymatically active 1,4-NADH limit such an application.

Very recently, pristine g-C_3N_4 photocatalysts have shown a capability of synthesising a series of important azo- and azoxy-aromatic dyes from their corresponding nitro-aromatics with very high selectivity and conversion.[59] These dyes are economically much more valuable than amines, which are the fully reduced products of their corresponding nitro-aromatics. In comparison to other popular photocatalysts that normally show poor conversion or poor selectivity, g-C_3N_4 shows a full conversion over a series of nitro-aromatics. And, amazingly, the selectivity to either azo- or azoxy-compound can be switched by controlling the irradiation wavelength. While selective formation of the azoxy-aromatics can be achieved using blue light (450 nm) excitation, a purple (410 nm) excitation result in the formation of solely azo-aromatics. The photosynthesis of azoxybenzene remains highly efficient and selective

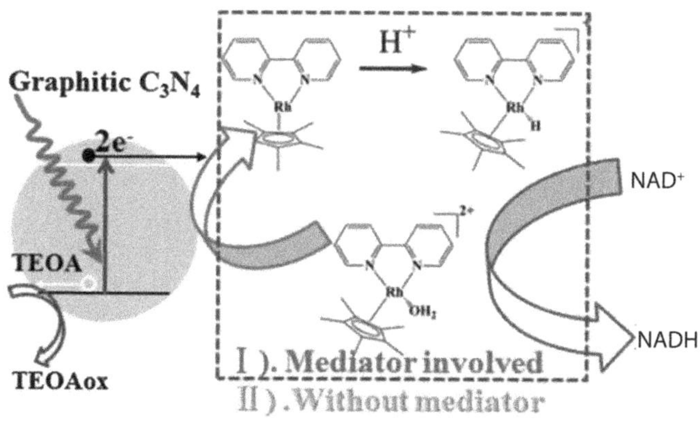

Figure 9.16 Scheme of the photocatalytic regeneration of NADH in the absence and presence of a mediator. Reproduced from ref. 58 with permission from The Royal Society of Chemistry.

even under solar irradiation conditions. The nitrobenzene can gradually convert into azoxybenzene with high selectivity. Although the conversion has not been optimised for scale-up experiments, the financial and environmental potential of such a process renders photocatalysis applications at an industrial level possible.

9.4 Concluding Remarks and Perspectives

In this chapter, we have gone through carbon-based photocatalyst materials from linear polymers, polymer networks, and covalent organic frameworks, to popular polymeric carbon nitrides and derivatives. The design strategy, synthesis methodology, and optimisation have been discussed in detail. The design and optimisation of carbon-based photocatalyst materials have mainly focused on (i) reducing the optical bandgaps, and (ii) engineering the electronic structure and geometric structure. Several strategies (*i.e.*, introducing dopant, surface engineering) have been successfully applied to improve the photocatalytic performance, however, it seems that the performance (especially for the HER) of carbon-based materials has reached a bottleneck. This requires an understanding of the whole photocatalytic process from light absorption, charge carrier generation and separation, radical formation, and charge recombination, to chemical reactions on the surface, to be improved further. Furthermore, the stability and failure mechanisms of carbon-based photocatalyst materials need to be explored.

Acknowledgements

I thank the NSFC (project numbers: 21503257 & 21601198) for financial support.

References

1. M. Qureshi and K. Takanabe, *Chem. Mater.*, 2017, **29**, 158.
2. R. Su, R. Bechstein, L. So, R. T. Vang, M. Sillassen, B. Esbjornsson, A. Palmqvist and F. Besenbacher, *J. Phys. Chem. C*, 2011, **115**, 24287.
3. D. O. Scanlon, C. W. Dunnill, J. Buckeridge, S. A. Shevlin, A. J. Logsdail, S. M. Woodley, C. R. A. Catlow, M. J. Powell, R. G. Palgrave, I. P. Parkin, G. W. Watson, T. W. Keal, P. Sherwood, A. Walsh and A. A. Sokol, *Nat. Mater.*, 2013, **12**, 798.
4. S. Yanagida, A. Kabumoto, K. Mizumoto, C. Pac and K. Yoshino, *Chem. Commun.*, 1985, 474.
5. T. Yamamoto, Y. Hayashi and A. Yamamoto, *Bull. Chem. Soc. Jpn.*, 1978, **51**, 2091.
6. T. Shibata, A. Kabumoto, T. Shiragami, O. Ishitani, C. Pac and S. Yanagida, *J. Phys. Chem.*, 1990, **94**, 2068.

7. T. Yamamoto, T. Maruyama, Z.-H. Zhou, T. Ito, T. Fukuda, Y. Yoneda, F. Begum, T. Ikeda and S. Sasaki, *J. Am. Chem. Soc.*, 1994, **116**, 4832.
8. R. S. Sprick, B. Bonillo, R. Clowes, P. Guiglion, N. J. Brownbill, B. J. Slater, F. Blanc, M. A. Zwijnenburg, D. J. Adams and A. I. Cooper, *Angew. Chem., Int. Ed.*, 2016, **55**, 1792.
9. M. G. Schwab, M. Hamburger, X. Feng, J. Shu, H. W. Spiess, X. Wang, M. Antonietti and K. Mullen, *Chem. Commun.*, 2010, **46**, 8932.
10. R. S. Sprick, J.-X. Jiang, B. Bonillo, S. Ren, T. Ratvijitvech, P. Guiglion, M. A. Zwijnenburg, D. J. Adams and A. I. Cooper, *J. Am. Chem. Soc.*, 2015, **137**, 3265.
11. J. H. Park, K. C. Ko, N. Park, H.-W. Shin, E. Kim, N. Kang, J. Hong Ko, S. M. Lee, H. J. Kim, T. K. Ahn, J. Y. Lee and S. U. Son, *J. Mater. Chem. A*, 2014, **2**, 7656.
12. R. Su, F. Besenbacher and G. Hutchings, in *Heterogeneous Photocatalysis: From Fundamentals to Green Applications*, ed. J. C. Colmenares and Y.-J. Xu, Springer Berlin Heidelberg, Berlin, Heidelberg, 2016, p. 109.
13. S. Wan, J. Guo, J. Kim, H. Ihee and D. Jiang, *Angew. Chem., Int. Ed.*, 2008, **47**, 8826.
14. A. P. Côté, A. I. Benin, N. W. Ockwig, M. O'Keeffe, A. J. Matzger and O. M. Yaghi, *Science*, 2005, **310**, 1166.
15. P. Kuhn, M. Antonietti and A. Thomas, *Angew. Chem., Int. Ed.*, 2008, **47**, 3450.
16. F. J. Uribe-Romo, J. R. Hunt, H. Furukawa, C. Klöck, M. O'Keeffe and O. M. Yaghi, *J. Am. Chem. Soc.*, 2009, **131**, 4570.
17. L. Stegbauer, K. Schwinghammer and B. V. Lotsch, *Chem. Sci.*, 2014, **5**, 2789.
18. J. Thote, H. B. Aiyappa, A. Deshpande, D. Díaz Díaz, S. Kurungot and R. Banerjee, *Chem.–Eur. J.*, 2014, **20**, 15961.
19. V. S. Vyas, F. Haase, L. Stegbauer, G. Savasci, F. Podjaski, C. Ochsenfeld and B. V. Lotsch, *Nat. Commun.*, 2015, **6**, 8508.
20. J. Bi, W. Fang, L. Li, J. Wang, S. Liang, Y. He, M. Liu and L. Wu, *Macromol. Rapid Commun.*, 2015, **36**, 1799.
21. K. Schwinghammer, S. Hug, M. B. Mesch, J. Senker and B. V. Lotsch, *Energy Environ. Sci.*, 2015, **8**, 3345.
22. X. Wang, K. Maeda, A. Thomas, K. Takanabe, G. Xin, J. M. Carlsson, K. Domen and M. Antonietti, *Nat. Mater.*, 2009, **8**, 76.
23. J. Liebig, *Ann. Pharm.*, 1834, **10**, 1.
24. G. Algara-Siller, N. Severin, S. Y. Chong, T. Björkman, R. G. Palgrave, A. Laybourn, M. Antonietti, Y. Z. Khimyak, A. V. Krasheninnikov, J. P. Rabe, U. Kaiser, A. I. Cooper, A. Thomas and M. J. Bojdys, *Angew. Chem., Int. Ed.*, 2014, **53**, 7450.
25. L. Lin, H. Ou, Y. Zhang and X. Wang, *ACS Catal.*, 2016, **6**, 3921.
26. J. Liu, T. Zhang, Z. Wang, G. Dawson and W. Chen, *J. Mater. Chem.*, 2011, **21**, 14398.
27. G. Zhang, J. Zhang, M. Zhang and X. Wang, *J. Mater. Chem.*, 2012, **22**, 8083.

28. P. Wu, J. Wang, J. Zhao, L. Guo and F. E. Osterloh, *J. Mater. Chem. A*, 2014, **2**, 20338.
29. Y. Kang, Y. Yang, L.-C. Yin, X. Kang, G. Liu and H.-M. Cheng, *Adv. Mater.*, 2015, **27**, 4572.
30. C. Li, C.-B. Cao and H.-S. Zhu, *Mater. Lett.*, 2004, **58**, 1903.
31. R. C. Dante, P. Martín-Ramos, A. Correa-Guimaraes and J. Martín-Gil, *Mater. Chem. Phys.*, 2011, **130**, 1094.
32. F. Goettmann, A. Fischer, M. Antonietti and A. Thomas, *Angew. Chem., Int. Ed.*, 2006, **45**, 4467.
33. J. Zhang, F. Guo and X. Wang, *Adv. Funct. Mater.*, 2013, **23**, 3008.
34. F. He, G. Chen, Y. Zhou, Y. Yu, Y. Zheng and S. Hao, *Chem. Commun.*, 2015, **51**, 16244.
35. J. Sun, J. Zhang, M. Zhang, M. Antonietti, X. Fu and X. Wang, *Nat. Commun.*, 2012, **3**, 1139.
36. W. Wu, J. Zhang, W. Fan, Z. Li, L. Wang, X. Li, Y. Wang, R. Wang, J. Zheng, M. Wu and H. Zeng, *ACS Catal.*, 2016, **6**, 3365.
37. Y.-S. Jun, J. Park, S. U. Lee, A. Thomas, W. H. Hong and G. D. Stucky, *Angew. Chem., Int. Ed.*, 2013, **52**, 11083.
38. P. Niu, L. Zhang, G. Liu and H.-M. Cheng, *Adv. Funct. Mater.*, 2012, **22**, 4763.
39. J. Tong, L. Zhang, F. Li, M. Li and S. Cao, *Phys. Chem. Chem. Phys.*, 2015, **17**, 23532.
40. Q. Han, B. Wang, J. Gao, Z. Cheng, Y. Zhao, Z. Zhang and L. Qu, *ACS Nano*, 2016, **10**, 2745.
41. S. Wang, C. Li, T. Wang, P. Zhang, A. Li and J. Gong, *J. Mater. Chem. A*, 2014, **2**, 2885.
42. J. Zhang, M. Zhang, C. Yang and X. Wang, *Adv. Mater.*, 2014, **26**, 4121.
43. J. Ran, T. Y. Ma, G. Gao, X.-W. Du and S. Z. Qiao, *Energy Environ. Sci.*, 2015, **8**, 3708.
44. G. Gao, Y. Jiao, F. Ma, Y. Jiao, E. Waclawik and A. Du, *J. Catal.*, 2015, **332**, 149.
45. T. Xiong, W. Cen, Y. Zhang and F. Dong, *ACS Catal.*, 2016, **6**, 2462.
46. Y. Hou, A. B. Laursen, J. Zhang, G. Zhang, Y. Zhu, X. Wang, S. Dahl and I. Chorkendorff, *Angew. Chem., Int. Ed.*, 2013, **52**, 3621.
47. Y. Sui, J. Liu, Y. Zhang, X. Tian and W. Chen, *Nanoscale*, 2013, **5**, 9150.
48. Q. Han, B. Wang, J. Gao and L. Qu, *Angew. Chem., Int. Ed.*, 2016, **55**, 10849.
49. J. Liu, Y. Liu, N. Liu, Y. Han, X. Zhang, H. Huang, Y. Lifshitz, S.-T. Lee, J. Zhong and Z. Kang, *Science*, 2015, **347**, 970.
50. C. Han, L. Ge, C. Chen, Y. Li, X. Xiao, Y. Zhang and L. Guo, *Appl. Catal., B*, 2014, **147**, 546.
51. X. Zhang, T. Peng, L. Yu, R. Li, Q. Li and Z. Li, *ACS Catal.*, 2015, **5**, 504.
52. E. Wirnhier, M. Döblinger, D. Gunzelmann, J. Senker, B. V. Lotsch and W. Schnick, *Chem.–Eur. J.*, 2011, **17**, 3213.
53. K. Schwinghammer, B. Tuffy, M. B. Mesch, E. Wirnhier, C. Martineau, F. Taulelle, W. Schnick, J. Senker and B. V. Lotsch, *Angew. Chem., Int. Ed.*, 2013, **52**, 2435.

54. S. Chu, Y. Wang, Y. Guo, J. Feng, C. Wang, W. Luo, X. Fan and Z. Zou, *ACS Catal.*, 2013, **3**, 912.
55. G. Dong, W. Ho, Y. Li and L. Zhang, *Appl. Catal., B*, 2015, **174**, 477.
56. Y. He, Y. Wang, L. Zhang, B. Teng and M. Fan, *Appl. Catal., B*, 2015, **168**, 1.
57. X. Chen, J. Zhang, X. Fu, M. Antonietti and X. Wang, *J. Am. Chem. Soc.*, 2009, **131**, 11658.
58. J. Liu and M. Antonietti, *Energy Environ. Sci.*, 2013, **6**, 1486.
59. Y. Dai, C. Li, Y. Shen, T. Lim, J. Xu, Y. Li, H. Niemantsverdriet, F. Besenbacher, N. Lock and R. Su, *Nature Commun.*, 2018, **9**, 60.

CHAPTER 10

Metal-free Doped Carbons for Electroanalytical Sensors

V. PIFFERI, A. TESTOLIN AND L. FALCIOLA*

Università degli Studi di Milano, Dipartimento di Chimica, via Golgi 19, Milano, 20133, Italy
*E-mail: luigi.falciola@unimi.it

10.1 Introduction

Carbon materials have been used extensively in the electroanalytical field in the development of modified electrodes to be employed as sensor platforms for different analytical targets.

The advantages of these materials are, primarily, their very low cost and extensive availability. Moreover, carbon-based materials are adaptable and versatile. In fact, the possibility to prepare, dope and functionalize them with different surface modifiers and in various ways allows the creation of materials with different characteristics and properties according to the needs of the analytical target to be determined.

The main advantages brought by the use of these materials for electroanalytical purposes are: a high surface–volume ratio, wide potential windows in aqueous and non-aqueous media, chemical inertness, miniaturization, efficient mass transport, stable and low background currents,

Catalysis Series No. 31
Metal-free Functionalized Carbons in Catalysis: Synthesis, Characterization and Applications
Edited by Alberto Villa and Nikolaos Dimitratos

Published by the Royal Society of Chemistry, www.rsc.org

accuracy (precision and trueness[1]), selectivity, sensitivity, reproducibility, low limits of detection, long lifetimes for real-time detection in real matrices, fast response, low fabrication costs, robustness, and portability. Moreover, the great simplicity in the preparation of electrodes and sensors allows their use in different fields with no need for expertise or lengthy training, as is the case with screen-printed electrodes (SPE), which have completely changed electroanalytical potentialities.[2–8] The principal methodologies of deposition and preparation of carbon-based modified electrodes are described in Section 10.2 starting from simpler ones (drop casting, spin coating) to the more complex (CVD, pyrolysis, screen printing) to reach the final challenge of recent years: the preparation of sensors on paper supports.[6,9–14]

Carbon materials for the modification of electrodes can be produced in different forms (carbon paste, carbon fibers, carbon dots, graphite (pyrolytic, highly ordered, ...), carbon nanotubes, graphene, glassy carbon, boron doped diamond, ...), and carbon-based sensors have been used for a plethora of analytical targets, also in trace electroanalysis,[15–18] from metals to organic pollutants, to biological analytes with excellent results in terms of high sensitivities and very low detection limits, and also in real-world matrices.

Due to the extensive use of these materials over the last few decades, the literature in the field has exploded and many authoritative reviews have been produced.[19–30]

In this chapter, according to the book's philosophy, we would like to focus our attention on carbonaceous materials doped with non-metal heteroatoms, although controlled doping with these atoms—B, N, O, P, S, ...—together with surface functionalization procedures, are technical challenges that are yet to be optimized,[19] thus providing results that are often only preliminary evidences.

However, it is important to immediately underline that the presence of metal impurities is not completely avoidable, since such a doping is difficult to eliminate, often deriving from the synthetic procedure. In this context, a debate on the role of metal impurities in the electroanalytical performances of these materials has held court for a long time, as shown by the literature in the field.[31–38] A definitive response is still to be found, although it is clear that metallic impurities strongly influence the redox properties and are responsible for a large proportion of the observed "electrocatalytic" behaviour.[37] Furthermore, it is important to consider that the basis of the electrocatalytic nature, electron transfer and chemical reactivity of carbon-based materials, which are responsible for the improved electroanalytical performances of such materials, resides in edge-plane defect electroactive sites[32,38–41] (see Figure 10.1) or in graphitic and amorphous carbon impurities.[37,38,42–44]

In this context, in some cases, for electroanalytical purposes, it is more advantageous to use low-cost alternatives (carbon black, nano-carbon black) rather than more complex and expensive (carbon nanotubes, graphene, ...) carbon materials.[45–48]

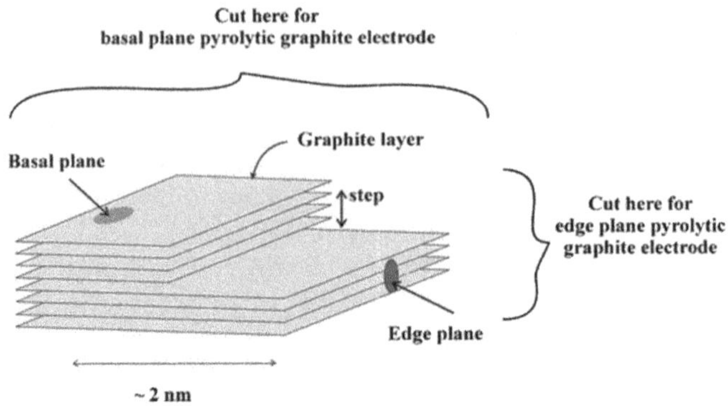

Figure 10.1 Schematic representation of basal-plane- and edge-plane-like-sites/defects. Reproduced from ref. 41 with permission from The Royal Society of Chemistry.

10.2 Modified Electrode Fabrication

It is noticeable that there are so many known, well-developed and currently explored electrode fabrication techniques that are not exclusively used to produce electroanalytical sensors.[49] In fact, the majority of these methods are often employed for solar cell production, catalysis, and coating material production. Since the employment of one technique usually results in completely different performances in comparison with the others, electrode fabrication represents a key parameter for the production of the final electroanalytical device. In this section, syntheses of the carbon materials are not taken into consideration; we will focus our attention on the different deposition methods for a large variety of supports. In some cases, the procedures are very simple and are based on simple mixing with a solvent, while in other cases, more complicated instrumentation or the formulation of inks are needed.

The casting method[49,50] is probably the simplest film-forming technique with the advantage that the only equipment needed is a horizontal work surface, where a solution of the chosen material in an appropriate solvent is casted and left to dry. The technique suffers from a lack of film thickness control and from possible inhomogeneous drying, due to the surface tension of the liquid and interactions of the carbon material with the solvent.

The spin coating procedure[49] involves the casting of a liquid on a support, followed by its acceleration to a chosen rotational speed. Most of the applied liquid is ejected from the substrate, due to its angular velocity and only a thin film remains and is left to dry. The advantages of this technique are the high reproducibility and the homogeneity of the deposition, since the thickness, the morphology and the surface topography are strongly dependent on the rotational speed, viscosity, volatility, diffusivity, molecular weight and concentration of the solutes, less on the amount of solution deposited, the rate of deposition and the spinning time. In contrast, the patterning is not allowed and the use of a relatively high quantity of solution is needed.

Chemical vapour deposition (CVD)[50–53] is based on the adsorption, desorption, evolution, and incorporation of vapour species at a film surface, traditionally using heat as the main energy source for the reaction (typically from 400 to 1000 °C for carbon materials). The process usually requires the use of a catalyst, assisting in the decomposition of vapour species. By tuning the deposition parameters, the type of catalyst and the vapour species, all the different carbon structures can be produced. The principal drawbacks are the presence of metal particles (from the catalyst), which can alter the electrochemical performances (see Introduction), and the necessity, in many cases, to remove the carbon material from the synthetic support (for some materials, the scotch tape procedure is often employed[54]) to be transferred to the surface of characterization or utilization.

Pyrolysis[30,55] (heating of a carbon surface at high temperature) can, in principle, produce various carbon materials, but it is principally applied to the preparation of conductive thin carbon films. The type of precursor allows us to distinguish six categories of pyrolysis: low molecular weight aliphatic or aromatic gas source, liquid polymers, electrografted layers derived from diazonium, amorphous carbon films and graphite oxides.

Sputtering[30] represents a powerful method to produce ultrasmooth, not amorphous, carbon films on a large variety of supports, with the temperature usually maintained at 200 °C. The principle consists of erosion of a target material (in this case carbon) and deposition of eroded particles on a chosen substrate. The main advantage is the production of a film with a narrow size particle distribution, easily reaching a few nanometres, and a tunable density.

Template synthesis[20] allows us to obtain carbon materials characterized by a chosen porosity (micro-, meso- or macropores) with a process named template-carbonization route, which can employ two different types of templates: hard or soft. Most procedures are based on the use of silica as a hard template, firstly preparing the silica structure by using surfactants and substituting them with carbon precursors (by wet impregnation or chemical vapour deposition) and secondly, after polymerization and carbonization, removing the silica template. The resulting carbon structure is strictly dependent on the initial silica template, since the space occupied by silica represents the pores in the carbon material. In the case of soft templates, supramolecular aggregates are used, with the advantages of reducing synthetic steps and avoiding silica preformation.

Carbon paste[24–26] electrodes probably represent the first example of formulation of a carbon-based device. Their production is based on mixing a carbon powder, in higher quantities, with a pasting liquid used as a binder. Suitable carbonaceous materials usually have a narrow micrometric distribution of particles, high chemical purity and low adsorption capability. Carbon sources are mainly represented by spectroscopic graphite, but many other carbon-based materials can be employed as powders. For this reason, paste mixtures are prepared also from soot and charcoal, acetylene black, glassy carbon powders, pulverized diamond, template carbon, porous carbon, carbon nanospheres, fullerenes, carbon nanofibers, carbon nanotubes and graphene. The choice of binder is also crucial, since this component can determine the electrode properties. The binder is required to be chemically and electrochemically inert, at high viscosity and low volatility, with a minimal solubility in aqueous solutions and immiscibility in organic solvents. The most popular pasting liquid is mineral oil (paraffin), but hydrocarbons, silicone oils, organic esters and ionic liquids are also employed.

The pencil graphite[56,57] electrode is composed of graphite and clay. As in the case of pencils used to draw or write, a higher quantity of graphite makes a soft pencil, while a higher quantity of clay makes a hard pencil. Pencil graphite electrodes can be divided into the conventional classes used for conventional pencils: from 9 H (the hardest) to 8 B (the softest). Among the numerous advantages of this type of electrode, such as wide availability, low technology, low cost, good mechanical rigidity, chemical inertness, low background currents, wide potential window, ease of modification and miniaturization, the most important feature is the renewability of the electrode surface, simply obtained by a fast polishing procedure (as in the case of glassy carbon electrodes).

The screen-printing technique[7,8,30,49,58] involves printing a viscous fluid through a mesh screen, appropriately designed depending on the final exigencies, onto a desired substrate (see Figure 10.2(A)).

Then, the substrate is cured at 60 °C for usually 30 minutes. The viscous fluid is generally composed of conductive materials (carbon, gold, silver) and binders, polymers, plasticizers, solvents and additives, all components of the ink. Generally, the entire electrochemical cell can be printed on the same substrate, using different meshes and inks for each electrode. In the case of carbon screen-printed electrodes, almost all types of carbon-based materials can be used in the formulation of the ink. However, the roles of the other ink components have to be taken into account in view of the final electrochemical performances. In fact, the type of plasticizer can affect the electrochemical behaviour of the electrode, as its quantity is inversely proportional to the film thickness. The type of binder is also very important, since the interaction with the plasticizer has to be evaluated together with its quantity, resulting in an increase in the ohmic resistance and film thickness. Lastly, the boiling point and the viscosity of the solvents represent key points in the formulation, to assure ink homogeneity, good adhesion to the support and efficient drying after printing. The main advantages of this printing technique are reliability, portability (the entire electrochemical cell is on the same support of a few square centimetres), modest cost, mass production and tunable design, with the drawbacks

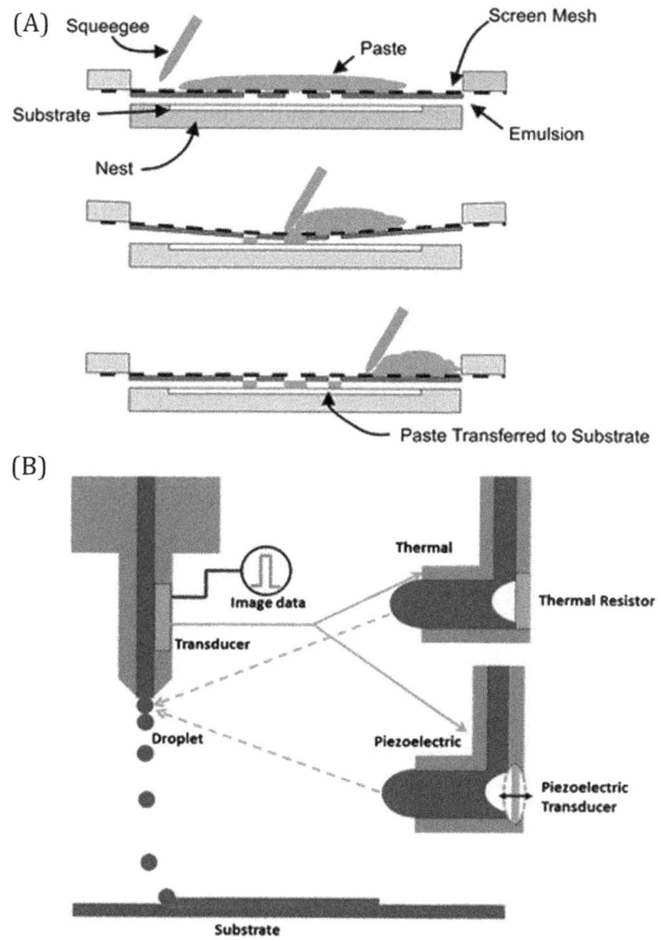

Figure 10.2 Schematic representations of two methods for the preparation of modified electrodes: (A) screen printed manufacturing and (B) droplet formation process in a thermal and piezoelectric inkjet nozzle. Reproduced from ref. 58 and 59 with permission from The Royal Society of Chemistry.

being that a high quantity of ink is needed and the relatively high film thickness.

Inkjet printing[49,52,59] is a relatively new technique, derived from the use of the well-known inkjet home/office printers. It is based on the production of small droplets by mechanical compression of ink in a nozzle or by heating and on the acceleration of these electrostatically charged droplets towards the substrate by an electric field (see Figure 10.2(B)). In the formulation of the ink, every type of carbon material can be used, paying attention to the viscosity, the electrostatic charge and to the surface tension, primarily derived from the solvent and additives employed. The main advantage of this printing method is the quite high resolution, together with the availability of a wide range of solvents and the tunability of the source image (entirely digital).

The fabrication methods presented in this chapter are the most employed techniques described in the literature to produce carbon-based electrodes. The choice of the support is also a key point for the final application, obviously considering the limits or requisites of the technique employed. Many examples of substrates can be found in the literature, from the conventional to the more exotic. Conventional supports are represented by metals, glasses, conductive glasses (indium tin oxide (ITO) or fluorine-doped tin oxide (FTO)), plastics and ceramics, while the most innovative are paper, polymer films, flexible and elastic clothing.[8,30,49]

10.3 Carbon Materials for the Preparation of Modified Electrodes

10.3.1 Carbon Paste and Ordered Carbons

Carbon paste materials were discovered accidentally and thus developed by Ralph Norman Adams and his coworkers in the late 1950s using graphite and a suitable binder, while testing a new "dropping carbon material" to be used in place of the conventional DME (dropping mercury electrode) in the anodic oxidation of organic compounds. Although the expected experiment failed, the authors discovered a new electrode material (carbon paste) characterized by interesting properties. The history of carbon paste and its development in the last decades is well described in a comprehensive review by Švancara *et al.*,[25] which also contains detailed information on the different analytical targets.

In general, carbon paste consists of a mixture of carbon powder and a binder. Over the last decades, a lot of different compositions have been developed and carbon paste represents one of the most popular and versatile electrode materials used for the construction of various electrochemical sensors and detectors, finding a wide range of applications.[25,60]

The most common carbonaceous material used in the mixture is spectroscopic graphite, with particles usually in the low micrometric scale, even though a lot of different carbon structures have been employed during the last few years: acetylene black, glassy carbon powders, graphene, diamond (synthetic or natural), fullerenes, carbon nanofibers and nanotubes.[25] The binders, generally, are either paraffin oils or silicone fluids.[26]

Carbon paste is intrinsically a metal-free carbon material, and, to the best of our knowledge, there are only a few examples in the electroanalytical field of doped carbon paste electrodes. One example was reported by Liu's group,[16] who worked with a nanocarbon paste electrode modified with N-doped graphene for the determination of traces of lead and cadmium. The paste was created by mixing a certain portion of nano-graphite powder with N-doped graphene and paraffin oil as a binder. The choice of nitrogen-doped graphene as the material for the construction of the sensor is due to the intrinsic advantages brought by the N atoms in the graphene structure (see Section 10.3.4). Another example is the use of a boron-doped diamond paste electrode,[9] prepared from a mixture of BDD powder and mineral oil, for the detection of biological species like norepinephrine and serotonin. The exemplary electrochemical properties of BDD electrodes are now combined with a microfluidic paper-based analytical device, to create a low-cost and high-performance electroanalytical sensor, overcoming the disadvantages of BDD electrodes of being expensive and not so available. In the new millennium, new technologies have been developed to modify carbon paste with complexants and redox electrocatalysts and mediators,[25] according to the need of the analytical target to be detected.

More recently, ordered mesoporous and nanoporous carbons have been used to design nanostructured electrodes for sensing applications due to the possibility to enhance the surface–volume ratio, improving interactions with analytes and also ensuring fast mass transport due to the 3D mechanically stable network.[20,61] Their electrocatalytic properties, which allow us to build more sensitive sensors, are mainly related to the presence of edge-plane graphite sites, since the chemical modification of these systems is difficult, due to carbon's low reactivity. Hence, these objects are used much more as supports for small objects' entrapment such as inorganic and organic redox mediators, metal oxides, organometallic catalysts, and metal nanoparticles.[20,61] However, some examples of oxygen-, aryl-diazonium- or fluorine-doping can be found with interesting analytical results.[20]

10.3.2 Glassy Carbon (GC), Boron-doped Diamond (BDD) and Tetrahedral Amorphous Carbon (ta-C)

In the panorama of metal-free carbon-based electrodes, the most popular and used device for electroanalytical applications is glassy carbon,[30,62] which is an important variant of the graphite structure. Heteroatoms are not present in this type of structure since its synthesis is based on the heat treatment of various polymers at high temperatures, where only carbon remains. The C–C bonds in the polymer are maintained and, for this reason, carbons can form only limited-size graphitic planes (30–70 Å) with an interplanar spacing higher than 3.6 Å (typical of HOPG—highly ordered pyrolytic graphite). The structure is generally described as a random distribution of graphitic planes with sp^2–sp^3 hybridization and presents a unique electroactivity, thanks to the presence of a high quantity of edge planes, which play a dominant role in electrochemical features (see Introduction). The advantages are represented by a wide potential window, a relative inert electrochemistry and a high stability in many solvents. For these reasons, this electrode finds application in the detection of a huge number of analytes, ranging from inorganic to organic compounds, and represents the most employed electrode for electroanalysis.[62] Surface modification by acidic functionalities (sulfonated species, carboxylic acids, boronic acids), attentively summarized in a review by Desimoni and Brunetti,[63] enhances the potentialities of this sensor. The most important drawback of glassy carbon and, in general, of all sp^2 hybridized carbon materials, is surface oxidation and fouling. In fact, the electrode surface reacts easily with oxygen and water, forming oxygen-containing functional groups, which are negatively charged, having a strong effect on adsorption, electron-transfer kinetics and electrocatalysis. For these reasons, considerable surface cleaning and conditioning are required.

Carbon materials with a completely sp^3 hybridized structure could overcome fouling and oxidation of the electrode surface. The principal representative element of this family is diamond,[64,65] which is perfect for electroanalytical purposes considering its hardness and high stability, but it is uninteresting as an electrode material due to its very low electrical conductivity. However, doping during the synthesis, especially with boron, can increase its conductivity, producing one of the most employed electrodes: boron-doped diamond[64] (BDD). Boron is electron-deficient in relation to carbon and represents, for diamond, a p-dopant, with a quite high doping level (10^{18}–10^{21} atoms cm^{-3} or B–C ratio from 10^{-5} to 10^{-3}). BDD is generally microcrystalline with randomly oriented crystallites of a few nanometres, forming a typical polycrystalline material. Its synthesis is based on CVD, with the gas phase consisting of a carbon source with boron and hydrogen as a carrier gas, and producing a H-terminated diamond surface, which reacts very slowly with oxygen. For this reason, BDD shows a hydrophobic nature and a high resistance to fouling processes that normally affect carbon electrode surfaces. The other characteristics that allowed BDD to be employed in many electroanalytical determinations of a vast range of compounds are: a wide potential window, a low background current over a wide potential range, a low electrostatic capacity, a high stability (electrical and mechanical), an insensitivity to dissolved oxygen, a chemical inertness, a high efficiency for oxidation and reduction and a good biocompatibility. Thanks to its versatility, several electroanalytical methods can be employed with BDD: classic voltammetries (cyclic, differential pulse and square-wave voltammetry), stripping techniques (both anodic and cathodic) and amperometric detection. The target analytes include both organic and inorganic species, in particular detection of heavy metal ions (Cd^{2+}, Ni^{2+}, Pb^{2+}, Hg^{2+}, As^{3+})[66] also at trace levels in complex matrices, and the detection of small biomolecules or redox proteins (*e.g.* dopamine, uric acid, glucose).[67–69]

In recent years, studies have begun on tetrahedral amorphous carbon (ta-C)—a mixture of sp^3 and sp^2 microstructured materials—as a possible substitute for BDD, due to its superior properties (low background currents, wide potential windows, fouling resistance, hardness, band gap, thermal stability, thermal and electrical conductivity) and electroanalytical performances.[19] This material could be conveniently prepared at room temperature and can be doped with nitrogen, which is able to control the sp^3–sp^2 ratio and the surface roughness, changing the final electroanalytical features for the detection of different analytes (serotonin, norepinephrine, propanolol, hydrochlorothiazide, glutathione). Notwithstanding the interesting applications, the role of nitrogen doping in these composites is yet to be clarified. In fact, according to the work of Niwa *et al.*,[70,71] the nitrogen doping increases the sp^3 fraction maintaining the surface roughness, while considering the work of Swain *et al.*,[72] the roughness and sp^2 carbon content increases with nitrogen incorporation.

10.3.3 Carbon Nanotubes (CNTs)

Among carbonaceous materials, carbon nanotubes (CNTs) stand out as the most promising materials. First discovered by Iijima in 1991,[73,74] they have received increasing attention over the last few decades, thanks to their unique characteristics. CNTs are composed of concentric cylinders of seamlessly closed rolled-up graphene sheets. They can be divided into single-walled carbon nanotubes (SWCNTs), composed of one graphene sheet, and multi-walled carbon nanotubes (MWCNTs), composed of different concentric graphene sheets. They represent an interesting materials that have had a profound impact on a wide range of applications.[27–29,37,75,76] In the electrochemistry and electroanalytical field, CNTs represent a crucial material, thanks to their demonstrated advantages such as increased voltammetric currents, increased heterogeneous electron-transfer rates, insignificant surface fouling and an apparent electrocatalytic effect toward the oxidation and reduction of a wide variety of compounds.[27–29,37,75,76]

CNTs are usually synthesized by CVD (chemical vapour deposition) or arch-evaporation, in which metallic nanoparticles act as catalysts. It has been demonstrated by Compton's group[33] that these metallic nanoparticles are responsible for the electrocatalytic effect of CNTs. It is quite impossible to think about CNTs without metal impurities, because they remain in the structure of carbon nanotubes, even after a strong washing treatment in nitric acid under elevated temperature.[77] Therefore, it is possible to explore the effect of different dopant atoms or functionalization, but the presence of metal impurities in the intimate structure of the nanotubes cannot be excluded.

10.3.3.1 Metal-free Functionalized Carbon Nanotubes

As previously mentioned, the unique properties of carbon nanotubes have made them attractive materials in many areas of science and technology, including electrochemistry and electroanalysis. In this particular case, it has been necessary to introduce some functionalizations in the carbon nanotubes' structures, by chemical or electrochemical modification, in order to improve their performances, for example enhancing their solubility in various solvents. The great use of carbon nanotubes in electrochemistry is due to their excellent electrical conductivity and mechanical strength, their chemical inertness in most electrolyte solutions, their high surface activity and a wide operational potential window,[24] all fundamental aspects for an electrode material. According to Compton's review,[24] the modification of carbon nanotubes can occur in three different ways: chemisorption, physisorption of the modifier or miscellaneous methods (incorporation of the nanotubes in a film and subsequent intercalation of the modifier).

An example of a strategy for modifying CNTs is the use of radical species generated by electrochemical reduction of aryldiazonium salts:[78,79] dinitrogen and free-radical species are generated and the radical is covalently attached to the surface of the nanotube. Depending on the type of diazonium salt, it could be possible to obtain different substituent on the CNTs surface and subsequent polymerization reactions could create a polymer-modified carbon nanotube, now soluble in different organic solvents and more suitable for the construction of sensors for analytical applications.[24] Moreover, different electropolymerization modification techniques can be found (electrochemical reduction of acrylonitrile, oxidative polymerization of aniline, electropolymerization of 1,8-diaminonaphthalene), in order to create different high-quality conducting polymer films on the surface of the nanotubes, to finally construct a sensor.[27,78,80,81] Electropolymerized polypyrrole coatings doped with nucleic acid probes showed great potential in the detection of complementary oligonucleotide sequences, representing an example of the applicability of CNT-based electrodes in the biomedical field[82] also (together with a lot of examples of carbon nanotubes onto which specific enzymes for specific targets are immobilized). Another interesting modification of CNTs is with a Nafion® cationic exchanger, a perfluorocarbon polymer-sensing membrane.[83] Such modified electrodes showed interesting results in the detection of different analytes like dopamine and carbohydrates.

Furthermore, the π-stacking interaction between aromatic compounds and the aromatic structure of carbon nanotubes yields to the possibility to adsorb such compounds on the tubes *via* non-covalent or covalent functionalization, allowing us to obtain novel nanostructures with new properties.[29] Covalent or non-covalent interactions may also be used to integrate biomaterials (proteins, enzymes, antigens, antibodies, DNA), producing composites that combine the conductive properties of CNTs with the high recognition selectivity of the biomolecules.[29]

10.3.3.2 Oxygen-doped CNTs

An important modification in the CNTs' structure is introduced by carboxylic moieties. It is still unclear if this functionalization has to be taken into account as "native" or as "modification", but the role of oxygen-containing groups on the electrochemical properties (in both positive and/or negative aspects) is undoubted.[32,37,75,84] The so-called "open-ended CNTs" (meaning that they do not end with a fullerene cap) normally have terminal oxygenated groups (hydroxyl, quinonyl and carboxyl moieties), which are responsible (together with metallic and nanographite impurities) for the properties of these materials.[37] However, since in some cases they are present at too low a concentration to be significant, some modification methods to introduce carboxylic groups have also been developed.[24] The final modified CNTs found important applications in electroanalysis and particularly in bioelectroanalysis, being used in the detection of dopamine, epinephrine and ascorbic acid.[85]

10.3.3.3 Nitrogen-doped CNTs

Nitrogen is one of the most suitable elements for doping carbon nanotubes, because of the similar dimension and proximity in the periodic table of C and N atoms. Different kinds of N-doped CNTs can be obtained:[86]

- *graphitic* nitrogen (or *substitutional*), where N atoms directly substitute a graphitic carbon atom being bonded to three C atoms in an sp^2 structure;
- *pyridine-like* nitrogen, where N is bound to two coordinated bonds in an sp^2 structure;
- *pyrrole-like* N, where nitrogen sits in an sp^3 coordination mode in a five-fold ring;
- *molecular* nitrogen, in which N_2 molecules are trapped within CNTs.

Depending on the concentration and on the final form of the modified material, nitrogen can act both as a p- or n-type dopant. The most suitable technique to understand the changes in the doping effect is Raman spectroscopy, investigating the changes in D–G mode ratios. Generally, by increasing the N concentration, the D–G band intensity ratio increases.[87]

N-doped carbon nanotubes find applications in gas sensors,[88] as matrix fillers in composite materials or in field emission devices,[89] since they are expected to emit with high current densities. Moreover, they are involved in Li-ion batteries[90] or, in general, in semiconductor technology. Finally, they can be used in sensors: the adsorption or desorption of different functional groups on the nanotubes can vary the electrical conductivity of the material.[86]

10.3.3.4 Boron-doped CNTs

Generally, as Ayala's group reported,[91] B is added to CNTs' structure in order to obtain a p-doping effect on the nanotubes to be used in semiconductor technology. It has been demonstrated, in fact, that there is a possibility to transform semiconducting tubes into metallic tubes by lowering the Fermi level into the valence band. Moreover, some studies reported a clear improvement in the conductivity of B-doped tubes. To the best of our knowledge, no works related to the use of such modified carbon nanotubes as electroanalytical sensors are present in the literature.

10.3.3.5 Phosphorus-doped CNTs

Scarce information about the doping effect of phosphorus can be found. Cruz-Silva *et al.* reported a study[92] in which the electronic transport and mechanical properties of P- and PN-doped CNTs were analyzed, but even in this case, there is no information on their final use as sensors.

10.3.4 Graphene

10.3.4.1 Advantages of the Use of Graphene

According to the IUPAC definition, graphene is a "single layer of the graphite structure, describing its nature by analogy to a polycyclic aromatic hydrocarbon of quasi-infinite size".[93] Thanks to its unique physicochemical properties, such as high surface area, excellent thermal and electrical conductivity, high mechanical strength and ease of functionalization and mass production, graphene has become, over the last few decades, one of the most promising materials for the construction of analytical and electroanalytical sensors.[94] However, the practical applications of graphene are limited because of the scarce catalytic activity related to the lack of an intrinsic band gap (graphene is a zero-band gap semiconductor). Therefore, an easy way to improve, or simply alter, the performances of pristine graphene is to dope or functionalize it with atoms and molecules. In the following sections, we will focus the attention on non-metal-doped graphene materials, in particular the ones suitable for analytical and electroanalytical sensors.

10.3.4.2 Metal-free Graphene-based Materials

An effective method to modify the electronic, mechanical and electrochemical properties of graphene is to dope it with different heteroatoms. This effect is due to a change of the electronic density within the carbonaceous sheets.[95,96] The heteroatoms introduce a huge amount of defects in graphene, which can act as active sites for the adsorption and activation of target analytes facilitating charge transfer processes, a fundamental aspect in the construction of electroanalytical sensors.[97] Generally, the dopant atom should have an electronegativity and atomic structure similar to carbon. For this reason, boron and nitrogen represent the most suitable candidates.

Heteroatom-doped graphene materials can be synthesized in different ways with *in situ* and *post-treatment* approaches. The former achieve graphene synthesis and heteroatom doping simultaneously. They include CVD (chemical vapour deposition), ball milling and bottom up methods. The latter include photo-chemistry, thermal annealing of graphite oxide, wet chemical methods, arc-discharge and plasma approaches.[96]

10.3.4.3 Boron-doped Graphene

Doping graphene with boron atoms does not change its excellent mechanical properties, although it does affect its thermal conductivity.[98] Regarding the structural properties, not many variations on the planar graphene structure are introduced, because *in-plane* doping is preferred (in terms of stability) to an *out-of-plane* one, due to the tendency of boron atoms to create sp^2 hybridization in the carbon lattice.[96] Only a slight alteration in the lattice parameters can be highlighted because of the different lengths of C–B bonds with respect to the C–C bonds (1.50 Å for the former and 1.40–1.42 Å for the latter).[99] Therefore, the predominant effect due to the doping approach is the charge polarization between C atoms and the electron-deficient neighbouring B atoms. As a matter of fact, when boron atoms are added to a graphene structure, a higher number of the original defects are created. In particular, B atoms induce a p-doping effect and a downshift of the Fermi level of graphene. This difference in the energy of the Fermi level is strongly related to the quantity of the dopant atoms (for example, it has been calculated that doping with 2% of boron moves the Fermi level to a value of approximatively 0.65 eV[100]).

Boron-doped graphene electrodes show promising features both as catalysts and sensors, with broad application fields. For example, Yeh *et al.*[101] demonstrated the high activity of B-doped graphene in the reduction and electrochemical detection of H_2O_2. They showed how the sensitivity towards the HPRR (hydrogen peroxide reduction reaction) increased using a modified graphene electrode instead of bare graphene. Zhang's group, instead, pointed out a facile synthetic route for the creation of B-doped graphene electrodes as electroactive tools for the simultaneous detection of hydroquinone and catechol.[102] This material showed higher activity and sensitivity with respect to pristine graphene towards the electrochemical reduction of those analytes, thus representing a powerful material to be used as a sensor in wastewater analysis. B-doped graphene was also used as an electrochemical sensor for the detection of a high melting explosive (HMX), a powerful tool in antiterrorism and related fields.[103] Some works are related to the potential application of B-doped graphene-modified electrodes for the detection of biomolecules[104] and DNA analysis.[105] Moreover, B-doped graphene has been used in gas sensors for the detection of CO, NO and NO_2.[106]

10.3.4.4 Nitrogen-doped Graphene

Doping graphene with electron-rich (with respect to the Pauling scale) neighbouring nitrogen atoms is the most studied approach. In this case, three different bonding configurations can be reached: graphitic, pyridinic and pyrrolic (see Figure 10.3).

The most stable ones are the graphitic and pyridinic, because in these cases, the structure of graphene is preserved, while the instauration of sp^3 C

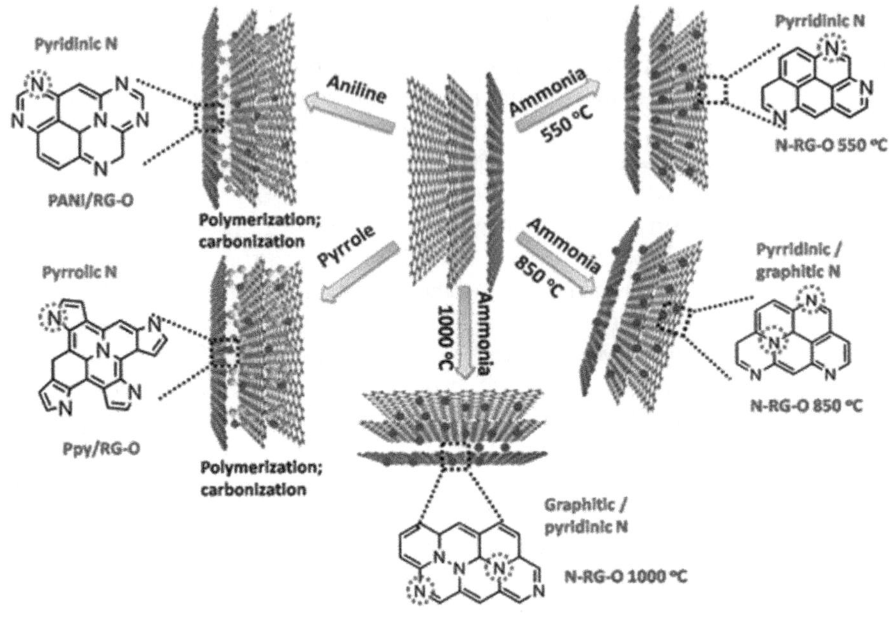

Figure 10.3 Schematic diagram of N-doped graphene with different N states. Reproduced from ref. 95 with permission from The Royal Society of Chemistry.

in the pyrrolic structure drastically changes the main features of bare graphene, breaking its planar structure. Pyrrolic and pyridinic ones usually locate at the edges or defects, while graphitic simply replaces a C atom in the basal plane of graphene.[107] These three different configurations also affect the electronic behaviour of graphene, which now behaves as a semiconductor (the N-doping introduces a band gap near the Dirac point). If it is in the pyridinic form, it can be considered as an n-type semiconductor (N atoms give electrons to the graphitic network),[108] while in the form of graphitic and pyrrolic, it is a p-type semiconductor (N atoms in defects keep electrons from the graphitic network, creating holes).[109]

Among the final applications of N-doped graphene, its great activity in the ORR (oxygen reduction reaction) is evident.[97,110] For this reason, it finds applications, especially, in fuel cells as a powerful substitute of the currently used Pt-based materials. Other applications of N-doped graphene are the activation of C–H bonds, the oxidation of benzylic alcohols and the reduction of nitro compounds.[95] In the electroanalytical field, thanks to its good biocompatibility, N-graphene can be used as an electrochemical sensor for small biomolecules such as dopamine (DA), uric acid (UA) and ascorbic acid (AA). These molecules play a fundamental role in several physiological functions of organisms and their determination is important because anomalous levels of these analytes may cause severe disease such as Parkinson's, cancer or cardiovascular diseases. In this context, electrochemistry could act as a simple, low cost and effective method of detection. The problem is that, in real matrices using conventional electrodes, DA, UA and AA oxidation potentials coexist. N-doped graphene electrodes, characterized by their electrocatalytic activity towards DA, AA and UA, allow their simultaneous detection.[111]

10.3.4.5 Oxygen-doped Graphene

Regarding atoms from group VI of the periodic table, oxygen is the most electronegative. For this reason, when graphene is doped with oxygen, it is impossible to observe a complete C–O substitution in the planar structure; therefore, a passage from an sp^2 to an sp^3 conformation of C atoms is obtained.[96] Oxygen is more or less always present in graphene, because most of the synthetic routes of graphene materials involve strong oxidizing agents.[95] In this way, so-called graphene oxide (GO) is produced and it can be easily ascribed to the O-doped graphene material. The oxygen functionalities in GO are in the form of epoxy and hydroxyl groups, located on the basal plane, or carboxyl, carbonyl, phenol, lactone or quinone groups on the sheet edges.[112] Compared to pristine graphene, GO has a lower electrical conductivity, which leads to limited applications. However, these functionalities (that cannot be completely removed even after a strong reductive treatment) also provide some advantages. For example, they make graphene more hydrophilic and suitable for processing and further functionalization, a powerful tool for the construction of electrode materials with excellent electrochemical properties. In fact, one of the most promising applications of O-doped graphene materials is electrochemical sensing, especially biosensing and gas sensing. Regarding the last application, GO can act as a sorption-based sensor, able to detect gas molecules even at very low concentrations, by observing a change in the conductance or capacitance values. Many works in the literature are related to the detection of gas molecules (NO_2, NO, NH_3, Cl_2) by GO-based sensors.[113–115] Other species can be electrochemically detected by GO sensors, for example hydrazine,[116] small organic molecules,[117] heavy metal ions,[118,119] drug molecules,[120,121] and pesticides.

10.3.4.6 Phosphorus-doped Graphene

Even though P belongs to the same group as N, changes induced by P-doping in graphene are larger than the corresponding N-doping. P, in fact, is bigger than N and when it is introduced into a graphene structure, it drastically changes the sp^2 nature of the C atoms to sp^3, and modifies the bond length. Moreover, the electronegativity of P atoms is lower than that of carbon, so the C–P polarity is the opposite with respect to the C–N one, resulting in a doping effect more similar to B-doping.[96] P-graphene materials find applications, especially, in catalysis. They are often used in supercapacitors or in fuel cells for the ORR. To the best of our knowledge, there is a lack of information about the use of P-graphene as a sensor.

10.3.4.7 Sulfur-doped Graphene

S-doped graphene can act as a powerful catalyst for the ORR, but, to the best of our knowledge, there are few examples of S-doped graphene sensors in the literature. Regarding the electronic and structural characteristics, S-doped graphene is, in general, more resistive than the pristine material and it can act as a small band gap semiconductor or as a metal, depending on the heteroatom doping level.[96]

10.3.4.8 Halogen (F, Cl, Br and I)-doped Graphene

Graphene can also be doped with halogen atoms, which introduce a strong distortion in the planar structure of the host material, together with a drastic change in the electronic properties. A lot of works are related to F- and Cl-doped graphene, while there is scarce information (both theoretical and experimental) about Br- and I-doped graphene. The doping effect of halogens is reflected in the formation of sp^3 C atoms, with completely different reactivity from the original sp^2 network. As far as we are concerned, no examples of halogen-doped materials used as electrochemical sensors are present in the literature.[96]

References

1. E. Desimoni and B. Brunetti, *Electroanalysis*, 2013, **25**, 1645–1651.
2. J. Wang, B. Tian, V. B. Nascimento and L. Angnes, *Electrochim. Acta*, 1998, **43**, 3459–3465.
3. F. Arduini, A. Amine, C. Majorani, F. Di Giorgio, D. De Felicis, F. Cataldo, D. Moscone and G. Palleschi, *Electrochem. Commun.*, 2010, **12**, 346–350.
4. P. Fanjul-Bolado, P. Queipo, P. J. Lamas-Ardisana and A. Costa-García, *Talanta*, 2007, **74**, 427–433.
5. P. Fanjul-Bolado, D. Hernández-Santos, P. J. Lamas-Ardisana, A. Martín-Pernía and A. Costa-García, *Electrochim. Acta*, 2008, **53**, 3635–3642.
6. S. Laschi, E. Bulukin, I. Palchetti, C. Cristea and M. Mascini, *IRBM*, 2008, **29**, 202–207.
7. M. Trojanowicz, *TrAC, Trends Anal. Chem.*, 2016, **84**, 22–47.
8. H. M. Mohamed, *TrAC, Trends Anal. Chem.*, 2016, **82**, 1–11.
9. S. Nantaphol, R. B. Channon, T. Kondo, W. Siangproh, O. Chailapakul and C. S. Henry, *Anal. Chem.*, 2017, **89**, 4100–4107.
10. S. Daniele, M. Baldo and C. Bragato, *Curr. Anal. Chem.*, 2008, **4**, 215–228.

11. C. Parolo, M. Medina-Sánchez, H. Montón, A. de la Escosura-Muñiz and A. Merkoçi, *Part. Part. Syst. Charact.*, 2013, **30**, 662–666.
12. B. Liu, D. Du, X. Hua, X.-Y. Yu and Y. Lin, *Electroanalysis*, 2014, **26**, 1214–1223.
13. R. Álvarez-Diduk, J. Orozco and A. Merkoçi, *Sci. Rep.*, 2017, **7**, 976.
14. A. M. López-Marzo and A. Merkoçi, *Lab Chip*, 2016, **16**, 3150–3176.
15. C. M. A. Brett and D. A. Fungaro, *J. Braz. Chem. Soc.*, 2000, **11**, 298–303.
16. X. Liu, Z. Li, R. Ding, B. Ren and Y. Li, *Microchim. Acta*, 2016, **183**, 709–714.
17. C. Gouveia-Caridade, R. Pauliukaite and C. M. A. Brett, *Electroanalysis*, 2006, **18**, 854–861.
18. J. Barek, J. Cvacka, A. Muck, V. Quaiserová and J. Zima, *Fresenius. J. Anal. Chem.*, 2001, **369**, 556–562.
19. N. Yang, G. M. Swain and X. Jiang, *Electroanalysis*, 2016, **28**, 27–34.
20. A. Walcarius, *TrAC, Trends Anal. Chem.*, 2012, **38**, 79–97.
21. D. W. H. Fam, A. Palaniappan, A. I. Y. Tok, B. Liedberg and S. M. Moochhala, *Sens. Actuators, B*, 2011, **157**, 1–7.
22. Y. Yulong and P. Xinsheng, *Analyst*, 2016, **141**, 2619–2628.
23. K. Szot and M. Opallo, *Electroanalysis*, 2016, **28**, 46–57.
24. G. G. Wildgoose, C. E. Banks, H. C. Leventis and R. G. Compton, *Microchim. Acta*, 2006, **152**, 187–214.
25. I. Švancara, A. Walcarius, K. Kalcher and K. Vytřas, *Cent. Eur. J. Chem.*, 2009, **7**, 598–656.
26. I. Švancara, K. Vytřas, K. Kalcher, A. Walcarius and J. Wang, *Electroanalysis*, 2009, **21**, 7–28.
27. L. Agüí, P. Yáñez-Sedeño and J. M. Pingarrón, *Anal. Chim. Acta*, 2008, **622**, 11–47.
28. J. J. Gooding, *Electrochim. Acta*, 2005, **50**, 3049–3060.
29. K. P. Gong, Y. M. Yan, M. N. Zhang, L. Su, S. X. Xiong and L. Q. Mao, *Anal. Sci.*, 2005, **21**, 1383–1393.
30. W. Zhang, S. Zhu, R. Luque, S. Han, L. Hu and G. Xu, *Chem. Soc. Rev.*, 2016, **45**, 715–752.
31. X. Dai, G. G. Wildgoose and R. G. Compton, *Analyst*, 2006, **131**, 901–906.
32. C. E. Banks, X. Ji, A. Crossley and R. G. Compton, *Electroanalysis*, 2006, **18**, 2137–2140.
33. C. E. Banks, A. Crossley, C. Salter, S. J. Wilkins and R. G. Compton, *Angew. Chem., Int. Ed. Engl.*, 2006, **45**, 2533–2537.
34. C. Batchelor-McAuley, G. G. Wildgoose, R. G. Compton, L. Shao and M. L. H. Green, *Sens. Actuators, B*, 2008, **132**, 356–360.
35. B. Sljukic, C. E. Banks and R. G. Compton, *Nano Lett.*, 2006, **6**, 1556–1558.
36. M. Pumera and H. Iwai, *J. Phys. Chem. C*, 2009, **113**, 4401–4405.
37. M. Pumera, *Chem. Rec.*, 2012, **12**, 201–213.
38. L. Wang and M. Pumera, *Appl. Mater. Today*, 2016, **5**, 134–141.
39. C. E. Banks, R. R. Moore, T. J. Davies and R. G. Compton, *Chem. Commun.*, 2004, **129**, 1804.

40. C. E. Banks, T. J. Davies, G. G. Wildgoose and R. G. Compton, *Chem. Commun.*, 2005, 829.
41. C. E. Banks and R. G. Compton, *Analyst*, 2006, **131**, 15.
42. A. Ambrosi and M. Pumera, *J. Phys. Chem. C*, 2011, **115**, 25281–25284.
43. L. Wang, A. Ambrosi and M. Pumera, *Anal. Chem.*, 2013, **85**, 6195–6197.
44. V. Pifferi, G. Facchinetti, A. Villa, L. Prati and L. Falciola, *Catal. Today*, 2015, **249**, 265–269.
45. T. W. B. Lo, L. Aldous and R. G. Compton, *Sens. Actuators, B*, 2012, **162**, 361–368.
46. F. Arduini, F. Di Nardo, A. Amine, L. Micheli, G. Palleschi and D. Moscone, *Electroanalysis*, 2012, **24**, 743–751.
47. F. Arduini, A. Amine, C. Majorani, F. Di Giorgio, D. De Felicis, F. Cataldo, D. Moscone and G. Palleschi, *Electrochem. Commun.*, 2010, **12**, 346–350.
48. F. Arduini, C. Majorani, A. Amine, D. Moscone and G. Palleschi, *Electrochim. Acta*, 2011, **56**, 4209–4215.
49. F. C. Krebs, *Sol. Energy Mater. Sol. Cells*, 2009, **93**, 394–412.
50. A. K. Wanekaya, *Analyst*, 2011, **136**, 4383.
51. N. O. Weiss, H. Zhou, L. Liao, Y. Liu, S. Jiang, Y. Huang and X. Duan, *Adv. Mater.*, 2012, **24**, 5782–5825.
52. S. Park, M. Vosguerichian and Z. Bao, *Nanoscale*, 2013, **5**, 1727.
53. A. V. Melechko, V. I. Merkulov, T. E. McKnight, M. A. Guillorn, K. L. Klein, D. H. Lowndes and M. L. Simpson, *J. Appl. Phys.*, 2005, **97**, 41301.
54. C. Ingrosso, G. V. Bianco, V. Pifferi, P. Guffanti, F. Petronella, R. Comparelli, A. Agostiano, M. Striccoli, I. Palchetti, L. Falciola, M. L. Curri and G. Bruno, *J. Mater. Chem. A*, 2017, **5**, 9307–9315.
55. A. Mardegan, V. Pifferi, E. Pontoglio, L. Falciola, P. Scopece and L. M. Moretto, *Electrochem. Commun.*, 2014, **48**, 13–16.
56. A.-N. Kawde, N. Baig and M. Sajid, *RSC Adv.*, 2016, **6**, 91325–91340.
57. M. R. Akanda, M. Sohail, M. A. Aziz and A.-N. Kawde, *Electroanalysis*, 2016, **28**, 408–424.
58. J. P. Metters, R. O. Kadara and C. E. Banks, *Analyst*, 2011, **136**, 1067.
59. Z. Zhan, J. An, Y. Wei, V. T. Tran and H. Du, *Nanoscale*, 2017, **9**, 965–993.
60. I. Švancara, K. Vytřas, K. Kalcher, A. Walcarius and J. Wang, *Electroanalysis*, 2009, **21**, 7–28.
61. R. C. Rodriguez, A. B. Moncada, D. F. Acevedo, G. a. Planes, M. C. Miras and C. a. Barbero, *Faraday Discuss.*, 2013, **164**, 147.
62. R. L. McCreery, *Chem. Rev.*, 2008, **108**, 2646–2687.
63. E. Desimoni and B. Brunetti, *Electroanalysis*, 2012, **24**, 1481–1500.
64. J. H. T. Luong, K. B. Male and J. D. Glennon, *Analyst*, 2009, **134**, 1965.
65. A. Qureshi, W. P. Kang, J. L. Davidson and Y. Gurbuz, *Diamond Relat. Mater.*, 2009, **18**, 1401–1420.
66. S. M. Tan, H. L. Poh, Z. Sofer and M. Pumera, *Analyst*, 2013, **138**, 4885–4891.
67. M. Wei, L.-G. Sun, Z.-Y. Xie, J.-F. Zhii, A. Fujishima, Y. Einaga, D.-G. Fu, X.-M. Wang and Z.-Z. Gu, *Adv. Funct. Mater.*, 2008, **18**, 1414–1421.
68. J. Barek, K. Jandová, K. Pecková and J. Zima, *Talanta*, 2007, **74**, 421–426.

69. B. V. Sarada, T. N. Rao, D. A. Tryk and A. Fujishima, *Anal. Chem.*, 2000, **72**, 1632–1638.
70. T. Kamata, D. Kato, S. Hirono and O. Niwa, *Anal. Chem.*, 2013, **85**, 9845–9851.
71. T. Kamata, D. Kato, S. Umemura and O. Niwa, *Anal. Sci.*, 2015, **31**, 651–656.
72. X. Yang, L. Haubold, G. DeVivo and G. M. Swain, *Anal. Chem.*, 2012, **84**, 6240–6248.
73. S. Iijima, *Nature*, 1991, **354**, 56–58.
74. B. O'Regan and M. Grätzel, *Nature*, 1991, **353**, 737–740.
75. M. Pumera, *Chemistry*, 2009, **15**, 4970–4978.
76. J. S. Stefano, D. P. Rocha, R. M. Dornellas, L. C. D. Narciso, S. R. Krzyzaniak, P. A. Mello, E. Nossol, E. M. Richter and R. A. A. Munoz, *Electrochim. Acta*, 2017, **240**, 80–89.
77. M. Pumera, *Langmuir*, 2007, **23**, 6453–6458.
78. M. Knez, M. Sumser, A. M. Bittner, C. Wege, H. Jeske, S. Kooi, M. Burghard and K. Kern, *J. Electroanal. Chem.*, 2002, **522**, 70–74.
79. P. R. Marcoux, P. Hapiot, P. Batail and J. Pinson, *New J. Chem.*, 2004, **28**, 302–307.
80. M. Burghard, S. Kooi, U. Schlecht, K. Balasubramanian and K. Kern, *AIP Conf. Proc*, ed. H. Kuzmany, American Institute of Physics, New York, 2002, pp. 77.
81. E. Tamburri, S. Orlanducci, M. L. Terranova, F. Valentini, G. Palleschi, A. Curulli, F. Brunetti, D. Passeri, A. Alippi and M. Rossi, *Carbon*, 2005, **43**, 1213–1221.
82. Y. Xu, Y. Jiang, H. Cai, P.-G. He and Y.-Z. Fang, *Anal. Chim. Acta*, 2004, **516**, 19–27.
83. Y. Tsai, J. Chen, S. Li and F. Marken, *Electrochem. Commun.*, 2004, **6**, 917–922.
84. A. Chou, T. Bocking, N. K. Singh and J. J. Gooding, *Chem. Commun.*, 2005, 842–844.
85. H. Luo, Z. Shi, N. Li, Z. Gu and Q. Zhuang, *Anal. Chem.*, 2001, **73**, 915–920.
86. P. Ayala, R. Arenal, M. Rümmeli, A. Rubio and T. Pichler, *Carbon*, 2010, **48**, 575–586.
87. L. S. Panchakarla, A. Govindaraj and C. N. R. Rao, *Inorg. Chim. Acta*, 2010, **363**, 4163–4174.
88. F. Villalpando-Páez, A. H. Romero, E. Muñoz-Sandoval, L. M. Martínez, H. Terrones and M. Terrones, *Chem. Phys. Lett.*, 2004, **386**, 137–143.
89. S. K. Srivastava, V. D. Vankar, D. V. Sridhar Rao and V. Kumar, *Thin Solid Films*, 2006, **515**, 1851–1856.
90. Z. Zhou, X. Gao, J. Yan, D. Song and M. Morinaga, *Carbon*, 2004, **42**, 2677–2682.
91. P. Ayala, W. Plank, A. Grüneis, E. I. Kauppinen, M. H. Rümmeli, H. Kuzmany and T. Pichler, *J. Mater. Chem.*, 2008, **18**, 5676.
92. E. Cruz-Silva, F. Lo, E. Mun, B. G. Sumpter, H. Terrones, J. Charlier, V. Meunier and M. Terrones, *ACS Nano*, 2009, **3**, 1913–1921.

93. E. Fitzer, K. H. Kochling, H. P. Boehm and H. Marsh, *Pure Appl. Chem.*, 1995, **67**, 473–506.
94. Y. Shao, J. Wang, H. Wu, J. Liu, I. A. Aksay and Y. Lin, *Electroanalysis*, 2010, **22**, 1027–1036.
95. X.-K. Kong, C.-L. Chen and Q.-W. Chen, *Chem. Soc. Rev.*, 2014, **43**, 2841.
96. X. Wang, G. Sun, P. Routh, D.-H. Kim, W. Huang and P. Chen, *Chem. Soc. Rev.*, 2014, **43**, 7067–7098.
97. J. Xu, Y. Wang and S. Hu, *Microchim. Acta*, 2017, **184**, 1–44.
98. B. Mortazavi and S. Ahzi, *Solid State Commun.*, 2012, **152**, 1503–1507.
99. R. H. Miwa, T. B. Martins and A. Fazzio, *Nanotechnology*, 2008, **19**, 155708.
100. P. Rani and V. K. Jindal, *RSC Adv.*, 2013, **3**, 802–812.
101. M. H. Yeh, Y. S. Li, G. L. Chen, L. Y. Lin, T. J. Li, H. M. Chuang, C. Y. Hsieh, S. C. Lo, W. H. Chiang and K. C. Ho, *Electrochim. Acta*, 2015, **172**, 52–60.
102. Y. Zhang, R. Sun, B. Luo and L. Wang, *Electrochim. Acta*, 2015, **156**, 228–234.
103. Y. Xu, W. Lei, Z. Han, T. Wang, M. Xia and Q. Hao, *Electrochim. Acta*, 2016, **216**, 219–227.
104. H. Tian, Z. Sofer, M. Pumera and A. Bonanni, *Nanoscale*, 2017, **9**, 3530–3536.
105. H. Tian, L. Wang, Z. Sofer, M. Pumera and A. Bonanni, *Sci. Rep.*, 2016, **6**, 33046.
106. S. S. Varghese, S. Swaminathan, K. K. Singh and V. Mittal, *Comput. Condens. Matter*, 2016, **9**, 40–55.
107. M. Fan, Z. Q. Feng, C. Zhu, X. Chen, C. Chen, J. Yang and D. Sun, *J. Mater. Sci.*, 2016, **51**, 10323–10349.
108. S. Jalili and R. Vaziri, *Mol. Phys.*, 2011, **109**, 687–694.
109. T. Schiros, D. Nordlund, L. Pálová, D. Prezzi, L. Zhao, K. S. Kim, U. Wurstbauer, C. Gutiérrez, D. Delongchamp, C. Jaye, D. Fischer, H. Ogasawara, L. G. M. Pettersson, D. R. Reichman, P. Kim, M. S. Hybertsen and A. N. Pasupathy, *Nano Lett.*, 2012, **12**, 4025–4031.
110. L. Chao, Y. Qin, Y. Liu, Y. Kong and F. Chu, *J. Solid State Electrochem.*, 2017, **21**, 1287–1295.
111. Z.-H. Sheng, X.-Q. Zheng, J.-Y. Xu, W.-J. Bao, F.-B. Wang and X.-H. Xia, *Biosens. Bioelectron.*, 2012, **34**, 125–131.
112. D. Chen, H. Feng and J. Li, *Chem. Rev.*, 2004, **112**, 6027–6053.
113. S. R. Ng, C. X. Guo and C. M. Li, *Electroanalysis*, 2011, **23**, 442–448.
114. V. Dua, S. P. Surwade, S. Ammu, S. R. Agnihotra, S. Jain, K. E. Roberts, S. Park, R. S. Ruoff and S. K. Manohar, *Angew. Chem., Int. Ed.*, 2010, **49**, 2154–2157.
115. J. D. Fowler, M. J. Allen, V. C. Tung, Y. Yang, R. B. Kaner and B. H. Weiller, *ACS Nano*, 2009, **3**, 301–306.
116. C. Wang, L. Zhang, Z. Guo, J. Xu, H. Wang, K. Zhai and X. Zhuo, *Microchim. Acta*, 2010, **169**, 1–6.

117. L. Chen, Y. Tang, K. Wang, C. Liu and S. Luo, *Electrochem. Commun.*, 2011, **13**, 133–137.
118. J. Li, S. Guo, Y. Zhai and E. Wang, *Electrochem. Commun.*, 2009, **11**, 1085–1088.
119. B. Wang, Y.-H. Chang and L. Zhi, *New Carbon Mater.*, 2011, **26**, 31–35.
120. F. Zhao, F. Wang, W. Zhao, J. Zhou, Y. Liu, L. Zou and B. Ye, *Microchim. Acta*, 2011, **174**, 383–390.
121. X. Kang, J. Wang, H. Wu, J. Liu, I. A. Aksay and Y. Lin, *Talanta*, 2010, **81**, 754–759.

Subject Index